CAMBRIDGE TRACTS IN MATHEMATICS

General Editors

J. BERTOIN, B. BOLLOBÁS, W. FULTON, B. KRA, I. MOERDIJK,
C. PRAEGER, P. SARNAK, B. SIMON, B. TOTARO

**232   The Art of Working with the Mathieu Group $M_{24}$**

# CAMBRIDGE TRACTS IN MATHEMATICS

## GENERAL EDITORS

J. BERTOIN, B. BOLLOBÁS, W. FULTON, B. KRA, I. MOERDIJK,
C. PRAEGER, P. SARNAK, B. SIMON, B. TOTARO

A complete list of books in the series can be found at www.cambridge.org/mathematics. Recent titles include the following:

200. Singularities of the Minimal Model Program. By J. Kollár
201. Coherence in Three-Dimensional Category Theory. By N. Gurski
202. Canonical Ramsey Theory on Polish Spaces. By V. Kanovei, M. Sabok, and J. Zapletal
203. A Primer on the Dirichlet Space. By O. El-Fallah, K. Kellay, J. Mashreghi, and T. Ransford
204. Group Cohomology and Algebraic Cycles. By B. Totaro
205. Ridge Functions. By A. Pinkus
206. Probability on Real Lie Algebras. By U. Franz and N. Privault
207. Auxiliary Polynomials in Number Theory. By D. Masser
208. Representations of Elementary Abelian $p$-Groups and Vector Bundles. By D. J. Benson
209. Non-homogeneous Random Walks. By M. Menshikov, S. Popov, and A. Wade
210. Fourier Integrals in Classical Analysis (Second Edition). By C. D. Sogge
211. Eigenvalues, Multiplicities and Graphs. By C. R. Johnson and C. M. Saiago
212. Applications of Diophantine Approximation to Integral Points and Transcendence. By P. Corvaja and U. Zannier
213. Variations on a Theme of Borel. By S. Weinberger
214. The Mathieu Groups. By A. A. Ivanov
215. Slenderness I: Foundations. By R. Dimitric
216. Justification Logic. By S. Artemov and M. Fitting
217. Defocusing Nonlinear Schrödinger Equations. By B. Dodson
218. The Random Matrix Theory of the Classical Compact Groups. By E. S. Meckes
219. Operator Analysis. By J. Agler, J. E. McCarthy, and N. J. Young
220. Lectures on Contact 3-Manifolds, Holomorphic Curves and Intersection Theory. By C. Wendl
221. Matrix Positivity. By C. R. Johnson, R. L. Smith, and M. J. Tsatsomeros
222. Assouad Dimension and Fractal Geometry. By J. M. Fraser
223. Coarse Geometry of Topological Groups. By C. Rosendal
224. Attractors of Hamiltonian Nonlinear Partial Differential Equations. By A. Komech and E. Kopylova
225. Noncommutative Function-Theoretic Operator Function and Applications. By J. A. Ball and V. Bolotnikov
226. The Mordell Conjecture. By A. Moriwaki, H. Ikoma, and S. Kawaguchi
227. Transcendence and Linear Relations of 1-Periods. By A. Huber and G. Wüstholz
228. Point-Counting and the Zilber–Pink Conjecture. By J. Pila
229. Large Deviations for Markov Chains. By A. D. de Acosta
230. Fractional Sobolev Spaces and Inequalities. By D. E. Edmunds and W. D. Evans
231. Families of Varieties of General Type. By J. Kollár
232. The Art of Working with the Mathieu Group $M_{24}$. By R. T. Curtis
233. Category and Measure. By N. H. Bingham and A. J. Ostaszewski
234. The Theory of Countable Borel Equivalence Relations. By A. S. Kechris

# The Art of Working with the Mathieu Group $M_{24}$

ROBERT T. CURTIS
*University of Birmingham*

CAMBRIDGE
UNIVERSITY PRESS

Shaftesbury Road, Cambridge CB2 8EA, United Kingdom

One Liberty Plaza, 20th Floor, New York, NY 10006, USA

477 Williamstown Road, Port Melbourne, VIC 3207, Australia

314–321, 3rd Floor, Plot 3, Splendor Forum, Jasola District Centre, New Delhi – 110025, India

103 Penang Road, #05–06/07, Visioncrest Commercial, Singapore 238467

Cambridge University Press is part of Cambridge University Press & Assessment, a department of the University of Cambridge.

We share the University's mission to contribute to society through the pursuit of education, learning and research at the highest international levels of excellence.

www.cambridge.org
Information on this title: www.cambridge.org/9781009405676

DOI: 10.1017/9781009405683

© Robert T. Curtis 2025

This publication is in copyright. Subject to statutory exception and to the provisions of relevant collective licensing agreements, no reproduction of any part may take place without the written permission of Cambridge University Press & Assessment.

When citing this work, please include a reference to the DOI 10.1017/9781009405683

First published 2025

*A catalogue record for this publication is available from the British Library*

*A Cataloging-in-Publication data record for this book is available from the Library of Congress*

ISBN 978-1-009-40567-6 Hardback

Cambridge University Press & Assessment has no responsibility for the persistence or accuracy of URLs for external or third-party internet websites referred to in this publication and does not guarantee that any content on such websites is, or will remain, accurate or appropriate.

This book is dedicated to the memory of John Horton Conway (1937–2020) who was my teacher, inspiration and friend over many years.

# Contents

|   |   | | |
|---|---|---|---|
| | *List of Figures* | *page* x |
| | *List of Tables* | xiv |
| | *Preface* | xv |
| | *Acknowledgements* | xx |
| **1** | **Introduction** | 1 |
| | 1.1 Motivation | 1 |
| | 1.2 Other Constructions | 2 |
| | 1.3 The Construction of This Book | 4 |
| | 1.4 The Centrality of $M_{24}$ | 4 |
| **2** | **Steiner Systems** | 6 |
| | 2.1 Definition and Numerical Conditions | 6 |
| | 2.2 The Fano Plane | 7 |
| | 2.3 The 9-Point Affine Plane and Its Extensions | 10 |
| | 2.4 A Todd Triangle for the Steiner System S(5, 6, 12) | 13 |
| | 2.5 A Steiner System S(3, 4, 16) | 14 |
| | 2.6 The Projective Plane of Order 4 and Its Extensions | 15 |
| | 2.7 Block Designs in General | 16 |
| **3** | **The Miracle Octad Generator** | 17 |
| | 3.1 The Miracle Octad Generator or MOG | 18 |
| | 3.2 Mnemonic for the Standard MOG Labelling | 23 |
| | 3.3 Finding Octads Using the MOG | 24 |
| **4** | **The Binary Golay Code** | 28 |
| | 4.1 Creating a Code from a Block Design | 28 |
| | 4.2 The Binary Golay Code $\mathscr{C}$ | 30 |
| | 4.3 The Dual Space $\mathscr{C}^\star \cong P(\Omega)/\mathscr{C}$ | 34 |

viii                                      *Contents*

| | | | |
|---|---|---|---|
| **5** | **Uniqueness of the Steiner System $S(5, 8, 24)$ and the Group $M_{24}$** | | **36** |
| | 5.1 | Properties of the Sextets | 36 |
| | 5.2 | Uniqueness of the Steiner System $S(5, 8, 24)$ and the Order of Its Group of Symmetries | 37 |
| **6** | **The Hexacode** | | **42** |
| | 6.1 | The Hexacodewords | 42 |
| | 6.2 | Interpreting Hexacodewords as $\mathscr{C}$-sets | 43 |
| **7** | **Elements of the Mathieu Group $M_{24}$** | | **46** |
| | 7.1 | Eight Octads Suffice | 46 |
| | 7.2 | Using a Computer Package | 65 |
| **8** | **The Maximal Subgroups of $M_{24}$** | | **67** |
| | 8.1 | The Maximal Subgroups of $M_{24}$ | 67 |
| | 8.2 | Proof of Completeness of the List of Maximal Subgroups | 83 |
| | 8.3 | Other Interesting Subgroups | 94 |
| **9** | **The Mathieu Group $M_{12}$** | | **97** |
| | 9.1 | The Steiner System $S(5, 6, 12)$ | 97 |
| | 9.2 | $\Pi$, the Affine Plane of Order 3 | 98 |
| | 9.3 | Action of H on Subsets of Points of $\Pi$ | 99 |
| | 9.4 | The Kitten | 101 |
| | 9.5 | Recognizing Hexads in the Kitten | 102 |
| | 9.6 | Mnemonic for Reconstructing the Kitten | 103 |
| | 9.7 | Finding the Hexad Containing a Given 5 Points | 103 |
| | 9.8 | Useful Generators Using the Kitten | 104 |
| | 9.9 | A Class of Involutions in $M_{12}$ Using the Kitten | 105 |
| | 9.10 | $M_{12}$ as a Subgroup of $M_{24}$ | 105 |
| | 9.11 | Embedding the Kitten in the MOG | 106 |
| | 9.12 | The Outer Automorphism of $S_6$ | 108 |
| | 9.13 | More on the Outer Automorphism of $M_{12}$ | 114 |
| **10** | **The Leech Lattice** | | **117** |
| | 10.1 | Sphere-Packing and Lattices | 117 |
| | 10.2 | The Leech Lattice | 119 |
| **11** | **The Conway Group $\cdot O$** | | **131** |
| | 11.1 | The Conway Element $\xi_T$ | 132 |
| | 11.2 | Transitivity on Short Vectors | 133 |
| | 11.3 | The Order of $\cdot O$ | 135 |
| | 11.4 | Constructing $\cdot O$ Using MAGMA | 135 |
| | 11.5 | The Conway Group $Co_2$ | 138 |

|  |  |  |
|---|---|---|
| | 11.6 The Conway Group $Co_3$ | 143 |
| | 11.7 The Finite Simple Groups | 150 |
| | 11.8 The Classification of Finite Simple Groups | 163 |
| | 11.9 A Word about Moonshine | 163 |
| **12** | **Permutation Actions of $M_{24}$** | **165** |
| | 12.1 Background | 165 |
| | 12.2 Orbitals and Graphs | 169 |
| | 12.3 Graph Diagrams | 172 |
| | 12.4 Eigenvalues and Invariant Subspaces | 177 |
| **13** | **Natural Generators of the Mathieu Groups** | **187** |
| | 13.1 The Combinatorial Approach | 188 |
| | 13.2 The Mathieu Group $M_{12}$ | 200 |
| **14** | **Symmetric Generation Using $M_{24}$** | **209** |
| | 14.1 Introduction to Symmetric Generation | 209 |
| | 14.2 The Conway Group $\cdot O$ | 212 |
| | 14.3 The Janko Group $J_4$ | 222 |
| **15** | **The Thompson Chain of Subgroups of $Co_1$** | **234** |
| | 15.1 Introduction | 234 |
| | 15.2 The Progenitor | 235 |
| | 15.3 Possible Images of a Triangle | 236 |
| | 15.4 Extension to $K_n$ for Higher $n$ | 242 |
| | 15.5 The Progenitor $7^{\star \binom{n}{2}} : S_n$ | 245 |
| | 15.6 Identification of the Groups in the Chain | 246 |
| | 15.7 The Real Normed Division Algebras | 250 |
| | 15.8 The Group $K_n$ for $n > 7$ | 253 |
| | 15.9 Doubling Up | 254 |
| | 15.10 The Case $N \cong A_9$ | 255 |
| | 15.11 Properties of the Complete 9-Graph | 257 |
| | 15.12 Connection with the Leech Lattice | 258 |
| *Appendix* | MAGMA **Code for** $7^{\star 36} : A_9 \mapsto Co_1$ | 273 |
| | *References* | 276 |
| | *Index* | 281 |

# Figures

| | | |
|---|---|---|
| 2.1 | The Fano plane as 3-dimensional vectors over $\mathbb{Z}_2$ | *page* 8 |
| 2.2 | The Fano plane with numbered points and lines | 9 |
| 2.3 | The labelling of points and lines in terms of integers modulo 7 | 9 |
| 2.4 | The unique Steiner system S(3, 4, 8) | 10 |
| 2.5 | The lines of the 9-point affine plane | 11 |
| 2.6 | Intersection diagram for an S(5, 6, 12) | 14 |
| 2.7 | A Steiner system S(3, 4, 16) | 15 |
| 2.8 | Translational symmetries of this S(3, 4, 16) | 15 |
| 3.1 | Todd's triangle of octad intersections | 18 |
| 3.2 | The Miracle Octad Generator | 20 |
| 3.3 | Two sextets: the columns of the MOG and the rows of the complementary 16-ad | 21 |
| 3.4 | The $L_2(7)$ acting identically on the three bricks of the MOG | 23 |
| 3.5 | A mnemonic for recalling the standard MOG labelling | 24 |
| 3.6 | Some octads as they appear in the MOG | 25 |
| 4.1 | The Point space $\mathcal{P}$ and the Line space $\mathcal{L}$ | 30 |
| 4.2 | The Point and Line subspaces | 31 |
| 4.3 | The three bricks | 31 |
| 5.1 | Permutations preserving $S_\infty$, $S_0$ and $S_1$ | 39 |
| 6.1 | The odd and even interpretations of the hexacode | 44 |
| 7.1 | The even and odd lines of sextets | 47 |
| 7.2 | The fixed point free involution associated with the even line $\epsilon$ | 50 |
| 7.3 | Generators of the octern group, showing the eight terns | 52 |
| 7.4 | Orbit representatives of the 87 = 21 + 24 + 42 octads annihilated by Oct$_3$ | 53 |
| 7.5 | Pet$_3$, a purely sextet subspace with no even lines | 54 |
| 7.6 | Eight octads that define the S(5, 8, 24) and their extension to a basis of $\mathscr{C}$ | 54 |
| 7.7 | Pet$_4$, a purely sextet subspace with no even lines | 56 |
| 7.8 | The Pet$_4$ space octads | 57 |

## List of Figures

| | | |
|---|---|---|
| 7.9 | Constructing involutions of cycle shape $1^8.2^8$ | 59 |
| 7.10 | Constructing involutions of cycle shape $2^{12}$ | 60 |
| 7.11 | Constructing elements of cycle shape $1^6.3^6$ | 61 |
| 7.12 | The action of $M_{24}$ on $P(\Omega)$ | 62 |
| 7.13 | The canonical $U_8$ preserved by a subgroup of shape $2^4 : S_4$ | 64 |
| 7.14 | The canonical $T_{12}$ and its complement preserved by a subgroup of shape $(2^2 \times L_2(5)) : 2$ | 65 |
| 8.1 | The 21-point projective plane as it appears in the MOG | 69 |
| 8.2 | The correspondence between elements of $M_{21}$ and matrices of $SL_3(4)$ | 70 |
| 8.3 | The outer automorphisms of $M_{21}$ | 71 |
| 8.4 | Generators of the stabilizer of a sextet | 72 |
| 8.5 | The isomorphism $A_8 \cong L_4(2)$ | 74 |
| 8.6 | The stabilizer of an octad as matrices over $\mathbb{Z}_2$ | 75 |
| 8.7 | Generators for a Sylow 2-subgroup of $M_{24}$ as they appear in $L_5(2)$ | 76 |
| 8.8 | The elementary abelian $2^6$ in the trio group | 81 |
| 8.9 | The four umbral hexads $4 \times 6$ array | 87 |
| 8.10 | The dodecahedra defined on a duum by the blocks of type (vi) | 89 |
| 8.11 | The $3 \times 8$ brick arrangement | 90 |
| 8.12 | The blocks of size 3 in Case (v) | 91 |
| 8.13 | The pairing preserved by a subgroup isomorphic to $PGL_2(11)$ | 92 |
| 8.14 | An involution $\rho$ in $S_{24}$ but outside $M_{24}$ pairs of whose transpositions generate the Pet$_5$ subspace of $C^*$ | 94 |
| 8.15 | The canonical $T_{12}$ and its stabilizing subgroup | 96 |
| 9.1 | The 9-point affine plane | 98 |
| 9.2 | Elements generating the Hessian group $H \cong 3^2 : 2S_4$ | 99 |
| 9.3 | Three arrangements of the 9-point affine plane corresponding to $\infty$, 0 and 1 in the Kitten triangle | 100 |
| 9.4 | Quadrangles containing $\{2, 6, X\}$ | 100 |
| 9.5 | Crosses and squares: unions of two perpendicular lines in the 9-point plane and their complements | 101 |
| 9.6 | The Kitten – a device for recognizing the hexads of the Steiner system $S(5, 6, 12)$ | 102 |
| 9.7 | The hexads of the three examples as they appear in the Kitten | 104 |
| 9.8 | The canonical dodecad as the symmetric difference of two octads | 106 |
| 9.9 | Some octads having six points in the canonical dodecad | 106 |
| 9.10 | The non-permutation identical action of $S_6$ on two sets of size 6 | 107 |
| 9.11 | Embedding the Kitten in the MOG | 107 |
| 9.12 | Useful elements of $M_{12}$ in the MOG | 108 |
| 9.13 | The six synthematic totals | 108 |
| 9.14 | The isomorphism between $S_6$ and $P\Sigma L_2(9)$ | 110 |
| 9.15 | Permutations in $P\Gamma L_2(9)$ and Aut $A_6$ | 111 |

xii  *List of Figures*

| | | |
|---|---|---|
| 9.16 | Diagrams of the Hoffman–Singleton graph and elements of $M_{24}$ preserving it | 112 |
| 9.17 | Vertices of the Hoffman–Singleton graph in the MOG | 112 |
| 9.18 | The labelling of the vertices of the Hoffman–Singleton graph | 113 |
| 9.19 | The canonical duum and the non-permutation identical actions of $M_{12}$ on two sets of size 12 | 115 |
| 9.20 | The transitive copy of $L_2(11)$ in $M_{12}$ | 115 |
| 9.21 | A hexad in each dodecad of a duum, exhibiting the two Steiner systems $S(5, 6, 12)$ preserved by $L_2(11)$ | 116 |
| 10.1 | The square lattice in $\mathbb{R}^2$ illustrating the density of the associated sphere-packing | 118 |
| 10.2 | The hexagonal lattice in $\mathbb{R}^2$ illustrating the density of the associated sphere-packing | 119 |
| 10.3 | Orbits of vectors of types 2, 3 and 4 in $\Lambda$ under the action of $N$ | 123 |
| 10.4 | Calculating 4-vectors in involution type crosses | 128 |
| 11.1 | One of the 2300 hexagonal sublattices containing a given 2-vector $v$ | 139 |
| 11.2 | The action of $Co_2$ on the 2300 hexagonal lattices | 140 |
| 11.3 | The permutation $\pi$ | 141 |
| 11.4 | An element outside $M_{23}$ fixing our 3-element | 145 |
| 11.5 | Generators for the centralizer in $M_{23}$ of $\eta$ | 147 |
| 11.6 | The Dynkin diagrams | 160 |
| 12.1 | The class list for $M_{24}$ showing some permutation characters | 168 |
| 12.2 | The action of some maximal subgroups on octads | 173 |
| 12.3 | The action of some maximal subgroups on dua | 175 |
| 12.4 | The action of some maximal subgroups on sextets | 176 |
| 12.5 | The action of some maximal subgroups on trios | 177 |
| 13.1 | Generators for $M_{24}$ on the faces of the Klein map | 193 |
| 13.2 | A map requiring seven colours on a surface of genus 1 | 196 |
| 13.3 | Generators for $M_{12} : 2$ on the faces of two dodecahedra | 206 |
| 13.4 | The alternative colouring of the vertices of the two dodecahedra | 206 |
| 13.5 | Correspondence between the combinatorial and geometric approaches | 207 |
| 14.1 | Symmetric generators for the Conway group $\cdot O$ | 212 |
| 14.2 | Elements of $K$, the central element $\nu$ and an element $\rho$ commuting with $\nu$ | 213 |
| 14.3 | Some elements $\nu_{xyz}$ | 214 |
| 14.4 | Verification that the relation of Theorem 14.3 holds in $\cdot O$ | 215 |
| 14.5 | Octads demonstrating that reflection on a dodecad is well defined | 218 |
| 14.6 | The intersection matrices for two trios | 222 |
| 14.7 | The intersection matrices for two sextets | 223 |
| 14.8 | The five trios yielding the defining relation for the Janko group $J_4$ | 224 |

*List of Figures*

| | | |
|---|---|---|
| 14.9 | Generators for the subgroup of $M_{24}$ fixing two trios that intersect one another as in matrix $I_3$ | 225 |
| 14.10 | Involutions used to obtain the $J_4$ relations | 230 |
| 15.1 | The brick labelling of $\Omega$ | 259 |
| 15.2 | Elements of $\cdot O$ commuting with $z$ | 260 |
| 15.3 | Generators for the intersection of $Co_2$ with $2^{12} : M_{24}$ | 272 |

# Tables

| | | |
|---|---|---|
| 4.1 | The weights of codewords in $\mathscr{C}$ | page 33 |
| 10.1 | The orbits of $2^{12}{:}M_{24}$ on crosses | 125 |
| 11.1 | Calculation of 2-vectors having i.p. 16 with a fixed 2-vector | 138 |
| 11.2 | The 2300 sublattices permuted by $Co_2$ | 139 |
| 12.1 | Orbits of some maximal subgroups of $M_{24}$ acting on the cosets of one another | 174 |
| 13.1 | Labelling of the 24 7-cycles in an orbit of $L_3(2)$ with the points of $\Omega$ | 190 |
| 13.2 | The seven involutory generators for $M_{24}$ acting on 7-cycles in $L_3(2)$ | 190 |
| 13.3 | Labelling the 24 5-cycles with elements of $P_1(11)$ | 201 |
| 13.4 | Action of the five symmetric generators on $\Lambda \cup \bar{\Lambda}$ | 202 |
| 15.1 | Candidates for the $T_{ij}$ forming $t_{89}$, stored in **tt89ss** | 265 |

# Preface

In 1968, John Conway took me on as his research student, having previously been my undergraduate director of studies from 1964 to 1967. The first project he assigned me was to classify all the finite subloops of the Cayley algebra which is nowadays referred to as the set of *octonions*; they are known as *Moufang loops* after the geometer Ruth Moufang as they satisfy the Moufang identities and are mentioned briefly here in Chapter 15 in connection with the group $G_2(4)$. In recent years new sporadic groups had been emerging in various unrelated ways. Conway hoped that I would find a new loop, such that the permutation group generated by the right and left multiplication of the elements of this loop would be yet another new simple group. Sadly I proved that the list of known finite subloops of the Cayley algebra was complete. However, I was able to produce a way of constructing Moufang loops from any group that possesses an element in its centre that squares to the identity. This element could, of course, itself be the identity; see Curtis (2007a).

Conway had recently worked out the symmetries of the Leech lattice $\Lambda$ and discovered the three groups named after him, and so it was essential that I myself become familiar with $\Lambda$. Todd, whose lectures I had attended the previous year during Part III of the Tripos, was in the process of retiring and rarely came into the department. So Conway took me down to Todd's office and liberated a copy of his famous paper on $M_{24}$; see Todd (1966), which lists the 759 octads of $S$, the Steiner system S(5, 8, 24), at length over five pages. At that time working on the Conway group often entailed finding an octad with certain properties, so people would spend ages searching through the list. Having reached the bottom of the fifth page without finding the one they needed, they would simply start again. The frustration caused by this tedious process persuaded me to look for an arrangement of the 24 points in which the octads would be immediately recognizable, so I obtained sheets of graph paper and began to experiment. It was whilst enjoying a beer in a Cambridge pub called *The Cricketers* and at the same time playing around with diagrams on the graph paper that I realized that the patterns I was finding could be given a sixfold symmetry, which meant that just 35 diagrams would encapsulate all 759 of the octads.

The next day I presented these diagrams to Conway who immediately saw how this would revolutionize the way in which we worked in $\Lambda$. He and I stayed up late in the department drawing a definitive version. We needed a glass-topped table on which to place an inverted sheet of graph paper. With a light underneath the lines would guide our work but would not show up on the final version. Lacking such a table, we removed a sliding window from the librarian's office and placed it between two chairs; we then put an Anglepoise lamp in a litter bin underneath it. This worked perfectly, and we had made good progress when Richard Guy, who was visiting Cambridge at the time, came into the common room and suggested that we finish the job in his guest room in Gonville and Caius College, having picked up some wine on the way. This we were happy to do, and we finally got the job done in the early hours of the morning.

The next day I cycled around to a printer and asked him to make 200 copies of our original, shrunk down to a quarter of its size and printed on card. This reduction removed blemishes due to our shaky hands and general inebriation, and the end product looked quite professional. We discussed at great length what we should call the device and eventually came up with the *Miracle Octad Generator* or MOG, although we knew even at that stage that it could be used for far more than just finding octads. Indeed, it is in many ways a diagram of the Steiner system and the binary Golay code, and can thus be used to work out and exhibit elements of $M_{24}$ and $\cdot O$. Conway and I became very proficient at using the MOG, but we often wished we had a mnemonic that would obviate the need to carry the card with you; such a tool was provided when David Benson was working on the largest Janko group $J_4$ and made extensive use of the MOG; see Benson (1980). He and Conway together introduced the *hexacode*, see Chapter 6, which provided algebraic respectability for the MOG and simplified several proofs.

This book traces the manner in which the techniques afforded by the MOG were used to prove various important facts about $M_{24}$ and $\cdot O$, and how, more recently, familiarity with $M_{24}$ led to the notion of *symmetric generation* of groups, which has provided pleasing constructions of many of the other sporadic simple groups; see Curtis (2007b).

Chapter 1 contains a brief discussion of various ways in which $M_{24}$ may be constructed, and then in Chapter 2, we give a gentle introduction to the notion of Steiner systems, concentrating on those that occur as subsystems in the $S(5, 8, 24)$. In Chapter 3, we proceed to $S$ itself and explain how the MOG may

## Preface

be used to find an octad containing a given five points. Chapter 4 describes how the binary Golay code $\mathscr{C}$ is obtained from the Steiner system $S$ and introduces the dual space $\mathscr{C}^* \cong P(\Omega)/\mathscr{C}$. In Chapter 5, we prove that the octads exhibited in the MOG do indeed constitute a Steiner system S(5, 8, 24) and that $S$ is unique up to relabelling; we also use the MOG to calculate the order of the group of symmetries of $S$. These proofs appear in the author's PhD thesis; see Curtis (1972). The hexacode is introduced in Chapter 6 and used to count again the number of octads in the binary Golay code. In Chapter 7, we show how the MOG can be used to write down various elements of $M_{24}$ of cycle shapes $1^8.2^8$, $2^{12}$ and $1^6.3^6$ having certain designated properties. The orbits of $M_{24}$ on subsets of $\Omega$ were worked out by Todd and displayed by Conway in a useful diagram that has been reproduced here. Our ability to work in the dual space $P(\Omega)/\mathscr{C}$ is used to produce a minimal possible test to check whether a permutation in $S_{24}$ lies in our favoured copy of $M_{24}$.

In Chapter 8, we describe in some detail each of the maximal subgroups of $M_{24}$ and proceed to prove that this list of nine conjugacy classes of subgroups is indeed complete. This list consists of the eight maximal subgroups described in Todd (1966), supplemented by the maximal *octern* group that is isomorphic to $L_2(7)$ and which was discovered later by McKay and Wales. This short proof leans heavily on the techniques afforded by the MOG. Two further non-maximal but transitive subgroups, $Pet_5$ and $T_{12}$, are described.

Chapter 9 deals with the Mathieu group $M_{12}$ and introduces the so-called *Kitten* that plays the same role for $M_{12}$ as the MOG plays for $M_{24}$. We show how the Kitten may be embedded in the MOG and illustrate the outer automorphism of the symmetric group $S_6$ and the isomorphism Aut $S_6 \cong P\Gamma L_2(9)$.

In Chapter 10, we introduce the Leech lattice $\Lambda$ and explain how the elements of the factor space $\Lambda/2\Lambda$ correspond to the short vectors of $\Lambda$. Special attention is given to the *frames of reference* or *crosses* that are type 4-vectors modulo $2\Lambda$, which is to say sets of 24 mutually orthogonal 4-vectors and their negatives.

In Chapter 11, we describe explicitly the Conway element $\xi_T$ that Conway produced in order to demonstrate that the Leech lattice had far more symmetries than those of $M_{24}$ and the binary Golay code $\mathscr{C}$ that was used in its construction. The order of ·O is calculated, and the maximal sporadic simple groups $Co_2$ and $Co_3$ are described. The chapter ends with a brief description of how these groups fit into the family of all finite simple groups.

Chapter 12 gives an account of the connection between permutation representations of a group and characters. It applies this theory to the primitive actions of $M_{24}$ on octads, dua, trios and sextets and gives detailed information about the manner in which the stabilizer of a point in one action acts on the others.

In Chapter 13, we describe how familiarity with $M_{24}$ led to my finding that permutation generators for it can essentially be read off from $L_2(7)$, the second smallest (non-abelian) finite simple group, acting on 24 letters. The construction of these generators is given a combinatorial, geometric and algebraic interpretation. In particular, we show how these generators appear naturally in the Klein map. An analogous investigation is carried out for $M_{12}$ in which $L_2(7)$ is replaced by $L_2(5)$, and the Klein map by the regular dodecahedron.

Chapter 13 was the motivation for the concept of *symmetric generation of groups* as explored in Curtis (2007b). In these two cases, the *control groups* are $L_2(7)$ and $L_2(5)$ acting on 7 symmetric generators of order two and 5 symmetric generators of order three, respectively. In Chapter 14, we describe two examples of symmetric generation, namely the Conway group $\cdot$O and the Janko group $J_4$, in which $M_{24}$ itself is playing the role of the control subgroup.

The process culminates in Chapter 15 where we consider symmetric generators of order 7 corresponding to the edges of a complete graph on $n$ vertices. The control subgroup is the symmetric group $S_n$ permuting the vertices of the complete graph, a direct product with an element of order 3 that squares each of the symmetric generators. Consideration of what can be generated by a triangle of generators leads us to imposing a relation, $R = 1$, which forces it to be the unitary group $U_3(3)$. It then follows spontaneously that complete graphs with 4, 5, 6 and 7 vertices result in the Thompson chain of subgroups of $\cdot$O. If we reduce the control subgroup to the alternating group $A_n$ direct product with the cyclic group of order 3, then the process continues to a complete 9-graph, and we have

$$\frac{7^{\star \binom{9}{2}} : (A_9 \times 3)}{R = 1} \cong \mathrm{Co}_1.$$

In this book, we make extensive use of the computer algebra package MAGMA of which John Cannon was the initiator and driving force, and which has proved an invaluable tool for verifying claims made in the text. The MAGMA code is generally self-explanatory, but where this is not the case, a supplement to the code is provided.

The character table of a finite group contains an extraordinary amount of information about the group in a remarkably concise fashion. It is essentially a square $n \times n$ matrix, where $n$ is the number of conjugacy classes the group possesses, which equals the number of (inequivalent) irreducible complex representations. Thus, for instance, although $M_{24}$ has over 200 million elements, its character table is just a $26 \times 26$ matrix; and the Monster group M, whose order is around $8 \times 10^{53}$, has fewer than 200 irreducible characters. Indeed, when we were working on the ATLAS of Finite Groups in the early days, the phrase 'let's have a look at...' a certain group meant let's have a look at its character table. I regret that it has not been possible to include character tables in this book, but I would encourage the reader to have access to a copy of the ATLAS, see Conway et al. (1985), alongside this volume so that they can see how various claims are reflected in the tables. It is worth noting that Todd's paper, see Todd (1966), contains character tables of several subgroups of $M_{24}$.

Similarly, the description of all finite simple groups given here is rudimentary and concentrates, for the most part, on those groups that play a significant role in the text. The reader who wishes to know more about the infinite families and the various sporadic finite simple groups will find Wilson (2009a) immensely useful.

# Acknowledgements

The wonderful work of many mathematicians has, directly or indirectly, formed the foundation for this book; unfortunately, I can only mention a few of them here.

Firstly, of course, is Mathieu who discovered those truly remarkable groups that are named after him and which are at the heart of the present volume. Then comes Janko who in 1965 discovered a new simple group that did not fit into any of the known infinite families, and who thus launched the most exciting period in modern group theory, leading eventually to the classification of all finite simple groups. Pleasingly, he also discovered the last sporadic simple group which plays an important role here through its close connection to $M_{24}$. It was in 1964 that Leech exploited the combinatorial structure underlying $M_{24}$ to construct his famous lattice. He also deserves great credit for his contribution to the restoration of the paddle steamer Waverley on which he sadly but fittingly died in 1992. In 1967, Conway worked out the group of symmetries of the Leech lattice in a beautiful piece of work whose elegance was to become his trademark. Todd's list of the 759 octads of the Steiner system was invaluable in enabling me to construct the Miracle Octad Generator, which was to prove so useful in working with $M_{24}$, the Conway group, $J_4$ and many other groups.

The reader of this book will find it useful to have access to the *Atlas of Finite Groups* on which John Conway and I started work in 1976. I am immensely grateful to our co-authors, Simon Norton (1952–2019), Richard Parker (1953–2024) and Robert Wilson, for extending and completing this demanding project. I should particularly like to thank Wilson for writing his monumental book *The Finite Simple Groups*, which is a natural follow-up for any reader of this book who wishes to know more about these fascinating structures. Along with many mathematicians, I am indebted to John Cannon and his team for initiating and developing the computer program MAGMA which is used extensively here. I should also like to thank John Bray for implementing in MAGMA and extending the double-coset enumerator for symmetrically generated groups which I had developed for hand calculations. This, together with Bray's extension, proved invaluable in working with $\cdot$O and $J_4$ in Chapter 14.

I should like to thank David Tranah for encouraging me to write this book, and for urging me to extend its scope. I should also like to thank the anonymous referees for their positive comments and support. Anna Scriven and Clare Dennison were very helpful in the early stages and, latterly, Sunantha Ramamoorthy and Preethi Sekar have been immensely supportive as we have worked together to produce the final version.

# 1

# Introduction

## 1.1 Motivation

The Mathieu group $M_{24}$ together with its subgroups $M_{23}, M_{22}, M_{12}$ and $M_{11}$ are arguably the most famous, most studied and most beautiful groups that exist. Indeed $M_{24}$ and $M_{12}$ acting with degrees 24 and 12 respectively are the only quintuply transitive permutation groups other than the alternating and symmetric groups, and similarly their point stabilizing subgroups $M_{23}$ and $M_{11}$ are the only quadruply transitive groups. A permutation group $G$ acting on a set $X$ is said to be $n$-transitive if, and only if, given $n$ distinct points of $X$, $\{x_1, x_2, \ldots, x_n\}$ say, and any other $n$ distinct points $\{y_1, y_2, \ldots, y_n\}$, there is a permutation $\pi \in G$ such that $\pi(x_i) = y_i$ for $i = 1, \ldots, n$. In the case of $M_{12}$ with $n = 5$ this element $\pi$ is unique and we say that $M_{12}$ is *sharply* 5-transitive; similarly $M_{11}$ is sharply 4-transitive. This implies that their orders are given by

$$|M_{12}| = 12.11.10.9.8 = 95\,040 \text{ and } |M_{11}| = 11.10.9.8 = 7920.$$

The order of $M_{24}$ has the form

$$|M_{24}| = 24.23.22.21.20.|H|,$$

where $H$ denotes the stabilizer in $M_{24}$ of five points; it emerges naturally in our proof of the uniqueness of the Steiner system $S(5, 8, 24)$ in Chapter 3. The fact that the Mathieu groups are the only quintuply and quadruply transitive groups, other than the alternating and symmetric groups, is now known to hold as a consequence of the Classification of Finite Simple Groups, see Section 11.8. However, up to now no direct proof of this remarkable fact has been found. They were discovered by Emil Mathieu and their existence was announced in two papers (Mathieu, 1861, 1873). Not only are these groups of huge interest in their own right, but they are involved in many of the other sporadic simple groups and play important roles in coding theory, sphere packing and other combinatorial structures. They arise most naturally as groups of permutations preserving certain Steiner systems, see Chapter 2 of this book, in particular $S(5, 8, 24)$ and $S(5, 6, 12)$ and their subsystems, and it is through these block designs that they are best studied.

## 1.2 Other Constructions

As befits mathematical structures of such beauty and importance, the Mathieu groups have been constructed in many different ways. Before proceeding to the approach that is the subject of this book, I shall give a brief description of some, but far from all, of these alternatives, which are fascinating in their diversity.

Mathieu himself in the aforementioned papers constructed the groups by 'gluing together' copies of the projective special linear groups $L_2(11)$ and $L_2(23)$ acting on 12 and 24 points, respectively. Indeed, he believed his construction could be carried out for primes larger than 11 and 23. In fact, numerologically speaking $n = 48$ has much in common with 24 and 12: $n - 1$ is prime, $n - 2$ is twice a prime and $n - 5$ is also prime. But sadly $M_{48}$ does not exist! Of course Mathieu knew that the order of $M_{12}$ divides the order of $M_{24}$ but he was not aware that it is in fact a subgroup.

Witt's approach, see Witt (1938a,b), was to start with a well-known Steiner system $S(l, m, n)$ and build successive transitive extensions $S(l+1, m+1, n+1)$, $S(l+2, m+2, n+2), \ldots$. Thus he started with the projective plane of order 4, which is a Steiner system $S(2, 5, 21)$, and showed that it can be extended to an $S(3, 6, 22)$. He then proved that this new system had a triply transitive group of automorphisms, which is in fact a group of shape $M_{22} : 2$, the simple Mathieu group $M_{22}$ extended by an outer automorphism of order 2. This process could then be repeated to form an $S(4, 7, 23)$ preserved by the quadruply transitive simple Mathieu group $M_{23}$. Finally he extended this system to an $S(5, 8, 24)$ that has the magnificent, quintuply transitive simple group $M_{24}$ as its group of symmetries. This Steiner system cannot be extended to an $S(6, 9, 25)$ as the number of *nonads* in such a system would be

$$\binom{25}{6} / \binom{9}{6},$$

which is not an integer.

Todd's approach also focused on the Steiner system $S(5, 8, 24)$ and his paper (Todd, 1966) actually lists the 759 octads, but it is an alternative method that

## 1.2 Other Constructions

he lectured on in Cambridge in the late 1960s that I wish to mention here.[1] The symmetric group $S_6$ is exceptional in that it has an outer automorphism of order 2. If we think of $G \cong S_6$ as permuting the six points of the projective line $\{\infty, 0, \ldots, 4\}$ then the copy of $S_5$ fixing $\infty$ clearly has index 6 in $G$, but so does the transitive subgroup $H \cong \mathrm{PGL}_2(5)$, which also has index 6. Thus $G$ acts in a non-permutation identical way on two different sets of size 6. These two sets are interchanged by the outer automorphism. In Sylvester's terminology a partition of six letters into three pairs is known as a *syntheme* and a set of five synthemes that includes all 15 unordered pairs is a synthematic total or simply a *total*. It is easy to see that there are just six possible totals, one of which is

$$\infty/01234 \sim \begin{array}{c} \infty 0.14.23 \\ \infty 1.20.34 \\ \infty 2.31.40 \\ \infty 3.42.01 \\ \infty 4.03.12 \end{array}.$$

It is readily checked that this total is preserved by

$$\langle x \mapsto x+1 \cong (\infty)(0\ 1\ 2\ 3\ 4), x \mapsto 2/x \cong (\infty\ 0)(1\ 2)(3\ 4)\rangle \cong \mathrm{PGL}_2(5)$$

and so has just six images under the action of $S_6$. Todd demonstrated longhand that a transposition on the six points induces a permutation of cycle shape $2^3$ on the totals, and vice versa, and similarly a 3-cycle on one side has cycle shape $3^2$ on the other. These conjugacy classes of shapes $1^4.2/2^3$ and $1^3.3/3^2$ are then used to define the 132 hexads of a Steiner system $S(5, 6, 12)$ on the $6+6$ points and totals, namely: $90 = (\binom{6}{2} \times 3) \times 2$ hexads consisting of four fixed points on one side and a corresponding transposition on the other; $40 = \binom{6}{3} \times 2$ corresponding to three fixed points on one side and a corresponding 3-cycle on the other side; together with the set of six points and the set of six totals. Thus giving $90 + 40 + 2 = 132$. The automorphism group of this Steiner system is $M_{12}$ which in turn is shown to act non-permutation identically on two sets of size 12 and, in an analogous manner, the actions on the two 12s are used to define a Steiner system $S(5, 8, 24)$ on the $12+12$ points. The automorphism group of this system is, of course, $M_{24}$.

In his Three Lectures on Exceptional Groups (see Conway, 1971 or Conway and Sloane, 1988, Chapter 10) Conway directly constructs $M_{24}$ by extending $\mathrm{PSL}_2(23)$ acting on the 24-point projective line. He adjoins a permutation

$$\delta : x \mapsto x^3/9 \ (x \in Q) \quad \text{and} \quad x \mapsto 9x^3 \ (x \in N),$$

---

[1] Graham Higman was himself describing this method in lectures in Oxford at around the same time.

where $Q$ denotes the quadratic residues modulo 23 and $N$ denotes the non-residues, and proceeds to deduce that the resulting group has the familiar properties.

In Curtis (1989, 1990) the current author produces five elements of order 3 permuting the 12 pentagonal faces of a dodecahedron that together generate $M_{12}$, and seven involutions permuting the 24 heptagonal faces of the genus 3 Klein map that generate $M_{24}$. Algebraic and combinatorial explanations for these generators are also given. The details of these constructions are given in Chapter 13 of this book.

## 1.3 The Construction of This Book

The main properties of $M_{24}$ may be deduced from each of the above constructions, and from many others not mentioned here. However, none of them helps us to actually work within the group itself, to recognize when a permutation on 24 letters is in our chosen copy of $M_{24}$, or to write down a permutation of the group having certain desirable properties. Of course modern algebra packages such as GAP and MAGMA are wonderfully efficient for working with permutation groups of such low degree, and it may seem indulgent to develop techniques that are only relevant to a particular small family of groups. However, I would claim that the Mathieu groups are so exceptional, as has been demonstrated earlier, and so intimately involved in other mathematical structures that a dedicated theory is justified. Moreover, it is our contention that a deeper understanding of the intricacies and sheer beauty of these remarkable structures is afforded by the approach described later.

## 1.4 The Centrality of $M_{24}$

Besides being an extraordinary structure in its own right, the group $M_{24}$ plays a central role within the sporadic simple groups and hence within wider mathematics. Firstly, of course, the Conway group ·O that Conway himself often called '$M_{24}$ writ large', grows out of $M_{24}$ and the binary Golay code $\mathscr{C}$ by way of the Leech lattice $\Lambda$. In his book entitled *Twelve Sporadic Groups* Robert Griess (1998) describes the sporadic groups that are visible within ·O as the 'first generation'. Due to their connection to $\Lambda$, see Conway et al. (1982), in working with these groups many people have found MOG techniques useful.

The involvement of $M_{24}$ does not stop with the first generation though, as the largest Fischer group $Fi_{24}$ contains maximal subgroups of shape $2^{11}$. $M_{24}$

## 1.4 The Centrality of $M_{24}$

in an analogous way to $Co_1$. However, in $Co_1$ the elementary abelian group of order $2^{11}$ is isomorphic to $\mathscr{C}$ factored by the all 1s vector and the extension is a semidirect product, that is to say it is a *split* extension as is indicated by the colon in $2^{11} : M_{24}$. In the Fischer group the elementary abelian group is isomorphic to the even part of the dual code $\mathscr{C}^*$, see Chapter 4, and the extension is *non-split*, which is indicated by the 'upper dot' in $2^{11\cdot}M_{24}$. So in the latter case this affine subgroup contains no copy of $M_{24}$; however, the techniques of this book have still proved useful, see Conway (1973) and Rowley and Walker (2012, 2021). Moreover, the Conway group $Co_1$ is involved in the Monster group M, since the centralizer of an involution of ATLAS class $2B$ has shape $2_+^{1+24}.Co_1$. Accordingly, they have made extensive use of the MOG in their investigations of both M and the Baby Monster B, see Rowley (2005); Rowley and Walker (2004a,b).

The last sporadic simple group to be discovered was the Janko group $J_4$ and it too contains a subgroup of shape $2^{11} : M_{24}$, see Section 14.3. Indeed, it was intensive use of the MOG by Benson and others, see Benson (1980), in their work on $J_4$ that led to the hexacode, see Chapter 6.

Besides its pivotal role in finite groups, $M_{24}$ and the underlying combinatorial structures crop up in unexpected places. For instance, in Berlekamp et al. (1982, page 436) the Miracle Octad Generator is reproduced in connection with the game *Mogul*.

A deep connection between the algebraic structures dealt with here and number theory was discovered when John McKay noticed some intriguing numerological coincidences that are explained briefly in Section 11.9; the resulting investigations were christened *Monstrous Moonshine* by Conway; see Conway and Norton (1979).

These structures and the techniques for working with them have recently become of great interest to Theoretical Particle Physicists working in String Theory. In Monstrous Moonshine the degrees of the irreducible complex representations of the Monster group M are related to the modular function as explained in Section 11.9; here it is those of $M_{24}$ that are related to an object they call a 'mock modular form', see Taormina and Wendland (2013, 2015a,b).

# 2

# Steiner Systems

## 2.1 Definition and Numerical Conditions

A Steiner system $S(l, m, n)$ is a collection of $m$-element subsets of an $n$-element set $\Omega$ with the property that any $l$-element subset is contained in precisely one of them. The $m$-element subsets are the *blocks* of the system. So each $l$-element subset of $\Omega$ determines a block that in turn is determined by each of its own $l$-element subsets. Thus the number of blocks in the system is given by

$$\binom{n}{l} \bigg/ \binom{m}{l},$$

which must, of course, be an integer. Given a Steiner system $S(l, m, n)$, if we remove one of the $n$ points of $\Omega$, $x$ say, and consider the $(m-1)$-element subsets obtained by removing $x$ from those blocks that contain it, we are left with a Steiner system $S(l-1, m-1, n-1)$ on $\Omega \setminus \{x\}$, which will have

$$\binom{n-1}{l-1} \bigg/ \binom{m-1}{l-1}$$

blocks of size $m - 1$. In this manner, we obtain a series of numerical conditions on the parameters $l, m, n$ and eventually arrive at a rather trivial Steiner system $S(1, m - l + 1, n - l + 1)$, which is simply a partition of the remaining $n - l + 1$ points into blocks of size $m - l + 1$. In particular, $m - l + 1$ must divide $n - l + 1$.

On the other hand, it may be possible to adjoin a point $y$ to $\Omega$ to obtain a Steiner system $S(l + 1, m + 1, n + 1)$ on $\Omega \cup \{y\}$ some of whose blocks are the subsets of form $B \cup \{y\}$ for $B$ a block of the original system. Note that such an *extension* is certainly not possible unless

$$\binom{n+1}{l+1} \bigg/ \binom{m+1}{l+1}$$

is an integer.

A Steiner system $S(2, 3, n)$ is referred to as a *Steiner triple system* or STS. It is clear that for an STS to exist we must have $n - 1$ even and that, since the number of triples will be

$$\binom{n}{2} \Big/ \binom{3}{2},$$

we must have $n \equiv 0$ or 1 modulo 3. So $n$ must be congruent to 1 or 3 modulo 6. It is known that this condition is both necessary and sufficient and an elegant proof that at least one STS $S(2, 3, n)$ exists whenever this congruency condition is satisfied is presented in Cameron (1994). The Steiner systems that are at the core of this book are $S(5, 6, 12)$ and $S(5, 8, 24)$; however, in order to get a feeling for these intriguing structures, we shall investigate some of the smaller systems in some detail. These will reappear frequently in what follows.

## 2.2 The Fano Plane

From Section 2.1, we see that the smallest $n$ for which there will exist an STS is $n = 7$, and it is an easy exercise to show that, up to relabelling the $n$ points, there is a unique such system; it is the famous *Fano plane* whose points are the seven non-zero vectors of a 3-dimensional vector space over, $\mathbb{Z}_2$, the integers modulo 2, and whose blocks or *lines* are the seven 2-dimensional subspaces. If the points are denoted by vectors $(x, y, z)$, with not all $x, y, z$ being zero, then the lines are all given by equations of the form $ax + by + cz = 0$ as shown in Figure 2.1. Note that the 'circular' line corresponds to the subspace defined by $x + y + z = 0$.

It is clear from the the definition of the Fano plane in terms of subspaces of a vector space $V$ that all automorphisms of $V$ preserve the Steiner system. That is to say, they permute the points (the vectors) whilst preserving the lines (the 2-dimensional subspaces). Thus the system is preserved by the linear group $L_3(2)$ of order $(2^3 - 1)(2^3 - 2)(2^3 - 2^2) = 168$, consisting of all non-singular $3 \times 3$ matrices over $\mathbb{Z}_2$, see Section 11.7.2. However, it is more convenient when working with permutations to label the points and lines with single numbers as shown in Figure 2.2, where the quadratic residues in $\mathbb{Z}_7$, the integers modulo 7, namely $\{1, 2, 4\}$, together with cycles of this set may be taken as the lines. Thus the labelling of the lines becomes the one that is shown in Figure 2.3.

The Fano plane is the smallest member of the family of classical projective planes: Take a field $\mathbb{F}$ of order $q = p^m$, see Section 11.7.2, for some prime number $p$, and let $V$ be a 3-dimensional vector space over $\mathbb{F}$; then $V$ possesses $(q^3 - 1)/(q - 1) = q^2 + q + 1$ 1-dimensional subspaces that we designate

8                         Steiner Systems

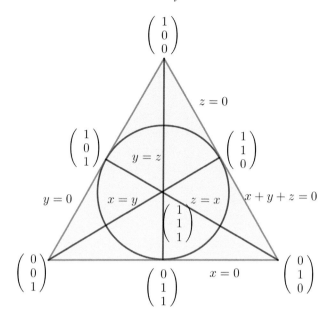

Figure 2.1 The Fano plane as 3-dimensional vectors over $\mathbb{Z}_2$

the *points* of the plane, and $(q^3 - 1)(q^3 - q)/(q^2 - 1)(q^2 - q) = q^2 + q + 1$ 2-dimensional subspaces that form the *lines*. Then any two distinct points lie together on a unique line. Moreover, since $\dim(U_1 + U_2) = \dim U_1 + \dim U_2 - \dim(U_1 \cap U_2)$, we see that any two distinct lines intersect at a unique point. Thus the lines form the blocks of a Steiner system $S(2, q+1, q^2+q+1)$. Indeed, we see that if we interchange the roles of points and lines, then we obtain a *dual* Steiner system with the same parameters. The Fano plane, of course, corresponds to the case when $q = 2$.

Note that the permutation (0 1 2 3 4 5 6) of the points preserves the set of lines and permutes them as (6 5 4 3 2 1 0). The permutation (2 4)(5 6) on points similarly has action (2 4)(5 6) on the lines. Together these two permutations generate the group $L_3(2)$ of order 168 consisting of all permutations of the points that preserve the lines. Incidentally, the permutation

$$(0\ 0)(0\ 1)(2\ 2)(3\ 3)(4\ 4)(5\ 5)(6\ 6)$$

interchanges points and lines whilst preserving point–line incidence, thus producing an outer automorphism of $L_3(2)$ and exhibiting the point–line duality.

The necessary numerical condition for there to exist an extension to a Steiner

## 2.2 The Fano Plane

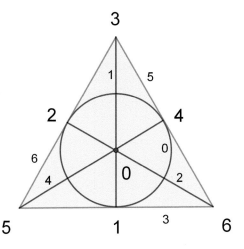

Figure 2.2 The Fano plane with numbered points and lines

$$\begin{array}{ccccccc} 124 & 235 & 346 & 450 & 561 & 602 & 013 \\ 0 & 6 & 5 & 4 & 3 & 2 & 1 \end{array}$$

Figure 2.3 The labelling of points and lines in terms of integers modulo 7

system S(3, 4, 8), namely that

$$\binom{4}{3} \text{ divides } \binom{8}{3},$$

holds. In fact, there must be

$$\binom{8}{3} \Big/ \binom{4}{3} = 14$$

special tetrads, of which we already have seven. Suppose that we take $\Omega = \mathbb{Z}_7 \cup \{\infty\}$ as our 8-element set; then the seven tetrads containing $\infty$ are known, namely $\infty$ together with one of the lines from Figure 2.2. We shall show that the complements of these tetrads must also be in the system. For, in an S(3, 4, 8), a pair of points would lie together in precisely $(8 - 2)/2 = 3$ tetrads; so, given a tetrad $T = \{x_1, x_2, x_3, x_4\}$, there are $\binom{4}{2} \times 2 = 12$ tetrads intersecting $T$ in two points. There is also the tetrad $T$ itself and so these account for all $3 \times 2 + 1 = 7$ tetrads containing, $x_i$ say. Thus the remaining tetrad must be disjoint from $T$. So the S(3, 4, 8) is uniquely defined and contains the tetrads shown in Figure 2.4. In order to verify that this is indeed a Steiner system

```
∞124    0365
∞235    1406
∞346    2510
∞450    3621
∞561    4032
∞602    5143
∞013    6254
```

Figure 2.4 The unique Steiner system S(3, 4, 8)

S(3, 4, 8), we make use of the group preserving it. First recall that the group $L_3(2)$ acts doubly transitively on the seven vectors, and then observe further that the permutation $e = (\infty\ 0)(1\ 3)(2\ 6)(4\ 5)$ preserves the set of 14 tetrads in Figure 2.4, either fixing or interchanging every pair of complementary tetrads. So the group preserving this set of tetrads acts transitively, and indeed triply transitively, on the eight points with order $8 \times 168 = 1344$. The seven images of $e$ under conjugation by elements of $L_3(2)$ form an elementary abelian normal subgroup of order 8 fixing the seven pairs of a tetrad and its complement, whilst the $L_3(2)$ permutes the pairs. Thus the group has shape

$$2^3 : L_3(2).$$

But since it acts triply transitively, every triple of points must lie in the same number of tetrads. The subset $\{\infty, 0, 1\}$ lies in $\{\infty, 0, 1, 3\}$ and no other tetrad, and so the tetrads shown in Figure 2.4 do indeed form an S(3, 4, 8).

This system cannot be extended to an S(4, 5, 9) as 5 does not divide $\binom{9}{4}$.

## 2.3 The 9-Point Affine Plane and Its Extensions

The next smallest $n$ for which an STS S(2,3,$n$) can exist is 9, and a short argument shows that, as with $n = 7$, such a Steiner system is unique:

**Lemma 2.1** *Up to relabelling, there is a unique Steiner system S(2, 3, 9).*

*Proof* Suppose the nine points are $a, b, \ldots, i$, and, without loss of generality, suppose that the pairs that together with $a$ make triples of the system are

$$a : bc,\ de,\ fg,\ hi.$$

Then we may assume that $bdf$ is a triple when *beg cannot* be a triple or we should be left with $bhi$, giving two triples containing $hi$. Thus we may assume

## 2.3 The 9-Point Affine Plane and Its Extensions

that the triples containing $b$ are given by

$$b : ca, df, eh, gi.$$

Now consider triples containing $d$. We have seen $dae$ and $dbf$; but $hi$ and $gi$ have already occurred together, so we must have $dci$ as a triple. So we have

$$d : ae, bf, ci, gh.$$

We now have three triples containing $g$, and so $gce$ has to be a triple. This in turn gives three triples containing $c$, and so $cfh$ is forced to be a triple. All other triples readily follow, and we see that, if we write the nine points as a $3 \times 3$ array, then the $\binom{9}{2}/\binom{3}{2} = 12$ triples of the system will be the rows, columns and generalized diagonals of a noughts and crosses or a tic-tac-toe board, see Figure 2.5:

| a | b | c |
|---|---|---|
| d | g | h |
| e | i | f |

□

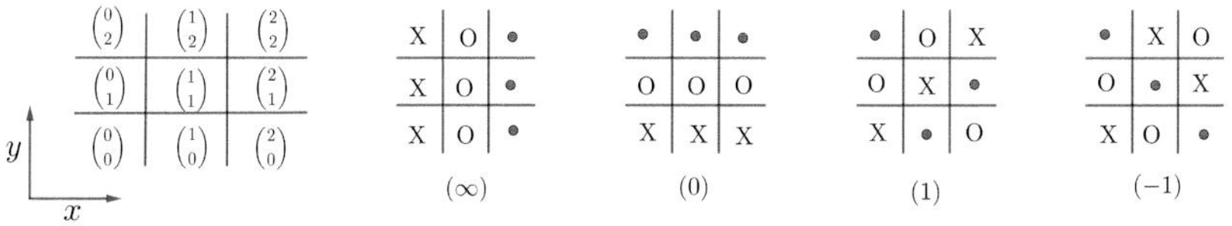

Figure 2.5 The lines of the 9-point affine plane

This system is also classical. For let $W$ be a 2-dimensional vector space over the field $\mathbb{F}$ of order $q = p^m$; take its $q^2$ vectors as our points and its

$$(q^2 - 1)/(q - 1) \times q = q^2 + q$$

cosets of 1-dimensional subspaces as our lines. Then any two vectors, $u$ and $v$, are contained together in a unique coset, namely

$$\langle u - v \rangle + v = \{\lambda u + \mu v \mid \lambda, \mu \in \mathbb{F}, \lambda + \mu = 1\}.$$

We thus obtain a Steiner system $S(2, q, q^2)$. Note that we no longer have the projective property that distinct lines intersect at a point; they form an *affine* plane in which the lines fall into sets of $q$ parallel lines, the $q$ cosets of a 1-dimensional subspace. The Steiner system $S(2, 3, 9)$ corresponds, of course, to the case $q = 3$ and in Figure 2.5 we show the $3^2 + 3 = 12$ lines partitioned into four sets of parallel lines with slopes $\infty, 0, 1$ and $-1$. Every line takes the form $x = c$ or $y = mx + c$ and so, for instance $y = x - 1$ is represented by the 'dots' in the third array, see also Section 9.2.

The numerical condition for there to exist an extension to a Steiner system S(3, 4, 10) is satisfied, and there is indeed a unique such system. It must have

$$\binom{10}{3} \bigg/ \binom{4}{3} = 30$$

(special) tetrads. We let

$$\mathbb{F}_9 = \{0, \pm 1, \pm i, \pm 1 \pm i \mid 1 + 1 + 1 = 0, i^2 = -1\}$$

be the field of order 9, when the 10-point projective line is $P_1(9) = \{\infty\} \cup \mathbb{F}_9$. It is acted on by the group $G \cong P\Gamma L_2(9)$ consisting of all linear fractional maps, see Section 11.7.2, of the form

$$x \mapsto \frac{ax + b}{cx + d} \quad ad - bc \neq 0$$

together with the field automorphism that interchanges $i$ and $-i$. It thus has order $10 \cdot 9 \cdot 8 \times 2 = 1440$. The subset $T = \{\infty, 0, 1, -1\}$ is fixed by the subgroup $H \cong PGL_2(3)$, consisting of all linear fractional maps with coefficients in the subfield of order 3, together with the field automorphism that clearly fixes every point of $T$. This subgroup has order $(3^2 - 1)(3^2 - 3)/2 \times 2 = 48$, and so $T$ has $1440/48 = 30$ images under the action of $G$. Denote this set of images of $T$ by $S$. Since $G$ acts triply transitively, every triple of points must be contained in the same number of tetrads in $S$. But the number of triples contained in tetrads of $S$ is $30 \times 4 = 120$, which is the total number of triples of elements in $P_1(9)$. So every triple is contained in a unique tetrad of $S$ which thus defines a Steiner system S(3, 4, 10) on the points of $P_1(9)$.

The numerical condition for there to exist an extension of this S(3, 4, 10) to an S(4, 5, 11) again holds, and we find that the number of *pentads* required is given by

$$\binom{11}{4} \bigg/ \binom{5}{4} = 66.$$

Furthermore, we have that

$$\binom{12}{5} \bigg/ \binom{6}{5} = 132,$$

and so a Steiner system S(5, 6, 12) would consist of 132 *hexads* with the property that any 5 of the 12 points are contained in precisely one of them. Each of these Steiner systems S($2 + i, 3 + i, 9 + i$), for $i = 0, \ldots, 3$ exists and is unique up to relabelling the points. However, note that if our affine plane S(2, 3, 9) defined on a set $\Lambda$ with $|\Lambda| = 9$ is extended to a Steiner system S(5, 6, 12) defined on

$\Lambda \cup \{a, b, c\}$, then the hexads containing $\{b, c\}$ must form a system $S(3, 4, 10)$ on $\Lambda \cup \{a\}$; the hexads containing $\{c, a\}$ must form a system $S(3, 4, 10)$ on $\Lambda \cup \{b\}$ and the hexads containing $\{a, b\}$ must form a system $S(3, 4, 10)$ on $\Lambda \cup \{c\}$. The tetrads containing none of $\{a, b, c\}$ must be different in the three cases, and so there are at least three ways in which the affine plane can be extended to an $S(3, 4, 10)$. These three systems can be mapped into one another by permutations preserving the $S(5, 6, 12)$, but the extension is not unique in the way it was for our extension $S(2, 3, 7) \mapsto S(3, 4, 8)$. In Chapter 9, we show how these three extensions may be glued together to obtain a device called the *Kitten* that exhibits the 132 hexads of the $S(5, 6, 12)$ in a revealing manner.

Note that
$$7 \text{ does not divide } \binom{13}{6},$$
and so extension to an $S(6, 7, 13)$ is not possible.

## 2.4 A Todd Triangle for the Steiner System S(5, 6, 12)

The remarkable Steiner system $S(5, 6, 12)$ will be explored in more detail in Chapter 9. In the meantime, we explore some of its numerical properties. As we have seen, the hexads containing a fixed point, $x_1$ say, must form a system $S(4, 5, 11)$, and so there are 66 of them. Thus the remaining $132 - 66 = 66$ do not contain that fixed point. The hexads containing a further fixed point $x_2$ form an $S(3, 4, 10)$, and so there are 30 of them, and there must be $66 - 30 = 36$ hexads containing $x_1$ but not $x_2$ and $66 - 36 = 30$ containing neither. Continuing in this way, we see that there are 12 hexads containing $x_1, x_2$ and $x_3$; $30 - 12 = 18$ containing $x_1$ and $x_2$ but not $x_3$; $36 - 18 = 18$ containing $x_1$ but neither $x_2$ nor $x_3$; and finally $30 - 18 = 12$ containing none of $x_1, x_2$ or $x_3$. Proceeding in this manner, we obtain Figure 2.6, the bottom line of which refers to a subset $\{x_1, x_2, \ldots, x_6\}$, which is a hexad of the system. So we see that there are three hexads that contain $\{x_1, x_2, x_3, x_4\}$ but neither of $x_5$ or $x_6$; two hexads that contain $\{x_1, x_2, x_3\}$ but none of $\{x_4, x_5, x_6\}$ and so on. In particular, we note that there is one hexad disjoint from $\{x_1, x_2, x_3, x_4, x_5, x_6\}$. In other words, the complement of a hexad is a hexad of the system; so the hexads come in 66 complementary pairs. Note that

$$1 + \binom{6}{2} \times 3 + \binom{6}{3} \times 2 + \binom{6}{4} \times 3 + 1 = 1 + 45 + 40 + 45 + 1 = 132.$$

Figure 2.6 is named after J. A. Todd who first produced a similar diagram for the Steiner system $S(5, 8, 24)$.

|   |   |   |   | 132 |   |   |   |   | S(5, 6, 12) |
|---|---|---|---|---|---|---|---|---|---|
|   |   |   | 66 |   | 66 |   |   |   | S(4, 5, 11) |
|   |   | 30 |   | 36 |   | 30 |   |   | S(3, 4, 10) |
|   | 12 |   | 18 |   | 18 |   | 12 |   | S(2, 3, 9) |
| 4 |   | 8 |   | 10 |   | 8 |   | 4 |   | S(1, 2, 8) |
| 1 | 3 |   | 5 |   | 5 |   | 3 | 1 |   |
| 1 | 0 | 3 |   | 2 |   | 3 | 0 | 1 | hexad |

Figure 2.6 Intersection diagram for an S(5, 6, 12)

## 2.5 A Steiner System S(3, 4, 16)

A further block design that will play an important part in what follows is a Steiner system S(3, 4, 16). First note that such a system, which is not unique, will consist of

$$\binom{16}{3} \bigg/ \binom{4}{3} = 140$$

tetrads. Let $V$ be a vector space of dimension 4 over the field $\mathbb{Z}_2$ of integers modulo 2, and note that $V$ possesses

$$(2^4 - 1)(2^4 - 2)/(2^2 - 1)(2^2 - 2) = 35$$

2-dimensional subspaces, and thus $35 \times 4 = 140$ cosets of 2-dimensional subspaces. If $u, v$ and $w$ are three distinct vectors in $V$, then they lie in

$$\{u, v, w, u + v + w\} = \langle u + w, v + w \rangle + w,$$

and this is the only coset of a 2-dimensional subspace of $V$ in which they all lie. Thus these 140 cosets do indeed define a Steiner system S(3, 4, 16) on the 16 vectors of $V$. We let

$$V = \{(x, y, z, t) \mid x, y, z, t \in \mathbb{Z}_2\}.$$

Then in Figure 2.7, we exhibit the 16 vectors as a $4 \times 4$ array in which the top row is the $xy$-subspace and the first column is the $zt$-subspace. Every other entry is the sum of the vectors in the first row and the first column of the row and column in which it lies. With this arrangement of the vectors, the tetrads are immediately recognizable as those sets of four points that intersect every row with the same parity and intersect every column with the same parity. From the geometric definition of the tetrads, it is clear that this system is preserved by the linear group $L_4(2)$. But it is clear that translations of the form $v \mapsto v + a$ also preserve the set of 140 cosets whilst acting transitively on the vectors as illustrated in Figure 2.8. These translations form an elementary abelian normal

## 2.6 The Projective Plane of Order 4 and Its Extensions

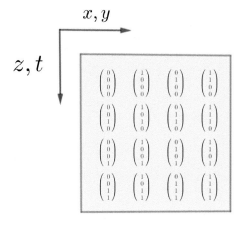

Figure 2.7 A Steiner system S(3, 4, 16)

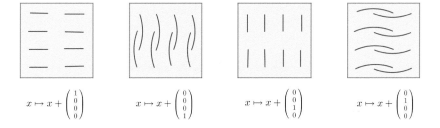

Figure 2.8 Translational symmetries of this S(3, 4, 16)

subgroup of the automorphism group of the Steiner system of order 16, and this system is preserved by a group of shape

$$2^4 : L_4(2).$$

## 2.6 The Projective Plane of Order 4 and Its Extensions

In Section 2.2, we saw how a finite field of order $q = p^n$ can be used to construct a Steiner system $S(2, q + 1, q^2 + q + 1)$ known as a projective plane, and we investigated the case $q = 2$ that gives rise to the Fano plane. As we shall see, the case $q = 2^2$ is of fundamental importance to the subject of this book.

The projective plane of order 4 has $4^2 + 4 + 1 = 21$ points and 21 lines;

every line contains 5 points, and every point has 5 lines passing through it. These 21 lines thus form the blocks of a Steiner system S(2, 5, 21) that is preserved by the projective general linear group $PGL_3(4)$ extended by the field automorphism that squares every element of the underlying field and thus interchanges $\omega \leftrightarrow \omega^2$:

$$\mathbb{F}_4 = \{0, 1, \omega, \omega^2 \mid 1 + 1 = 1 + \omega + \omega^2 = 0\},$$

obtained by extending $\mathbb{Z}_2$, the field of order 2, by a root of the irreducible quadratic $x^2 + x + 1$.

As was mentioned briefly in Chapter 1, this system possesses a sequence of extensions:

$$S(2, 5, 21) \mapsto S(3, 6, 22) \mapsto S(4, 7, 23) \mapsto S(5, 8, 24),$$

and it was Witt's approach to construct this sequence and show that each extension was unique and possessed a transitive group of automorphisms. The numerical properties that we have presented in Figure 2.6 for the Steiner system S(5, 6, 12) are given for the S(5, 8, 24) in Figure 3.1.

## 2.7 Block Designs in General

The Steiner systems described in this chapter are examples of *t-designs*: *t-*$(v, k, \lambda)$ block designs, where $v$ denotes the total number of points, $k$ the number of points in each block and $\lambda$ the number of blocks in which each subset of $t$ points is contained. Thus Steiner systems have $\lambda = 1$, with a Steiner system $S(l, m, n)$ being an $l$-$(n, m, 1)$ block design. The Steiner systems $S(5, 6, 12)$ and $S(5, 8, 24)$ are thus 5-designs, with parameters $(12, 6, 1)$ and $(24, 8, 1)$, respectively, whose groups of automorphisms act 5-transitively on the points. The first 6-designs were discovered by Leavitt and Magliveras (1984); however, the CFSG, see Section 11.8, tells us that a non-trivial 6-design cannot admit a 6-transitive group of automorphisms.

In this notation, the 11-point biplane mentioned in Section 11.7.2 is a 2-$(11, 5, 2)$ design.

# 3

# The Miracle Octad Generator

We are now in a position to introduce the Steiner system, which is at the heart of this book, namely the majestic S(5, 8, 24). From Chapter 2 we see that, if one exists, it will consist of

$$\binom{24}{5} \Big/ \binom{8}{5} = 759$$

8-element subsets of a 24-element set $\Omega$ such that any five points of $\Omega$ lie together in precisely one of them. These 8-element subsets are the *special octads* or simply *octads* of the system. As we shall see later, such a system is unique up to relabelling the points of $\Omega$ and the group of permutations of $\Omega$ preserving the set of octads acts quintuply transitively and has order 244 823 040; it is the famous Mathieu group $M_{24}$. $M_{24}$ contains copies of the linear group $L_2(23)$ as linear fractional maps on the points of the projective line $P_1(23)$ and so we usually label the 24 points $\Omega = \{\infty, 0, 1, \ldots, 22\}$. It is possible to say a great deal about such a Steiner system before we know if one actually exists. For instance, in his PhD thesis, Curtis (1972) proves that if an S(5, 8, 24) exists then it is unique, and its group of symmetries acts quintuply transitively on the 24 points with order 244 823 040. Much of this proof is reproduced in MacWilliams and Sloane (1977). J. A. Todd (1966, 1970) worked out the number of octads that can intersect a subset of $l$ points in a given $k$ points for all $k \leq l \leq 5$ and for $6 \leq l \leq 8$ provided the $l$ points lie in an octad. He exhibited this information in the form of a triangle, see Figure 3.1. If we fix a point $x_1$ then the octads that contain it must form an S(4, 7, 23) on the remaining 23 points and so there are

$$\binom{23}{4} \Big/ \binom{7}{4} = 253$$

octads containing $x_1$, and $759 - 253 = 506$ not containing $x_1$. Again, if we fix a further point $x_2$ then the octads containing both $x_1$ and $x_2$ form an S(3, 6, 22)

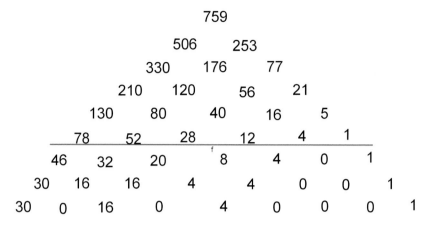

Figure 3.1 Todd's triangle of octad intersections

on the remaining 22 points, and so there are

$$\binom{22}{3} \bigg/ \binom{6}{3} = 77$$

octads containing $x_1$ and $x_2$, $253 - 77 = 176$ containing $x_1$ but not $x_2$, and $506 - 176 = 330$ containing neither, and so on. Below the horizontal line in Figure 3.1 we assume that the $l$ points lie in an octad. So if $\{x_1, x_2, \ldots, x_6\}$ is contained in an octad, then there is one octad containing all of them, zero octads containing $\{x_1, x_2, \ldots, x_5\}$ but not $x_6$, four containing $\{x_1, x_2, \ldots, x_4\}$ but not $x_5$ or $x_6$, and so on. The last line is particularly informative. It tells us that if $\{x_1, x_2, \ldots, x_8\}$ is an octad, then there are no octads intersecting that octad in 1, 3, 5 or 7 points. In fact, it tells us that one octad contains all of its points; $\binom{8}{4} \times 4 = 280$ octads intersect it in four points; $\binom{8}{2} \times 16 = 480$ intersect it in two points and 30 octads are disjoint from it. Note that

$$1 + 280 + 448 + 30 = 759.$$

## 3.1 The Miracle Octad Generator or MOG

The Miracle Octad Generator or MOG is essentially a diagram of the Steiner system S(5, 8, 24), see Curtis (1972, 1976) and Conway and Sloane (1988). In it the octads take a particularly recognizable form and, given any five of the 24 points, it enables us rapidly to obtain the remaining three points that complete it to an octad. The MOG can also be used to write down permutations

## 3.1 The Miracle Octad Generator or MOG

of $M_{24}$ having certain desired properties and to work within the Leech lattice, see Chapter 10.

As we see from Todd's triangle, if $U$ is a fixed octad then through any four points of $U$ there are four octads intersecting U in precisely those four points. These four octads must intersect $\Omega \setminus U$, the *complementary 16-ad*, in four sets of size four called *special tetrads* that partition the 16-ad, otherwise we should have two octads intersecting in more than four points.

**Lemma 3.1** (Todd)   *The symmetric difference of two octads that intersect in four points is another octad of the system.*

*Proof*   Suppose

$$A = \{a_1, a_2, a_3, a_4, a_5, a_6, a_7, a_8\} \text{ and } B = \{a_1, a_2, a_3, a_4, b_5, b_6, b_7, b_8\}$$

are two distinct octads of the system, and that $\{a_5, a_6, a_7, a_8, b_5, b_6, b_7, b_8\}$ is *not* an octad. Then the octad $C$ containing $\{a_5, a_6, a_7, a_8, b_5\}$ must contain a further element of $B$ that cannot be an $a_i$ and so we may assume it is $b_6$. Similarly the octad $D$ containing $\{a_5, a_6, a_7, a_8, b_7\}$ must contain $b_8$. Consider now the octad containing $\{a_5, a_6, a_7, b_5, b_7\}$. It must contain a further $a_i$ that cannot be $a_8$ and so we may take it to be $a_1$. But now it must contain a further element of $B$ that cannot be an $a_i$ or we should have five points of $A$. So it must be $b_6$ or $b_8$. In either case we should have five points in common with $C$ or $D$, which is impossible unless the original assumption was false. □

So the six tetrads obtained by a partition of a given octad into two tetrads, together with the four tetrads that partition the complementary 16-ad as above have the property that the union of any two of them is an octad. Such a system of six complementary tetrads is known as a *sextet*. Now there are $\binom{8}{4}/2 = 35$ partitions of our fixed octad into two tetrads, and the $35 \times 4 = 140$ *special tetrads* in the complementary 16-ad must be distinct or we should have two octads with more than four points in common. Suppose that a given three points of the 16-ad, $x_1, x_2, x_3$ say, lie in two distinct special tetrads $T_1 = \{x_1, x_2, x_3, x_4\}$ and $T_2 = \{x_1, x_2, x_3, y_4\}$ and let the partitions of $U$ into two tetrads that give rise to $T_1$ and $T_2$ respectively be $A_1 \cup A_2$ and $B_1 \cup B_2$, where $A_1 \cup A_2 = B_1 \cup B_2 = U$ and $|A_i| = |B_i| = 4$ for $i = 1, 2$. Then for some $k$ and $l$ we have $|A_k \cap B_l| \geq 2$ and $T_1 \cup A_k$ and $T_2 \cup B_l$ would be two distinct octads with at least five points in common. So three points of the 16-ad can be contained in at most one special tetrad. But there are $140 \times 4 = 560$ triples contained in special tetrads, and there are in total $\binom{16}{3} = 560$ subsets of size three in the 16-ad, and so every triple of points in the 16-ad is contained in a unique special tetrad. That is to say they define a Steiner system S(3, 4, 16).

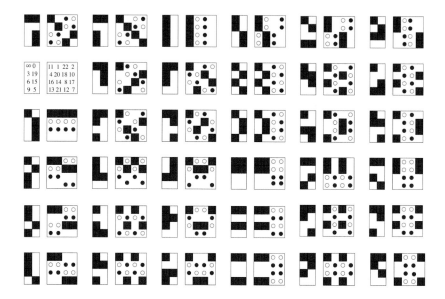

Figure 3.2 The Miracle Octad Generator

We can arrange the 24 points of $\Omega$ into a $4 \times 6$ array so that the columns form a sextet as described earlier, with the first two columns being the octad $U$. Since the Todd triangle tells us that any two octads intersect one another evenly, we see that any octad must cut the six tetrads of a sextet as $4^2.0^4$, $3.1^5$ or $2^4.0^2$; indeed we may readily count that the number of octads in each of these three categories is 15, 384 and 360, respectively. Of course,

$$15 + 384 + 360 = 759.$$

If we consider the sextet defined by the tetrad consisting of the top point in the first column, and all but the top point in the second, the partition of the 16-ad $\Omega \setminus U$ must consist of tetrads with a point in each of the last four columns. The points in each column can be arranged so that these tetrads form the four rows of the 16-ad, see Figure 3.3. Since an octad must intersect every tetrad of a sextet evenly or every tetrad oddly, we see that any special tetrad in the 16-ad $\Omega \setminus U$ must intersect every column of $\Omega \setminus U$ with the same parity and every row with the same parity. Any three points of $\Omega \setminus U$ must intersect the columns as $(\underline{3}, 0, 0, 0)$, $(2, \underline{1}, 0, 0)$ or $(1, 1, 1, \underline{0})$ where we have underlined the column with the exceptional parity, which is where the 4th point of the special tetrad containing them must lie. The same applies to the rows, and so

## 3.1 The Miracle Octad Generator or MOG

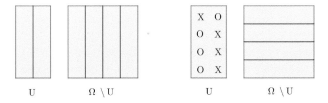

Figure 3.3 Two sextets: the columns of the MOG and the rows of the complementary 16-ad

the row and column in which the 4th point of the special tetrad must lie are determined. In fact, the 16 points of $\Omega \setminus U$ may be labelled with the vectors of $V$, a 4-dimensional vector space over $\mathbb{Z}_2$ as shown in Figure 2.7

$$V = \{(x, y, z, t) \mid x, y, z, t, \in \mathbb{Z}_2\}$$

with the horizontal axis corresponding to the $xy$-subspace and the vertical axis corresponding to the $zt$-subspace. The special tetrads are simply the cosets of the $(2^4 - 1)(2^4 - 2)/(2^2 - 1)(2^2 - 2) = 35$ 2-dimensional subspaces of $V$, as described in Chapter 2. From the Todd triangle we see that there are 30 octads disjoint from our given octad $U$ and so 30 octads contained in $\Omega \setminus U$. Since we know that the columns form four tetrads of a sextet, as do the rows, and that the symmetric difference of any two octads that intersect in four points is itself an octad, we see that these 30 consist of two columns (six such), two rows (six such) and the symmetric difference of two rows and two columns ($6 \times 6/2 = 18$ such, since each of these occurs in two ways). These are precisely the 3-dimensional subspaces of $V$ in Figure 2.7, together with their complements, and so we see that each 2-dimensional subspace together with its cosets forms four tetrads of a sextet. So the 35 partitions of $\Omega \setminus U$ into four tetrads of a sextet, each corresponding to a 2-dimensional subspace of $V$ and its cosets, are in one-to-one correspondence with the 35 partitions of the eight points of $U$ into two tetrads.

It is this correspondence that is exhibited in the MOG, see Figure 3.2, which consists of 36 diagrams displayed as a $6 \times 6$ array. One of them, the one in the 2nd row and 1st column (2R1C), shows the labelling of the 24 points of $\Omega$ with the elements of the projective line $P_1(23)$. This correspondence between two sets of size 35 would be of limited value were it only valid for the particular octad $U$. However, this is where the full beauty of the Steiner system comes into play: It is possible to partition the six columns of our $4 \times 6$ array into three pairs, in other words into three mutually disjoint octads, so that the same correspondence holds for any one of the three. For us this will of course be the

1st and 2nd, the 3rd and 4th, and the 5th and 6th columns; we refer to these three octads as the *bricks* and denote them by $\Lambda_1$, $\Lambda_2$ and $\Lambda_3$. In fact, with the correspondence as in the MOG, Figure 3.2, the Mathieu group $M_{24}$ contains a subgroup isomorphic to the symmetric group $S_3$ bodily permuting these three bricks that commutes with a copy of the linear group $L_2(7)$ acting identically in the three bricks. Elements of this latter subgroup are shown in Figure 3.4, where each brick is labelled with the points of the 8-point projective line $P_1(7)$ and the permutations are given as linear fractional maps.

Each of the other 35 diagrams consists of an octad, which can be any one of the three bricks, together with the complementary 16-ad. The octad is partitioned into two subsets of size four coloured black and white, respectively; the complementary 16-ad is partitioned into four special tetrads. The union of either the black or the white subset in the brick together with one of the corresponding special tetrads in the 16-ad forms an octad of the Steiner system, as indeed does the union of two of the special tetrads. Thus each of the 35 diagrams in the MOG in effect represents three sextets.

The 35 diagrams in the MOG have been arranged in such a way that one can soon become familiar with where each special tetrad will occur. Thus, special tetrads that are contained in the first two columns or the last two columns of the 16-ad are in the 4th column of the MOG; those that are contained in the 1st and 3rd or the 2nd and 4th columns are in the fifth column of the MOG; and those that are contained in the 1st and 4th or the 2nd and 3rd columns are in the sixth column of the MOG. Similarly those special tetrads that are contained in the first two rows or the last two rows of the 16-ad appear in the fourth row of the MOG; those that are contained in the 1st and 3rd or the 2nd and 4th rows appear in the fifth row of the MOG; and those that are contained in the 1st and 4th or 2nd and 3rd rows appear in the sixth row of the MOG. The remaining eight types of special tetrad, which is to say the columns of the 16-ad, the rows of the 16-ad and those six types that intersect both rows and columns oddly, appear in the first three rows and columns of the MOG.

The stabilizer of an octad in $M_{24}$ is a subgroup of shape $2^4 : A_8$ where the elementary abelian normal subgroup of order 16 consists of permutations that fix every point of the octad and act as translations on the vector space $V$ as illustrated in Figure 2.8; factoring out this normal sugroup we obtain the group $A_8 \cong L_4(2)$ acting on the eight points of the octad and on the 15 non-zero vectors of V. Indeed the first two elements in Figure 3.4 correspond to matrices

$$\begin{pmatrix} 0 & 0 & 0 & 1 \\ 0 & 1 & 0 & 0 \\ 1 & 0 & 0 & 1 \\ 0 & 0 & 1 & 0 \end{pmatrix}, \quad \begin{pmatrix} 1 & 0 & 0 & 0 \\ 0 & 1 & 0 & 0 \\ 0 & 0 & 0 & 1 \\ 0 & 0 & 1 & 1 \end{pmatrix}$$

respectively and the correspondence between even permutations of the eight points of $U$ and $4 \times 4$ matrices of $L_4(2)$ can easily be read off from the MOG.

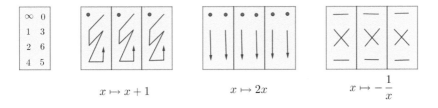

Figure 3.4 The $L_2(7)$ acting identically on the three bricks of the MOG

## 3.2 Mnemonic for the Standard MOG Labelling

Of course we could permute the labelling of the points of $\Omega$ as they appear in Figure 3.2 using any permutation of $M_{24}$ and we should obtain a valid version of the MOG. However, the ordering used here can be recovered rapidly using the following algorithm due to John Conway.

(A) Insert the first 12 numerals of $\mathbb{Z}_{23}$ as shown in Figure 3.5A.
(B) Negate the non-squares to obtain Figure 3.5B.
(C) Complete the labelling letting the permutation in Figure 3.5C correspond to

$$\gamma : x \mapsto -\frac{1}{x} \equiv \begin{matrix} (\infty\ 0)(1\ 22)(2\ 11)(3\ 15)(4\ 17)(5\ 9) \\ (6\ 19)(7\ 13)(8\ 20)(10\ 16)(12\ 21)(14\ 18) \end{matrix}.$$

(D) To obtain the standard labelling Figure 3.5D.

**Warning:** The observant reader may have found a slightly different version of the MOG appearing in print, see for instance Conway and Sloane (1988, page 309). This occurred because John Conway felt at one time that the mirror image of the original version had certain advantages. This is equivalent, modulo elements of the Mathieu group $M_{24}$, to interchanging the 5th and 6th columns of the MOG whilst preserving the rows. However, Conway himself later recanted his heresy and embraced orthodoxy!

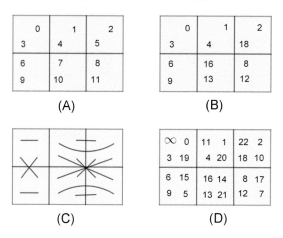

Figure 3.5 A mnemonic for recalling the standard MOG labelling

## 3.3 Finding Octads Using the MOG

Since octads must intersect evenly it is clear that any octad, other than the three bricks themselves, must have four points in one of the bricks and a special tetrad in the complementary 16-ad. As can be seen in Figure 3.2, just 14 special tetrads lie completely in one of the bricks (i.e., they satisfy the condition of intersecting every row with the same parity and have an even number of points in each of the two columns). They are the two columns of the MOG in a brick and the 12 tetrads exhibited in the 4th column of the MOG. So there are $3 \times 14 \times 2 = 84$ octads cutting the three bricks as $4^2.0$. The remainder must cut the bricks as $4.2^2$ and there are thus $3 \times 28 \times 2 \times 4 = 672$ of these. Note that $3 + 84 + 672 = 759$. So, given five points of $\Omega$, finding the unique octad containing them invariably consists of identifying the brick that contains four points, the so-called *heavy brick*. In Figure 3.6 we give some examples of octads together with the MOG diagram in which they can be seen; thus 3R5C means that the octad occurs in the third row and the 5th column of the MOG. The vector with entries from the field of order four, $\mathbb{F}_4 = \{0, 1, \omega, \bar{\omega}\}$, refer to the *hexacode* notation that is explained fully in Chapter 6.

**Example 3.1** Find the octad containing $\{1, 2, 3, 4, 5\}$.

## 3.3 Finding Octads Using the MOG

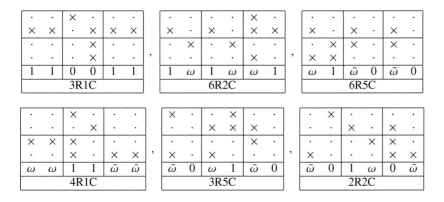

Figure 3.6 Some octads as they appear in the MOG

These five points appear in the MOG as

$$\begin{array}{|cc|cc|cc|}\hline \cdot & \cdot & \cdot & \times & \cdot & \times \\ \times & \cdot & \times & \cdot & \cdot & \cdot \\ \hline \cdot & \cdot & \cdot & \cdot & \cdot & \cdot \\ \cdot & \times & \cdot & \cdot & \cdot & \cdot \\ \hline \end{array},$$

cutting the bricks $2.2.1$. Now the four points in $\Lambda_1 \cup \Lambda_2$ visibly do not form a special tetrad, and so the special tetrad lies in $\Lambda_1 \cup \Lambda_3$ or $\Lambda_2 \cup \Lambda_3$. So the desired octad will contain either

$$\begin{array}{|cc|cc|cc|}\hline \cdot & \cdot & \cdot & \times & \cdot & \times \\ \times & \cdot & \times & \cdot & \cdot & \cdot \\ \hline \cdot & \cdot & \cdot & \cdot & o & \cdot \\ \cdot & \times & \cdot & \cdot & \cdot & \cdot \\ \hline \end{array} \quad \text{or} \quad \begin{array}{|cc|cc|cc|}\hline \cdot & \cdot & \cdot & \times & \cdot & \times \\ \times & \cdot & \times & \cdot & o & \cdot \\ \hline \cdot & \cdot & \cdot & \cdot & \cdot & \cdot \\ \cdot & \times & \cdot & \cdot & \cdot & \cdot \\ \hline \end{array}.$$

The special tetrad in the first case appears in the 3R2C (third row, second column) diagram in the MOG when we see that 1 and 4 appear with the same colour (both black) in the complementary octad $\Lambda_2$. Thus the desired octad is

$$\begin{array}{|cc|cc|cc|}\hline \cdot & \cdot & \times & \times & \cdot & \times \\ \times & \cdot & \times & \cdot & \cdot & \cdot \\ \hline \cdot & \cdot & \cdot & \cdot & \times & \cdot \\ \cdot & \times & \times & \cdot & \cdot & \cdot \\ \hline \end{array}.$$

Note that the special tetrad in the second case appears as the circles in the 4R2C diagram, when 3 and 5 have opposite colours in the complementary octad $\Lambda_1$.

**Example 3.2** Find the octad containing $\{3, 5, 7, 11, 13\}$.

These five points appear in the MOG as

|   | . | × | . | . | . |
|---|---|---|---|---|---|
| × | . | . | . | . | . |
| . | . | . | . | . | . |
| . | × | × | . | . | × |

,

again cutting the bricks 2.2.1, and again the four points in $\Lambda_1 \cup \Lambda_2$ visibly do not form a special tetrad. So the special tetrad lies in $\Lambda_1 \cup \Lambda_3$ or $\Lambda_2 \cup \Lambda_3$. So the desired octad will contain either

|   | . | × | . | . | . |      |   | . | × | . | . | o |
|---|---|---|---|---|---|------|---|---|---|---|---|---|
| × | . | . | . | o | . |  or  | × | . | . | . | . | . |
| . | . | . | . | . | . |      | . | . | . | . | . | . |
| . | × | × | . | . | × |      | . | × | × | . | . | × |

.

The special tetrad in the first case appears as circles in the 5R2C diagram in the MOG when we see that 11 and 13 appear with the same colour (both black) in the complementary octad $\Lambda_2$. Thus the desired octad is

|   | . | × | . | . | . |
|---|---|---|---|---|---|
| × | . | . | . | × | . |
| . | . | × | . | . | . |
| . | × | × | × | . | × |

.

Note that the special tetrad in the second case appears as the black squares in the 6R6C diagram, when 3 and 5 again have opposite colours in the complementary octad $\Lambda_1$.

If the five points we are given have four points in one of the bricks, then this is clearly the heavy brick and the relevant diagram is immediate. If the five points have three points in one of the bricks and two in the complementary 16-ad, then again the heavy brick is clear, but we must consider which of the five remaining points of that brick we must adjoin.

**Example 3.3**  Find the octad containing $\{0, 5, 10, 15, 20\}$.

The heavy brick is clearly $\Lambda_1$; nor can we adjoin 19 as the required octad clearly does not consist of two columns of the MOG, so there are four cases to consider: We must add $\infty$, 3, 6 or 9. Then $\{\infty, 0, 15, 5\}$ occurs in 1R1C and 20 and 10 are in different special tetrads; $\{3, 0, 15, 5\}$ occurs in 4R1C and again 20 and 10 are in different special tetrads; $\{6, 0, 15, 5\}$ occurs in 6R2C when 20 and 10 complete to a special tetrad along with 16 and 8. Thus the required octad is

$$\begin{array}{|cc|cc|cc|} \hline \cdot & \times & \cdot & \cdot & \cdot & \cdot \\ \cdot & \cdot & \cdot & \times & \cdot & \times \\ \hline \times & \times & \times & \cdot & \times & \cdot \\ \cdot & \times & \cdot & \cdot & \cdot & \cdot \\ \hline \end{array} = \{6, 0, 15, 5, 16, 20, 8, 10\}\,.$$

# 4

# The Binary Golay Code

## 4.1 Creating a Code from a Block Design

We may give any block design an algebraic structure by interpreting the blocks as vectors in a vector space over a finite field. In our case of course the design will be a Steiner system $\mathcal{S}$. Explicitly, if the design consists of subsets of a set $\Lambda$ containing $n$ points, then we define $V$ to be the $n$-dimensional vector space over the field $\mathbb{F}$, a basis for which is

$$\{v_i \mid i \in \Lambda\}.$$

Corresponding to each block $B$ is a vector

$$v_B = \sum_{i \in B} v_i,$$

and we may then consider the subspace of $V$ spanned by the vectors $v_B$. In our case the field will usually be $\mathbb{Z}_2$, the integers modulo 2, but we will also consider $\mathbb{Z}_3$ and $\mathbb{F}_4$, the Galois field of order 4.

A *linear code* is simply a subspace of a vector space over a finite field $\mathbb{F}$ together with a fixed ordered basis. If the dimension of the vector space is $n$, then its vectors or *codewords* may be written as $n$-tuples of elements from $\mathbb{F}$. The code is said to have *length n* and the number of non-zero entries in any codeword is said to be its *weight*. Of particular interest in coding theory is the *minimal weight* of a code, that is the smallest weight of a non-zero codeword. The *Hamming distance* between two codewords **u** and **v** is defined to be the number of positions in which they differ, which is to say the weight of **u-v**, and the *minimum distance d* of the code is the minimum distance between any two distinct codewords. For a linear code this is the same as the minimum weight, see for example MacWilliams and Sloane (1977, page 9). Suppose now that a codeword **u** has been altered in a small number of positions to give a vector

## 4.1 Creating a Code from a Block Design

**u**′ then, provided the minimal weight of the code is sufficiently large, we can deduce what **u** must have been. Indeed we have, see MacWilliams and Sloane (1977, page 10):

**Theorem 4.1** *A code with minimum distance d can correct $[\frac{1}{2}(d-1)]$ errors. If d is even then the code can correct $\frac{1}{2}(d-2)$ errors and detect $d/2$, where $[x]$ denotes the largest integer less than or equal to $x$.*

If the field is $\mathbb{Z}_2$, we say the linear code is *binary*. In this case, if $\Lambda$ denotes the set of $n$ indices of the fixed basis, and if $A$ and $B$ are subsets of $\Lambda$, then we have

$$v_A + v_B = v_C,$$

where $C$ is the *symmetric difference* of $A$ and $B$. We often write the symmetric difference of two sets $A$ and $B$ as $A + B$, thus

$$A + B = (A \setminus B) \cup (B \setminus A) = (A \cup B) \setminus (A \cap B)$$

since this binary operation satisfies the commutative and associative laws. So it is often convenient to think of the vector space itself as being $P(\Lambda)$, the power set of $\Lambda$, and the subsets of $\Lambda$ as the vectors or codewords. For further information on designs and codes, the reader is referred to Assmus and Key (1992) and MacWilliams and Sloane (1977).

If the block design in question is the unique Steiner system $S(3, 4, 8)$ described in Chapter 2, then the 14 tetrads are the 7 shown in the *Point Space* of Figure 4.1, together with their complements. The labelling of the eight points of $\Lambda$ is indicated in the left-hand diagram. These 14 tetrads may be conveniently described as those tetrads that intersect each column of the $4 \times 2$ array evenly and every row with the same parity. A helpful mnemonic for these tetrads is Adjacent, Broken, Central, Descending, Echelon, Flagged and Gibbous.[1] It is readily shown that any two distinct tetrads are either disjoint or intersect in two points, and that the symmetric difference of two non-trivially intersecting tetrads is again a tetrad of the system. As pointed out earlier, taking the symmetric difference is equivalent to adding the corresponding vectors over $\mathbb{Z}_2$. Thus, for example,

$$A + B = C \text{ or, more symmetrically, } A + B + C = 0,$$

where 0 denotes the empty set, the zero of this vector space.

The tetrads indicated by ×s in Figure 4.1 do not contain ∞ and so they span a subspace of dimension three, as shown in Figure 4.2. The Point space $\mathcal{P}$ is a

---

[1] This last is a bit of a cheat as, although the pattern is indeed moon shaped, it is *not* the shape of a gibbous moon!

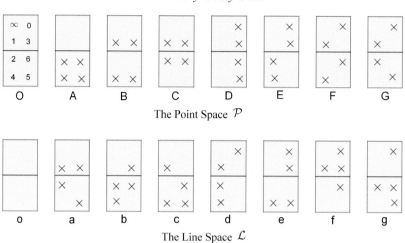

Figure 4.1 The Point space $\mathcal{P}$ and the Line space $\mathcal{L}$

copy of the *Hamming code* [7,4,3], which is to say it has length 7, dimension 3 and its 'shortest' vectors have 4 non-zero entries. Adjoining their complements, or equivalently the all 1s vector $v_\Lambda$, we obtain a 4-dimensional space $\mathcal{P}^e$, a copy of the *extended Hamming code* [8,4,4], see MacWilliams and Sloane (1977). Note that we denote the complement of the tetrad $X$ by $X'$.

Since the 14 tetrads of $\mathcal{P}$ form a Steiner system S(3, 4, 8), any subset of size 4 that is not one of these special tetrads must intersect four of them in three points, thus four of them in one point (the complements of the first four), and the remaining six in two points. These six fall into three pairs, a tetrad and its complement. The three of these that lie in $\mathcal{P}$, that is they do not contain $\infty$, must form a 2-dimensional subspace of $\mathcal{P}$ together with the zero vector.

The Line space $\mathcal{L}$ is a further 3-dimensional subspace of $V$ that consists of the zero vector together with seven such subsets of size four. Its non-zero vectors correspond to lines of $\mathcal{P}$, as shown in Figure 4.2. Note that, since $\mathcal{P}^e$ and $\mathcal{L}$ intersect in the zero vector, their sum $\mathcal{P}^e + \mathcal{L}$ has dimension $4 + 3 = 7$; it consists of all *even* subsets of $\Lambda$. That is to say, every even subset of $\Lambda$ can be written uniquely as the sum of an element of $\mathcal{P}^e$ and an element of $\mathcal{L}$.

## 4.2 The Binary Golay Code $\mathscr{C}$

We are now in a position to construct the binary Golay code $\mathscr{C}$; it will turn out the be a [24,12,8] code, which is to say it has length 24, dimension 12

## 4.2 The Binary Golay Code $\mathscr{C}$

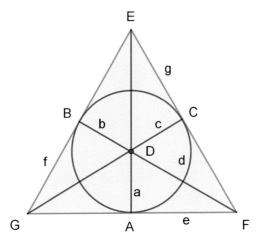

Figure 4.2 The Point and Line subspaces

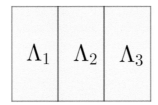

Figure 4.3 The three bricks

and the minimal length of non-trivial codewords will be 8. Thus $\mathscr{C}$ can correct $(8-2)/2 = 3$ errors. In fact, since it is so robust, $\mathscr{C}$ was used to send messages back to Earth from the Voyager 1 and 2 space missions in 1979–1981. We firstly take three copies of the space $\Lambda$ and lay them side by side as in Figure 4.3, when our 24-element set will be

$$\Omega = \Lambda_1 \cup \Lambda_2 \cup \Lambda_3,$$

the union of the three bricks. The space $V = \langle v_i \mid i \in \Omega \rangle = P(\Omega)$ will have dimension 24; we shall define a subspace of $P(\Omega)$ of dimension 12. To do this, we let $X, Y, Z$ be elements of $\mathcal{P}$ such that $X + Y + Z = 0$ ($X = Y, Z = 0$ or indeed $X = Y = Z = 0$ are permissible possibilities), and we let $t \in \mathcal{L}$. We then place

$$(X \text{ or } X') + t \text{ in } \Lambda_1; \quad (Y \text{ or } Y') + t \text{ in } \Lambda_2; \quad (Z \text{ or } Z') + t \text{ in } \Lambda_3.$$

Such an element will be denoted by

$$[X \text{ or } X', Y \text{ or } Y', Z \text{ or } Z']_t.$$

Thus for instance,

$[G, D', C]_a = $ (brick diagram) and $[0, 0, 0']_d = $ (brick diagram).

These are the *codewords* of $\mathscr{C}$. These two examples have weights 8 and 12, respectively. We now define

$$\mathscr{C} = \{[X \text{ or } X', Y \text{ or } Y', Z \text{ or } Z']_t \mid X, Y, Z \in \mathcal{P}, \ X + Y + Z = 0, \ t \in \mathcal{L}\}.$$

It is clear from the definition that $\mathscr{C}$ is closed under addition and so forms a subspace of $V$. We must now work out its dimension. We have seen already that the entry in $\Lambda_1$ can be any even subset and so there are $2^7$ possibilities, but now $t$ is fixed. However, any element of $\mathcal{P}^e$ is still available for $Y$ or $Y'$, so there are $2^4$ possibilities for the entry in $\Lambda_2$. Finally $Z$ and $t$ are fixed but we may still choose $Z$ or $Z'$ in $\Lambda_3$ so there are a further two choices. In total then we have

$$2^7 \cdot 2^4 \cdot 2 = 2^{12}$$

such elements and we see that $\mathscr{C}$ has dimension 12. This is the celebrated *binary Golay code*; we now investigate its *weight distribution*.

Notice that if the point $X$ is not on the line $t$, that is $X \notin t$, then $|X+t| = 2$ or 6 but if $X \in t$ then $|X+t| = 4$. We can only get the empty set in, say, $\Lambda_1$ if $X = t = 0$ when the entries in $\Lambda_2$ and $\Lambda_3$ are of size 4 (unless they are of weight 0 or 8). So there are $3 \times 7 \times 2 \times 2 = 84$ such codewords of weight 8, together with the the three bricks themselves that are

$$[0', 0, 0]_0, \ [0, 0', 0]_0 \text{ and } [0, 0, 0']_0.$$

If $t = 0$ but none of $X, Y$ or $Z$ is zero, then the codeword will have weight $4 + 4 + 4 = 12$.

The entry $(X \text{ or } X') + t$ can only have weight 2 if $X \notin t$ and so, since any two lines of the plane intersect one another non-trivially, we can never have weight 2 in all three bricks. Thus the minimal weight of codewords in $\mathscr{C}$ is 8. Moreover, the codeword $[X, Y, Z]_t$ for $t \neq 0$ can only have weight 8 if the line $t$ intersects the line $\{X, Y, Z\}$ in $X$, say, and we choose $Y$ or $Y'$, and $Z$ or $Z'$ so that entries in $\Lambda_2$ and $\Lambda_3$ have weight 2. There are thus $7 \cdot 6 \cdot 3 \cdot 2 \cdot 2 = 504$ octads of this type. In Table 4.1 it is implicitly understood that any of the points $X, Y, Z$ of the Point Space $\mathcal{P}$ may be replaced by its complement $X', Y'$ or $Z'$. In the Table 4.1 we

## 4.2 The Binary Golay Code $\mathscr{C}$

Table 4.1 The weights of codewords in $\mathscr{C}$

| Label | Number of octads | Description | Intersection with $\Lambda_1$ | $\Lambda_2$ | $\Lambda_3$ | Number |
|---|---|---|---|---|---|---|
| $[0,0,0]_0$ | 3 | the bricks | 0/8 | 0/8 | 0/8 | 8 |
| $[X,X,0]_0$ | 84 | $X \neq 0$ | 4/4 | 4/4 | 0/8 | 168 |
| $[X,Y,Z]_0$ |  | $X,Y,Z$ distinct | 4/4 | 4/4 | 4/4 | 336 |
| $[0,0,0]_t$ |  | $t \neq 0$ | 4/4 | 4/4 | 4/4 | 56 |
| $[X,X,0]_t$ |  | $0 \neq X \in t \neq 0$ | 4/4 | 4/4 | 4/4 | 504 |
| $[X,X,0]_t$ | 168 | $0 \neq X \notin t \neq 0$ | 2/6 | 2/6 | 4/4 | 672 |
| $[X,Y,Z]_t$ |  | $\{X,Y,Z\} = t \neq 0$ | 4/4 | 4/4 | 4/4 | 336 |
| $[X,Y,Z]_t$ | 504 | $\{X,Y,Z\} \cap t = \{X\}$ | 4/4 | 2/6 | 2/6 | 2016 |
| Total |  759 |  |  |  |  | 4096 |

record the weights of the various types of codeword and calculate how many of these have the minimal weight 8. We see that there are precisely 759 of these *octads* and the fact that there are no non-trivial codewords of weight less than 8 means that no two of these octads can intersect in more than four points. So a given five points can be contained in at most one octad. But the number of 5-element subsets of one of these 759 octads is

$$759 \times \binom{8}{5} = 42\,504 = \binom{24}{5},$$

the total number of 5-point subsets of $\Omega$. Thus every subset of five points of $\Omega$ is contained in precisely one of the octads in $\mathscr{C}$, and so these octads form a Steiner system S(5, 8, 24). It is, of course, the system given by the MOG. The whole set $\Omega$ is an element of $\mathscr{C}$, given by $[0',0',0']_0$, and so the 759 complements of octads are also in $\mathscr{C}$; they are known as *16-ads*. The remaining

$$2^{12} - (2 \times 759 + 2) = 2576$$

codewords have weight 12 and are known as *dodecads*. We denote the sets of octads, dodecads and 16-ads in $\mathscr{C}$ by $\mathscr{C}_8, \mathscr{C}_{12}$ and $\mathscr{C}_{16}$, respectively. Note that an octad cannot be contained in a dodecad as the sum of this octad and dodecad would be a codeword of weight 4. Let $D \in \mathscr{C}_{12}$, then any five points of $D$ are contained in a unique octad $O$ of $S$. But $|D \cap O| = 5, 7$ or 8 would imply $|D + O| = 10, 6$ or 4, respectively. Since there are no codewords of these weights in $\mathscr{C}$ we must have $|D \cap O| = 6$, and so every five points of $D$ are contained in a *special hexad* of $D$, which is the intersection of an octad with $D$. Thus these special hexads form a Steiner system S(5, 6, 12) on the points of

$D$, and there are

$$\binom{12}{5} \Big/ \binom{6}{5} = 132$$

of them. Note that the octad $D+O$ intersects $D$ in the complement of $D \cap O$, and so the complement in $D$ of a special hexad is a special hexad; so the 132 special hexads fall into 66 pairs of complementary hexads. Moreover, the two points of $O \setminus D$ cannot be contained in an octad $O'$ that intersects $D$ in six points, distinct from $O$ and $O + D$, as the sum $O + O'$ would be an octad contained in $D$, which as we have seen cannot occur. But $\Omega \setminus D$, the complement of $D$, contains

$$\binom{12}{2} = 66$$

pairs of points, and so every pair of points in $\Omega \setminus D$ defines a partition of $D$ into two complementary special hexads. The subgroup of $M_{24}$ fixing a dodecad is the Mathieu group $M_{12}$ that we shall investigate further in Chapter 9.

Note that the three bricks $\Lambda_1$, $\Lambda_2$ and $\Lambda_3$ form a partition of the 24 points of $\Omega$ into three disjoint octads; such a partition is known as a *trio*, see Section 8.1.7. The six tetrads of a sextet can be paired in 15 ways to form trios; these are known as the *coarsenings* of the sextet. There are $759 \cdot 15/3 = 3795$ trios and we know that there are 1771 sextets. Thus a trio may be *refined* to a sextet in $1771 \cdot 15/3795 = 7$ ways; in the case of the MOG trio these seven *refinements* correspond to the Point space above.

## 4.3 The Dual Space $\mathscr{C}^\star \cong P(\Omega)/\mathscr{C}$

The space $V = \langle v_i \mid i \in \Omega \rangle$ has a natural inner product defined on it, namely

$$\left( \sum_{i \in \Omega} \xi_i v_i, \sum_{j \in \Omega} \eta_j v_j \right) = \sum_{i \in \Omega} \xi_i \eta_i.$$

In the language of $P(\Omega)$ this becomes

$$(A, B) = \begin{cases} 0 & \text{if } |A \cap B| \in 2\mathbb{Z}, \\ 1 & \text{if } |A \cap B| \in 2\mathbb{Z} + 1 \end{cases}$$

for $A, B \subseteq \Omega$. Note that, since any two $\mathscr{C}$-sets intersect evenly, we have $(U, W) = 0$ for all $U, W \in \mathscr{C}$ and so $\mathscr{C} \subseteq \mathscr{C}^\perp$, the orthogonal complement of $\mathscr{C}$. Since $\mathscr{C}^\perp$ has dimension $24 - 12 = 12$, we see that $\mathscr{C}^\perp = \mathscr{C}$.

Now, for any $X \subset \Omega$ we may define $f_X : \mathscr{C} \mapsto \mathbb{Z}_2$ by

$$f_X(U) = (X, U),$$

## 4.3 The Dual Space $\mathscr{C}^\star \cong P(\Omega)/\mathscr{C}$

which is a linear functional on $\mathscr{C}$. If $\mathscr{C}^\star$ denotes the dual space of $\mathscr{C}$, then the function $f : X \mapsto f_X$ is a linear transformation from $P(\Omega)$ to $\mathscr{C}^\star$ and, since the kernel of this map is $\mathscr{C}^\perp = \mathscr{C}$, this map is *onto* and

$$P(\Omega)/\mathscr{C} \cong \mathscr{C}^\star.$$

But if $X \subset \Omega$ and $U$ is a $\mathscr{C}$-set, then

$$X + U \equiv X \text{ modulo } \mathscr{C},$$

and, since any five points of $\Omega$ are contained in an octad of $\mathscr{C}$, we may always take a representation of $\mathscr{C} + X$ that has at most four points in it. Clearly if that representative has 1, 2 or 3 points in it, that is to say it is a *monad, duad* or *triad* then it is the unique such representative. However, if that representative is a tetrad $T_1$, then any other tetrad that completes $T_1$ to an octad will be congruent to $T_1$ modulo $\mathscr{C}$. As we have seen from Todd's triangle, there are five tetrads $T_2, T_3, \ldots, T_6$ such that $(T_1 \cup T_i) \in \mathscr{C}_8$. Moreover, Todd's Lemma (Lemma 3.1) tells us that $(T_i \cup T_j) \in \mathscr{C}$ for all $1 \leq i < j \leq 6$. Such a partition of the 24 points of $\Omega$ into six tetrads so that the union of any two of them is an octad of the Steiner system is known as a *sextet*, as in Section 3.1,[2] and we may readily count that there are

$$\binom{24}{4}\bigg/6 = 1771$$

of them. So the elements of $P(\Omega)/\mathscr{C}$ may be taken as the monads, the duads, the triads and the sextets, and we may confirm that:

$$\binom{24}{1} + \binom{24}{2} + \binom{24}{3} + \binom{24}{4}\bigg/6 = 24 + 276 + 2024 + 1771 = 4096 = 2^{12}.$$

---

[2] Todd called such partitions 'sets of mutually complementary tetrads'; the term sextet, of course, stands for six tetrads.

# 5

# Uniqueness of the Steiner System S(5, 8, 24) and the Group $M_{24}$

In this chapter we shall show that there is a unique Steiner system S(5, 8, 24) up to relabelling of the 24 points of the set $\Omega$ on which the system is defined.[1] We shall do this by assuming such a system $S$ exists and forcing it to assume the form displayed in the MOG. In doing so, we shall show that the group of permutations of the 24 points of $\Omega$ that preserve $S$ acts quintuply transitively on $\Omega$ and has order $24.23.22.21.20.16.3 = 244\,823\,040$. It is the famous *Mathieu group* $M_{24}$. In order to do this, we shall investigate properties of the sextets as defined in Chapter 4.

## 5.1 Properties of the Sextets

Let $S$ be an S(5, 8, 24) defined on the set $\Omega$ and recall from Chapter 4 that a *sextet* is a partition of the 24 points of $\Omega$ into six tetrads $T_1, T_2, \ldots, T_6$ such that the union of any two of them is an octad. Our proof that the Steiner system S(5, 8, 24) is unique relies heavily on properties of these sextets and so we shall now investigate them further.

Since from Todd's triangle, Figure 3.1, we know that any two octads intersect evenly (in 0, 2, 4 or 8 points), we see that any octad must intersect each tetrad of a sextet evenly or every tetrad oddly; thus, it must cut across the six tetrads as $4^2.0^4$, $2^4.0^2$ or $3.1^5$. From this fact we readily obtain

**Lemma 5.1** (Intersection matrices for two sextets)   *The tetrads of any two*

---

[1] Other uniqueness proofs include Jónsson (1972).

*sextets have intersection matrix one of*

$$\begin{bmatrix} 4 & 0 & 0 & 0 & 0 & 0 \\ 0 & 4 & 0 & 0 & 0 & 0 \\ 0 & 0 & 4 & 0 & 0 & 0 \\ 0 & 0 & 0 & 4 & 0 & 0 \\ 0 & 0 & 0 & 0 & 4 & 0 \\ 0 & 0 & 0 & 0 & 0 & 4 \end{bmatrix} \begin{bmatrix} 2 & 2 & 0 & 0 & 0 & 0 \\ 2 & 2 & 0 & 0 & 0 & 0 \\ 0 & 0 & 2 & 2 & 0 & 0 \\ 0 & 0 & 2 & 2 & 0 & 0 \\ 0 & 0 & 0 & 0 & 2 & 2 \\ 0 & 0 & 0 & 0 & 2 & 2 \end{bmatrix}$$

$$\begin{bmatrix} 2 & 0 & 0 & 0 & 1 & 1 \\ 0 & 2 & 0 & 0 & 1 & 1 \\ 0 & 0 & 2 & 0 & 1 & 1 \\ 0 & 0 & 0 & 2 & 1 & 1 \\ 1 & 1 & 1 & 1 & 0 & 0 \\ 1 & 1 & 1 & 1 & 0 & 0 \end{bmatrix} \begin{bmatrix} 3 & 1 & 0 & 0 & 0 & 0 \\ 1 & 3 & 0 & 0 & 0 & 0 \\ 0 & 0 & 1 & 1 & 1 & 1 \\ 0 & 0 & 1 & 1 & 1 & 1 \\ 0 & 0 & 1 & 1 & 1 & 1 \\ 0 & 0 & 1 & 1 & 1 & 1 \end{bmatrix}$$

*where the rows correspond to the six tetrads of one sextet and the columns to the six tetrads of the other.*

## 5.2 Uniqueness of the Steiner System S(5, 8, 24) and the Order of Its Group of Symmetries

In order to prove the uniqueness theorem, we shall require a further lemma.

**Lemma 5.2** *Suppose every octad containing four points of a fixed octad U is known, then all octads of the system follow by symmetric differencing.*

*Proof* Let $x, y, z \in U$. Any octad containing $x, y$ and $z$ contains a further point of $U$ and thus is known. From Todd's triangle, Figure 3.1, there are 21 of these and, since any two of them must have four points in common, their symmetric difference is an octad. But there are $\binom{21}{2} = 210$ of these and they must be distinct since if $P, Q, R, S$ are octads containing $x, y, z$ such that $P + Q = R + S$ then 2 points of $P \setminus \{x, y, z\}$ are in $R$, say. Thus $P$ and $R$ have five points in common, and so $P = R$. From Figure 3.1 we see that this is all the octads disjoint from $\{x, y, z\}$. But every octad, apart from $U$ itself, is disjoint from some three points of $U$, and so all octads are known. □

We are now in a position to prove the following theorem. The proof here first appeared in Curtis (1972, 1976) and was reproduced in Conway and Sloane (1988).

**Theorem 5.3** *If $S$ is a Steiner system $S(5, 8, 24)$ defined on a 24-element set $\Omega$ then $S$ is unique up to relabelling the points of $\Omega$. The subgroup of the symmetric group $S_{24}$ acting on $\Omega$ that preserves $S$ is quintuply transitive and has order* $24.23.22.21.20.16.3 = 244\,823\,040$.

*Proof* Let $\{x_1, x_2, x_3, x_4, x_5, x_6\}$ be an ordered set of six points of an octad $U$ in $S$, a Steiner system $S(5, 8, 24)$ defined on a 24-element set $\Omega$, and let $x_7$ be a further point of $\Omega$ that is not contained in $U$. We shall show that the group of permutations of $\Omega$ that preserve $S$ acts sharply transitively on such ordered sets of seven points. We can choose the ordered quintuple $x_1, x_2, x_3, x_4, x_5$ in $24.23.22.21.20$ ways; there will then be three ways of choosing $x_6$ and 16 ways of choosing $x_7$. We shall show that the system is then completely determined and no further permutation is permitted. Thus the group preserving $S$ has order $24.23.22.21.20.16.3$.

Firstly we let $S_\infty$ denote the sextet defined by the tetrad $\{x_1, x_2, x_3, x_4\}$ and arrange the points of $\Omega$ as a $4 \times 6$ array whose columns are the tetrads of $S_\infty$. Thus

$$S_\infty = \begin{array}{|c|c|c|c|c|c|} \hline x_1 & x_5 & x_7 & \cdot & \cdot & \cdot \\ x_2 & x_6 & \cdot & \cdot & \cdot & \cdot \\ x_3 & \cdot & \cdot & \cdot & \cdot & \cdot \\ x_4 & \cdot & \cdot & \cdot & \cdot & \cdot \\ \hline \end{array},$$

where the octad $U$ is the union of the first two columns. The octad containing $\{x_2, x_3, x_4, x_5, x_7\}$ must cut $S_\infty$ as $3.1^5$ and so by rearranging the points in the last four columns if necessary we may assume that the sextet determined by $\{x_2, x_3, x_4, x_5\}$ is

$$S_0 = \begin{array}{|c|c|c|c|c|c|} \hline 0 & \times & 1 & 1 & 1 & 1 \\ \times & 0 & 2 & 2 & 2 & 2 \\ \times & 0 & 3 & 3 & 3 & 3 \\ \times & 0 & 4 & 4 & 4 & 4 \\ \hline \end{array}.$$

Note that at this stage we know all the octads in the 16-ad $\Omega \setminus U$. For the union of two columns of the 16-ad is an octad of the system, as is the union of two rows. Moreover, by Todd's lemma, the symmetric difference of two rows and two columns is also an octad. This gives us

$$2 \times \binom{4}{2} + \binom{4}{2} \cdot \binom{4}{2} / 2 = 12 + 18 = 30$$

octads disjoint from $U$, which Todd's triangle Figure 3.1 tells us is all of them.

Now the octad containing $\{x_1, x_3, x_4, x_5, x_7\}$ cuts both $S_\infty$ and $S_0$ as $3.1^5$ and so, by rearranging the bottom three rows of $\Omega \setminus U$ if necessary, we may assume

## 5.2 Uniqueness of S(5, 8, 24)

that

$$S_1 = \begin{array}{|cc|cccc|} \hline \times & \times & 1 & 2 & 4 & 3 \\ 0 & 0 & 2 & 1 & 3 & 4 \\ \times & 0 & 4 & 3 & 1 & 2 \\ \times & 0 & 3 & 4 & 2 & 1 \\ \hline \end{array} \text{ is a sextet.}$$

Note that once the tetrad indicated by the 1s is known, the other three indicated by 2s, 3s and 4s follow since the symmetric difference of any two rows and any two columns is an octad.

In Figure 5.1 we exhibit permutations of the unnamed 17 points that preserve the sextets $S_\infty, S_0$ and $S_1$, whilst possibly permuting their tetrads. The permutation $\pi$ is an element of order three consisting of five 3-cycles indicated by the arrows. The octad containing $\{x_1, x_2, x_3, x_5, x_7\}$ cannot contain points indicated by o in

$$\begin{array}{|cccccc|} \hline x_1 & x_5 & x_7 & o & o & o \\ x_2 & x_6 & o & o & \cdot & \cdot \\ x_3 & \cdot & o & \cdot & o & \cdot \\ x_4 & \cdot & o & \cdot & \cdot & o \\ \hline \end{array}$$

without having five points in common with a previously identified octad, and so it must be either

$$\begin{array}{|cccccc|} \hline \times & \times & \times & \cdot & \cdot & \cdot \\ \times & \cdot & \cdot & \cdot & \cdot & \times \\ \times & \cdot & \cdot & \times & \cdot & \cdot \\ \cdot & \cdot & \cdot & \cdot & \times & \cdot \\ \hline \end{array} \text{ or } \begin{array}{|cccccc|} \hline \times & \times & \times & \cdot & \cdot & \cdot \\ \times & \cdot & \cdot & \cdot & \times & \cdot \\ \times & \cdot & \cdot & \cdot & \cdot & \times \\ \cdot & \cdot & \cdot & \times & \cdot & \cdot \\ \hline \end{array}.$$

These two possibilities are interchanged by $\sigma$ and so we may assume that

$$S_2 = \begin{array}{|cc|cccc|} \hline \times & \times & 1 & 2 & 3 & 4 \\ \times & 0 & 3 & 4 & 1 & 2 \\ \times & 0 & 4 & 3 & 2 & 1 \\ 0 & 0 & 2 & 1 & 4 & 3 \\ \hline \end{array}$$

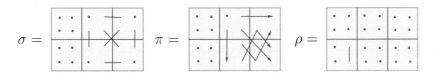

Figure 5.1 Permutations preserving $S_\infty, S_0$ and $S_1$

is a sextet, when

$$S_3 = \begin{array}{|cc|cccc|} \hline \times & \times & 1 & 2 & 3 & 4 \\ \times & 0 & 4 & 3 & 2 & 1 \\ 0 & 0 & 2 & 1 & 4 & 3 \\ \times & 0 & 3 & 4 & 1 & 2 \\ \hline \end{array}$$

is forced to be. The octad containing $\{x_1, x_2, x_5, x_6, x_7\}$ must cut $S_\infty$ $2^4.0^2$ and so, using the element $\pi$ of Figure 5.1, it may be assumed to have two points in each of the first four columns. But it also cuts $S_0$ as $2^4.0^2$ and so it must contain the top point of the 4th column. $S_1$ now forces it to be

$$\begin{array}{|cccc|cc|} \hline \times & \times & \times & \times & \cdot & \cdot \\ \times & \times & \times & \times & \cdot & \cdot \\ \cdot & \cdot & \cdot & \cdot & \cdot & \cdot \\ \cdot & \cdot & \cdot & \cdot & \cdot & \cdot \\ \hline \end{array} \quad \text{and so} \quad S_4 = \begin{array}{|cc|cccc|} \hline \times & \times & 1 & 1 & 2 & 2 \\ \times & \times & 1 & 1 & 2 & 2 \\ 0 & 0 & 3 & 3 & 4 & 4 \\ 0 & 0 & 3 & 3 & 4 & 4 \\ \hline \end{array} \quad \text{is a sextet.}$$

Since the octad containing

$$\begin{array}{|ccc|ccc|} \hline \times & \times & \times & \cdot & \cdot & \cdot \\ \cdot & \cdot & \cdot & \cdot & \cdot & \cdot \\ \times & \cdot & \times & \cdot & \cdot & \cdot \\ \cdot & \cdot & \cdot & \cdot & \cdot & \cdot \\ \hline \end{array} \quad \text{cuts} \quad S_\infty \quad \text{as} \quad 2^4.0^2$$

it must have an additional point in the second column, and so also cut $S_0$ as $2^4.0^2$. So the two remaining points must lie in the 1st and 3rd rows of one of the last three columns. But it also cuts $S_3$ as $2^4.0^2$, since it cannot have another point in the first column, and so these two points must be in the 4th column, giving

$$\begin{array}{|cccc|cc|} \hline \times & \times & \times & \times & \cdot & \cdot \\ \cdot & \cdot & \cdot & \cdot & \cdot & \cdot \\ \times & \cdot & \times & \times & \cdot & \cdot \\ \cdot & \cdot & \cdot & \cdot & \cdot & \cdot \\ \hline \end{array}.$$

Clearly the final point of this octad cannot be $x_6$ as this would be inconsistent with its intersection with $S_4$; and so, using the element $\rho$ of Figure 5.1, we may assume the octad is

$$\begin{array}{|cccc|cc|} \hline \times & \times & \times & \times & \cdot & \cdot \\ \cdot & \cdot & \cdot & \cdot & \cdot & \cdot \\ \times & \times & \times & \times & \cdot & \cdot \\ \cdot & \cdot & \cdot & \cdot & \cdot & \cdot \\ \hline \end{array} \quad \text{and so} \quad S_5 = \begin{array}{|cc|cccc|} \hline \times & \times & 1 & 1 & 3 & 3 \\ 0 & 0 & 2 & 2 & 4 & 4 \\ \times & \times & 1 & 1 & 3 & 3 \\ 0 & 0 & 2 & 2 & 4 & 4 \\ \hline \end{array} \quad \text{is a sextet.}$$

These seven sextets $S_\infty, S_0, S_1, \ldots, S_5$ may readily be used to generate all 35

## 5.2 Uniqueness of S(5, 8, 24)

sextets that have a tetrad in $U$. For, whenever two sextets intersect evenly (that is to say, a tetrad of one cuts the tetrads of the other $2^2.0^4$), we may take the symmetric difference of suitable octads to obtain a new octad and sextet. Thus, for example,

| × | × | × | × | . | . |
|---|---|---|---|---|---|
| × | × | × | × | . | . |
| . | . | . | . | . | . |
| . | . | . | . | . | . |

+

| × | × | × | × | . | . |
|---|---|---|---|---|---|
| . | . | . | . | . | . |
| × | × | × | × | . | . |
| . | . | . | . | . | . |

=

| . | . | . | . | . | . |
|---|---|---|---|---|---|
| × | × | × | × | . | . |
| × | × | × | × | . | . |
| . | . | . | . | . | . |

leading to the sextet:

| × | × | 1 | 1 | 3 | 3 |
|---|---|---|---|---|---|
| 0 | 0 | 2 | 2 | 4 | 4 |
| 0 | 0 | 2 | 2 | 4 | 4 |
| × | × | 1 | 1 | 3 | 3 |

.

In this way we readily obtain all sextets having a tetrad in $U$ and thus we know all octads that intersect $U$ in four points. Thus by Lemma 5.2 we have all the octads of the system.

To complete the proof of the theorem, note that all the permutations fixing the sextets $S_\infty, S_0$ and $S_1$ were used in the construction of $S_2, \ldots, S_5$, and so there is no permutation fixing $x_1, x_2, \ldots, x_7$ and preserving the Steiner system other than the identity element. Thus the group of automorphisms of the system acts sharply transitively on such 7-tuples and has the order stated. □

# 6

# The Hexacode

## 6.1 The Hexacodewords

The *hexacode* is a 3-dimensional, length 6 code over the Galois field

$$\mathbb{F}_4 = \{0, 1, \omega, \bar{\omega} \mid 1 + 1 = 0, \omega^2 = \bar{\omega}, \ 1 + \omega + \bar{\omega} = 0\},$$

which gives an immensely useful algebraic description of the codewords of $\mathscr{C}$ as they appear in the MOG. It was used extensively by Benson, Conway, Norton, Parker and Thackray (see Benson, 1980, Conway, 1981 and Norton, 1980) in constructing the largest Janko group $J_4$ using its subgroup ismomorphic to $M_{24}$, see also Curtis (2007b, p. 182) and Chapter 14 of this book. We may define[1] it as

$$\begin{aligned}\mathcal{H} &= \langle (0,0,1,1,1,1), (1,1,0,0,1,1), (\omega, \bar{\omega}, 1, 0, 0, 1) \rangle \\ &= \{(a, b, c, a+b+c, \omega a + \bar{\omega} b + c, \bar{\omega} a + \omega b + c) \mid a, b, c \in \mathbb{F}_4\}.\end{aligned}$$

It turns out that, if the six coordinate positions are labelled $1, \ldots, 6$, then the code is preserved by a group

$$\langle (1\ 2)(3\ 4), (1\ 3\ 5)(2\ 4\ 6), (3\ 5)(4\ 6) \rangle,$$

acting on the coordinates. That is to say, the coordinates have been paired as 12.34.56 and we may bodily permute the three pairs, and flip any two of them. Of course we may also multiply any vector by $\omega$. Altogether this gives a group of order $6 \times 4 \times 3 = 72$ preserving the code. Under the action of this group, we

---

[1] Following the warning issued in Section 3.5, the 'mirror-image' of the hexacode can be obtained by interchanging the 5th and 6th entries.

then have the following orbits:

| Shape | Subgroup order | Number |
|---|---|---|
| $(0,0 \mid 0,0 \mid 0,0)$ | 72 | 1 |
| $(0,0 \mid 1,1 \mid 1,1)$ | 8 | 3.3 = 9 |
| $(1,1 \mid \omega,\omega \mid \bar{\omega},\bar{\omega})$ | 12 | 3.2 = 6 |
| $(\omega,1 \mid \omega,1 \mid \omega,1)$ | 6 | 3.2.2 = 12 |
| $(\bar{\omega},\omega \mid 1,0 \mid 1,0)$ | 2 | 3.3.2.2 = 36 |
| | Total | 64 |

In order to gain the full benefit from the hexacode, we need to memorize these vectors.

## 6.2 Interpreting Hexacodewords as $\mathscr{C}$-sets

Each hexacodeword will represent $2^6$ $\mathscr{C}$-sets, that is to say words of the Golay code $\mathscr{C}$, as they appear in the MOG. To see this correspondence we label the four points in each column of the MOG with the elements of $\mathbb{F}_4$ as shown in the top-left diagram of Figure 6.1. The six coordinates of the hexacode will correspond to the columns of the MOG and the field element on a particular column will determine what subset of that column is included in the $\mathscr{C}$-set. There are, however, two interpretations depending on whether the $\mathscr{C}$-set intersects the MOG sextet evenly or oddly; they are the *odd* and *even* interpretations as illustrated in Figure 6.1. In the odd interpretation, if the hexacodeword is $(\xi_1, \xi_2 | \xi_3, \xi_4 | \xi_5, \xi_6)$ for $\xi_i \in \mathbb{F}_4$ then we either place a single point in the $\xi_i$-position in the $i$th column, or we place three points in the complement of the $\xi_i$th position. We have complete freedom to complement on any column subject to the condition that *the number of points in the top row of the MOG must be odd*. So corresponding to each hexacodeword there are $2^5$ $\mathscr{C}$-sets under the odd interpretation. Thus, for instance,

$(\omega, 1|\bar{\omega}, 0|\bar{\omega}, 0) \sim$ [MOG diagram] $\sim$ [MOG diagram].

As can be seen in Figure 6.1, the even interpretation is obtained by adding a point in the 0-position to the odd interpretation. Thus a 0 in the even interpretation represents the empty set or the whole set, a 1 represents points in the 0-position and the 1-position, or in the complement that is the $\omega$-position and the

44                           *The Hexacode*

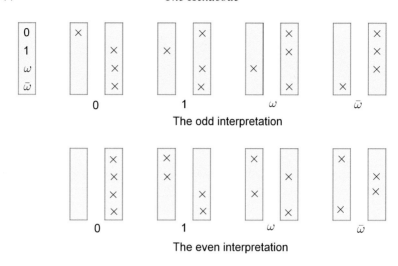

Figure 6.1  The odd and even interpretations of the hexacode

$\bar\omega$-position, and so on. As in the odd interpretation we have complete freedom to complement on any column subject to the condition that *the number of points in the top row of the* MOG *is even*. Thus, for instance,

$$(\omega, 1 | \bar\omega, 0 | \bar\omega, 0) \sim \begin{array}{|cc|cc|cc|} \hline \times & \times & \times & \cdot & \times & \cdot \\ \cdot & \times & \cdot & \cdot & \cdot & \cdot \\ \hline \times & \cdot & \cdot & \cdot & \cdot & \cdot \\ \cdot & \cdot & \times & \cdot & \times & \cdot \\ \hline \omega & 1 & \bar\omega & 0 & \bar\omega & 0 \end{array} \sim \begin{array}{|cc|cc|cc|} \hline \cdot & \times & \times & \cdot & \cdot & \cdot \\ \times & \times & \cdot & \cdot & \times & \cdot \\ \hline \cdot & \cdot & \cdot & \cdot & \times & \cdot \\ \times & \cdot & \times & \cdot & \cdot & \cdot \\ \hline \omega & 1 & \bar\omega & 0 & \bar\omega & 0 \end{array}.$$

The all 0s hexacodeword will represent the union of an even number of complete columns, in the even interpretation, or the symmetric difference of the top row of the MOG with an odd number of complete columns. Thus

$$(0, 0 | 0, 0 | 0, 0) \sim \begin{array}{|cc|cc|cc|} \hline \cdot & \cdot & \cdot & \times & \times & \cdot \\ \cdot & \cdot & \cdot & \times & \times & \cdot \\ \hline \cdot & \cdot & \cdot & \times & \times & \cdot \\ \cdot & \cdot & \cdot & \times & \times & \cdot \\ \hline 0 & 0 & 0 & 0 & 0 & 0 \end{array} \sim \begin{array}{|cc|cc|cc|} \hline \times & \cdot & \times & \cdot & \times & \cdot \\ \cdot & \times & \cdot & \times & \cdot & \times \\ \hline \cdot & \times & \cdot & \times & \cdot & \times \\ \cdot & \times & \cdot & \times & \cdot & \times \\ \hline 0 & 0 & 0 & 0 & 0 & 0 \end{array}.$$

## 6.2.1 Counting Octads in the Hexacode

It is particularly easy to count octads using the hexacode. For, in the odd interpretation, any octad will have three points in one column and one point in the other five. Thus for every hexacodeword there are six octads, giving a total of $2^6 \times 6 = 384$ octads. In the even interpretation we can only get an octad if the hexacodeword contains at least two 0s. If it is the all 0s hexacodeword then, as we have seen, we obtain the 15 octads consisting of the union of two columns. If the hexacodeword has just two 0s, then we can complement on any three of the non-zero columns to get eight octads, giving a total of $(36 + 9) \times 8 = 360$ octads. So altogether we obtain

$$384 + 15 + 360 = 759 \text{ octads.}$$

This provides a more concise proof that the octads form an $S(5, 8, 24)$ than the Point-Line space construction of Chapter 4, although it should be noted that the latter does afford a unique label for every $\mathscr{C}$-set.

# 7

# Elements of the Mathieu Group $M_{24}$

We have defined the group $M_{24}$ by constructing $S$, a Steiner system S(5, 8, 24) on a set $\Omega$, labelling $\Omega$ with $\{\infty, 0, 1, \ldots, 22\}$, the points of the projective line $P_1(23)$ and then setting

$$M_{24} := \{\pi \in S_{24} \mid S^\pi = S\}.$$

Our proof that the Steiner system is unique up to relabelling of its 24 points told us a great deal about $M_{24}$; in particular, we know that it acts quintuply transitively on the 24 points and has order 244 823 040. However, we know very little about the elements of the group. For instance, we have labelled $\Omega$ with $P_1(23)$ because we have *asserted* that $M_{24}$ contains subgroups isomorphic to the projective special linear group $L_2(23)$, and that linear fractional maps acting on the projective line preserve the Steiner system and hence lie in $M_{24}$. The question arises: How do we verify these assertions? In theory, if $\pi \in S_{24}$ in order to show that $\pi \in M_{24}$ as defined earlier, we need to show that the image of every octad of $S$ is in $S$. However, we can do much better than that! Recall that

$$\langle S \rangle = \mathscr{C},$$

a copy of the 12-dimensional binary Golay code. So, if we take 12 octads that form a basis for this space and check that their images under $\pi$ are all in $S$ then we shall know that $\pi \in M_{24}$. Indeed, the basis we choose could consist of the whole set $\Omega$ together with 11 octads and, since $\Omega$ is certainly fixed by $\pi$, we need only to check the images of these 11 octads.

## 7.1 Eight Octads Suffice

However, in Curtis (1984) we produced a set of just eight octads that can be extended to a copy of $S$ in a unique way, and so if the image under $\pi$ of each

## 7.1 Eight Octads Suffice

these eight octads lies in $S$ then $\pi \in M_{24}$. We proved moreover that any set of seven octads is contained in no copy of $S$, or more than one copy, and so this test is optimal.

Before proving the theorem we investigate the properties of the sextets in a little more detail. We have seen that the dual space $\mathscr{C}^*$ consists of 24 monads, 276 duads, 2024 triads and 1771 sextets. However, every *even* subset of $\Omega$ is congruent modulo $\mathscr{C}$ to a duad or a sextet, and so these $(276 + 1771) = 2047 = 2^{11} - 1$ elements together with the zero vector form an 11-dimensional subspace. From the sextet intersection matrices given in Lemma 5.1, page 36, we see that there are just two ways in which two sextets can intersect in such a way that their sum is another sextet. We refer to these as the *even* and *odd lines of sextets*. In Figure 7.1 we exhibit the canonical even and odd lines. Note, in passing, that the subgroup of $M_{24}$ fixing the even line must preserve the MOG trio as it is the only trio, see Section 8.1.7, of which each of the three sextets of the line is one of its seven refinements. Moreover the pairing as indicated in Figure 7.2 is invariant as these are the only pairs that occur together in a tetrad of each of the three sextets. So the group of the even line fixes a line of refinements of a fixed trio and that is the only restriction. It thus has index 7 in the trio group and has shape $2^6 : (S_3 \times S_4)$, where the $S_3$ permutes the three octads of the fixed trio, the $S_4$ acts on the seven refinements fixing a line, and the $2^6$ fixes the three octads and all refinements.

| × | × | 1 | 1 | 3 | 3 |
|---|---|---|---|---|---|
| × | × | 1 | 1 | 3 | 3 |
| o | o | 2 | 2 | 4 | 4 |
| o | o | 2 | 2 | 4 | 4 |

| × | × | 1 | 1 | 3 | 3 |
|---|---|---|---|---|---|
| o | o | 2 | 2 | 4 | 4 |
| × | × | 1 | 1 | 3 | 3 |
| o | o | 2 | 2 | 4 | 4 |

| × | × | 1 | 1 | 3 | 3 |
|---|---|---|---|---|---|
| o | o | 2 | 2 | 4 | 4 |
| o | o | 2 | 2 | 4 | 4 |
| × | × | 1 | 1 | 3 | 3 |

The canonical even line of sextets

| × | o | 1 | 1 | 1 | 1 |
|---|---|---|---|---|---|
| o | × | 2 | 2 | 2 | 2 |
| o | × | 3 | 3 | 3 | 3 |
| o | × | 4 | 4 | 4 | 4 |

| 1 | 1 | × | o | 1 | 1 |
|---|---|---|---|---|---|
| 2 | 2 | o | × | 2 | 2 |
| 3 | 3 | o | × | 3 | 3 |
| 4 | 4 | o | × | 4 | 4 |

| 1 | 1 | 1 | 1 | × | o |
|---|---|---|---|---|---|
| 2 | 2 | 2 | 2 | o | × |
| 3 | 3 | 3 | 3 | o | × |
| 4 | 4 | 4 | 4 | o | × |

The canonical odd line of sextets

Figure 7.1 The even and odd lines of sextets

The group of the odd line must also fix the MOG trio as the tetrads indicated by × and o in Figure 7.1 are the only tetrads that cut the other two sextets as $0^2.1^4$. Moreover the pairing of Figure 7.2 is also preserved as these pairs are the only ones that appear together in two of the three sextets; and so the group of the canonical odd line is contained in the group of the canonical even line. Now there are $4 \times 4 = 16$ ways in which we can choose special tetrads in the 16-ads $\Lambda_1 \cup \Lambda_2$ and $\Lambda_1 \cup \Lambda_3$, consisting of a pair as in Figure 7.2 in each of the two bricks (the special tetrad in $\Lambda_2 \cup \Lambda_3$ is then forced), and so the subgroup has index 16 in the group of the associated even line. It has shape $2^4 : (S_3 \times S_3)$ where the two copies of $S_3$ act independently on the three octads of the trio and the three refinements of the fixed line; the $2^4$ fixes both sets.

Now if $U \leq \mathscr{C}$ then we define

$$U^o = \{\xi \in \mathscr{C}^* \mid (u, \xi) = 0 \text{ for all } u \in U\},$$

the *annihilator* of $U$ in the dual space of $\mathscr{C}$. So $U^o$ will be a subspace of $\mathscr{C}^*$ of dimension $12 - \dim U$. In the following we shall take

$$U = \langle \Omega, O_1, O_2, \ldots, O_r \rangle,$$

where $\{O_1, O_2, \ldots, O_r\}$ is a set of linearly independent octads of $\mathcal{S}$, no sum of which is the whole of $\Omega$; and so $U^o$ will have dimension $11 - r$. Note that $U^o$ must consist entirely of duads and sextets, since monads and triads do not annihilate $\Omega$.

**Lemma 7.1** *If $U$ is as above and $d = \{i, j\}$ is a duad with $d \in U^o$, then $U$ extends to more than one copy of the Golay code.*

*Proof* The permutation $(i\ j)$ fixes all the generators of $U$ and so

$$U \subset \mathscr{C} \cap \mathscr{C}^{(i\ j)}.$$

However, $\mathscr{C}^{(i\ j)}$ is a copy of the binary Golay code different from $\mathscr{C}$, since $M_{24}$ can contain no transpositions. □

**Corollary** *If $U$ extends to a unique copy of the Golay code, then $U^o$ consists entirely of sextets.*

**Lemma 7.2** *If $U = \langle \Omega, O_1, O_2, \ldots, O_{10} \rangle$ is an 11-dimensional subspace of $\mathscr{C}$ such that $U^o = \langle \chi_0 \rangle$, where $\chi_0$ is a sextet, then $U$ extends to a unique copy of the binary Golay code, namely $\mathscr{C}$.*

*Proof* Naturally we take $\chi_0$ to be the columns of the MOG, so any octad that cuts the columns of the MOG evenly is in $U$. We now ask, what is the octad $O$

## 7.1 Eight Octads Suffice

containing

[grid figure]

Now $O$ must cut the sextet $\chi_0$ as $3.1^5$, and so has a further point in each of the last three columns. Each of the octads

[grid figure] and [grid figure]

of $U$ must contain a further point of $O$ that must lie in the 4th column and thus be the top point. Similarly,

[grid figure] and [grid figure]

and

[grid figure] and [grid figure]

force $O$ to contain the top points in the 5th and 6th columns. Thus

$$O = \text{[grid figure]}.$$

Thus $\langle U, O \rangle = \mathscr{C}$ and the result is proved. □

**Lemma 7.3** *If $U = \langle \Omega, O_1, O_2, \ldots, O_9 \rangle$ is a 10-dimensional subspace of $\mathscr{C}$ such that $U^o = \langle \chi_1, \chi_2 \rangle = \epsilon$ is an even line of sextets, then $U$ extends to more than one copy of the binary Golay code.*

*Proof* Associated with every even line $\epsilon$ is $\pi(\epsilon)$, a unique fixed point free involution of $S_{24}$ that fixes every tetrad of $\epsilon$. Note that this permutation does *not* belong to $M_{24}$. For the canonical even line of Figure 7.1, this permutation is exhibited in Figure 7.2 where we have included the permutation $\rho \in M_{24}$

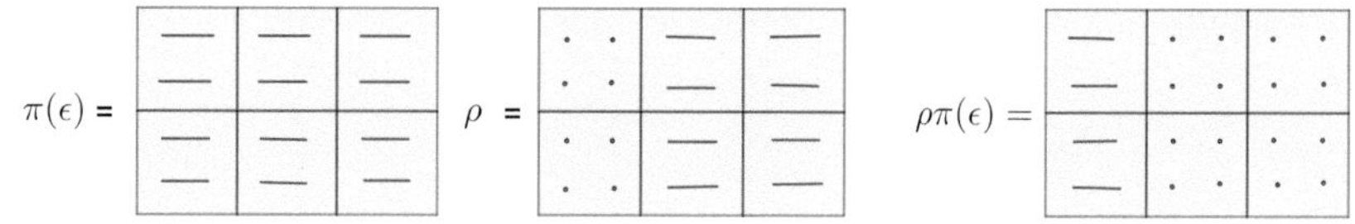

Figure 7.2 The fixed point free involution associated with the even line $\epsilon$

that is such that $\rho\pi(\epsilon)$ has cycle shape $2^4.1^{16}$; this will facilitate the necessary counting of octads. Since $\pi(\epsilon) \notin M_{24}$ we have $\mathscr{C}^{\pi(\epsilon)} \ne \mathscr{C}$; we shall show that

$$U \subset \mathscr{C}^{\pi(\epsilon)} \cap \mathscr{C} = \mathscr{C}^{\rho\pi(\epsilon)} \cap \mathscr{C}.$$

To do this we make use of a formula that may be deduced from the MacWilliams identities, see MacWilliams and Sloane (1977, p. 127): If $U$ is an $n$-dimensional subspace of $\mathscr{C}$ containing $\Omega$, $\delta$ is the number of duads in $U^o$, the annihilator of $U$, and $N$ is the number of octads in $U$ then we have

$$N = 2^{n-6}(12 + \delta) - 9. \tag{7.1}$$

In our case we have $n = 10$ and $\delta = 0$ and so $N = 183$. We shall show that this set of octads is fixed by $\rho\pi(\epsilon)$. However,

$$\{O \mid O \in \mathscr{C} \cap \mathscr{C}^{\rho\pi(\epsilon)}\} = \{O \in \mathscr{C} \mid O^{\rho\pi(\epsilon)} = O\} \cup \{O \in \mathscr{C} \mid O + O^{\rho\pi(\epsilon)} = \Lambda_1\}.$$

We may readily count that there are $30 + 4.16 + 6.4 + 1 = 119$ octads of the first kind, depending on whether they intersect $\Lambda_1$ in 0, 2, 4 or 8 points, and there are $2^4.4 = 64$ of the second type, making a total of 183 as required. Thus $U$ can be extended to distinct copies of the Golay code, either $\mathscr{C}$ or $\mathscr{C}^{\pi(\epsilon)}$. □

**Corollary** *The annihilator of a subspace $U = \langle \Omega, O_1, O_2, \ldots, O_r \rangle$ that extends to a unique copy of $\mathscr{C}$ must consist entirely of sextets, and all 2-dimensional subspaces must be odd lines.*

**Lemma 7.4** *If $U = \langle \Omega, O_1, O_2, \ldots, O_9 \rangle$ is a 10-dimensional subspace of $\mathscr{C}$ such that $U^o = \langle \chi_1, \chi_2 \rangle = \sigma$, an odd line of sextets, then $U$ extends to a unique copy of the binary Golay code, namely $\mathscr{C}$.*

*Proof* We shall, of course, choose $\sigma$ to be the canonical odd line of Figure 7.1 and label these three sextets $\chi_1, \chi_2$ and $\chi_3$ as they appear in Figure 7.1. Note that all octads that consist of two rows of one of the three 16-ads $\Omega \setminus \Lambda_i$ for

## 7.1 Eight Octads Suffice

$i \in \{1, 2, 3\}$ are annihilated by $\sigma$. We now ask what is the octad $O$ containing

$$X = \begin{array}{|cc|cc|cc|} \hline \times & \cdot & \times & \cdot & \cdot & \cdot \\ \times & \cdot & \cdot & \cdot & \cdot & \cdot \\ \hline \times & \cdot & \cdot & \cdot & \cdot & \cdot \\ \times & \cdot & \cdot & \cdot & \cdot & \cdot \\ \hline \end{array}.$$

This set cuts $\chi_1$ as $3.1^2.0^3$ and so $O$ must cut it as $3.1^5$; moreover $X$ cuts $\chi_2$ as $1^5.0^3$ and so $O$ must cut $\chi_2$ oddly as $3.1^5$. Since $O$ cuts $\chi_1$ and $\chi_2$ oddly, it must cut $\chi_3 = \chi_1 + \chi_2$ evenly as $2^4.0^2$ and so contain one further point from each of the 2nd, 3rd and 4th rows of $\Omega \setminus \Lambda_3$. These points cannot, of course, lie in the 2nd column of the MOG or we should have five points of $\Lambda_1$; thus they all lie in the 3rd or 4th columns. But, each of

$$\begin{array}{|cc|cc|cc|} \hline \times & \cdot & \cdot & \times & \cdot & \times \\ \times & \cdot & \times & \cdot & \times & \cdot \\ \hline \times & \times & \cdot & \cdot & \cdot & \cdot \\ \cdot & \cdot & \cdot & \cdot & \cdot & \cdot \\ \hline \end{array} \quad \begin{array}{|cc|cc|cc|} \hline \times & \cdot & \cdot & \times & \cdot & \times \\ \cdot & \cdot & \cdot & \cdot & \cdot & \cdot \\ \hline \times & \cdot & \times & \cdot & \times & \cdot \\ \times & \times & \cdot & \cdot & \cdot & \cdot \\ \hline \end{array} \text{ and } \begin{array}{|cc|cc|cc|} \hline \times & \cdot & \cdot & \times & \cdot & \times \\ \times & \times & \cdot & \cdot & \cdot & \cdot \\ \hline \cdot & \cdot & \cdot & \cdot & \cdot & \cdot \\ \times & \cdot & \times & \cdot & \times & \cdot \\ \hline \end{array}$$

is annihilated by $\sigma$ and contains three points of $X$; thus $O$ must contain a further point of each of them and so we see that

$$O = \begin{array}{|cc|cc|cc|} \hline \times & \cdot & \times & \cdot & \cdot & \cdot \\ \times & \cdot & \times & \cdot & \cdot & \cdot \\ \hline \times & \cdot & \times & \cdot & \cdot & \cdot \\ \times & \cdot & \times & \cdot & \cdot & \cdot \\ \hline \end{array}.$$

Then $\langle U, O \rangle$ is the 11-dimensional subspace annihilated by $\chi_3$, and we have seen that such a space extends uniquely to $\mathscr{C}$. □

We are now in a position to prove the main result in this section.

**Theorem 7.5** (Eight Octads Suffice) *There exist octads $\{O_1, O_2, \ldots, O_8\}$ that lie in one and only one Steiner system $S(5, 8, 24)$. Such a set of octads is spatially unique in the sense that if $\{U_1, U_2, \ldots, U_8\}$ is another set of octads with this property, then there exists a permutation $\pi \in M_{24}$ such that*

$$\langle \Omega, O_1, O_2, \ldots, O_8 \rangle^\pi = \langle \Omega, U_1, U_2, \ldots, U_8 \rangle.$$

*Any set of seven 8-element subsets of $\Omega$ is contained in no $S(5, 8, 24)$ or more than one.*

*Proof* The purely sextet subspaces of $\mathscr{C}^*$ were completely classified by hand in Curtis (1972). However, at this point we are only interested in those subspaces all of whose 2-dimensional subspaces are odd lines. It turns out that there are

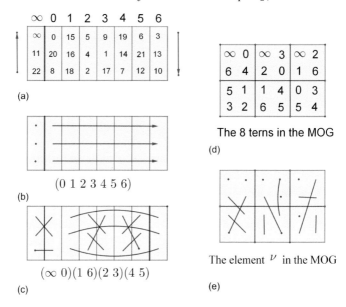

Figure 7.3 Generators of the octern group, showing the eight terns

just two orbits of $M_{24}$ on 3-dimensional subspaces of $\mathscr{C}^*$ with this property, as has been confirmed computationally using MAGMA, see Chapter 8. The first of these is known as $\text{Oct}_3$ as it is preserved by the *octern group* On, a maximal subgroup of $M_{24}$ isomorphic to $L_2(7)$, which is the centralizer in $M_{24}$ of $\tau$, a fixed point free permutation of order 3 outside $M_{24}$. We shall say a little more about this group in the section on maximal subgroups, see Section 8.1.9, but in Figure 7.3 we exhibit the 24 points of $\Omega$ arranged as a $3 \times 8$ array whose columns are preserved by the octern group; it is generated by the two permutations given here and $\tau$, in Figure 7.3(a), fixes all the columns, rotating the first column upwards and all other columns downwards. Now the normalizer of On in $S_{24}$ is a subgroup of shape $3 \times \text{PGL}_2(7)$ and if $v$ is an involution in this group but outside $M_{24}$ then $v$ fixes $U$, the 9-dimensional subspace of $\mathscr{C}$ annihilated by $\text{Oct}_3$. In fact using Equation 7.1 we see that the 9-dimensional subspace of $\mathscr{C}$ annihilated by On contains 87 octads. These split into three orbits of sizes 21, 24 and 42 under the action of $\text{Oct}_3$ and representatives of each orbit are exhibited in Figure 7.4. They consists of all octads that intersect each of the eight terns with the same parity. We may take

$$v = (1\ 14)(2\ 12)(3\ 15)(4\ 21)(5\ 6)(7\ 17)(9\ 19)(10\ 18)(13\ 16),$$

the element that acts as $x \mapsto -x \equiv (\infty)(0)(1\ 6)(2\ 5)(3\ 4)$ on the terns and preserves the rows of the octern arrangement of Figure 7.3. It stabilizes each of

## 7.1 Eight Octads Suffice

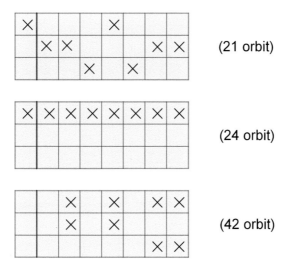

Figure 7.4 Orbit representatives of the 87 = 21 + 24 + 42 octads annihilated by $\mathrm{Oct}_3$

the three orbits whose representatives are given in Figure 7.4, but it certainly does not lie in $M_{24}$. So $\mathscr{C} \neq \mathscr{C}^{\nu}$ and we have

$$U \subseteq \mathscr{C} \cap \mathscr{C}^{\nu} \subsetneq \mathscr{C};$$

so $U$ can be extended to more than one copy of the binary Golay code and $\mathrm{Oct}_3$ will not afford the desired set of eight octads.

The other 3-dimensional purely sextet space all of whose 2-dimensional subspaces are odd lines is known as $\mathrm{Pet}_3$, see Figure 7.5, as it is contained in a larger space that is related to the Petersen graph. We obtain a copy of $\mathrm{Pet}_3$ by adjoining the sextet $\chi_4$ to the canonical odd line $\{\chi_1, \chi_2, \chi_3\}$. We see that the octads $\{O_1, O_2, \ldots, O_8\}$ in Figure 7.6 intersect each of $\chi_1, \chi_2$ and $\chi_4$ evenly, and so they intersect the whole $\mathrm{Pet}_3$ space evenly; however, the octads $\{O_9, O_{10}, O_{11}\}$ do not.

We shall show that the only Steiner system $S(5, 8, 24)$ containing the set of octads $\{O_1, O_2, \ldots, O_8\}$ is the one given by the MOG. First note that each of

| · | · | 1 | 1 | 2 | 2 |
|---|---|---|---|---|---|
| · | · | 1 | 1 | 2 | 2 |
| · | · | 3 | 3 | 4 | 4 |
| · | · | 3 | 3 | 4 | 4 |

and

| 1 | 1 | · | · | 2 | 2 |
|---|---|---|---|---|---|
| 3 | 3 | · | · | 4 | 4 |
| 1 | 1 | · | · | 2 | 2 |
| 3 | 3 | · | · | 4 | 4 |

forms four tetrads of a sextet, the union of any two of which is an octad in the

54                    *Elements of the Mathieu Group* $M_{24}$

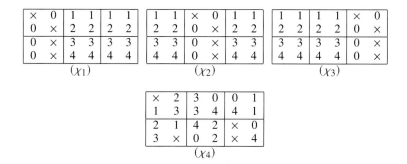

Figure 7.5 $Pet_3$, a purely sextet subspace with no even lines

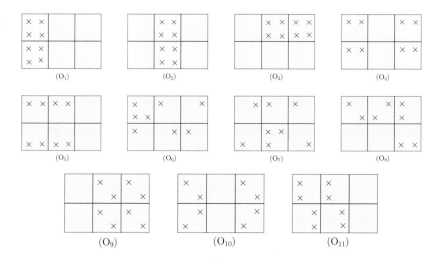

Figure 7.6 Eight octads that define the $S(5, 8, 24)$ and their extension to a basis of $\mathscr{C}$

$Pet_3$ subspace. We ask what are the octads containing

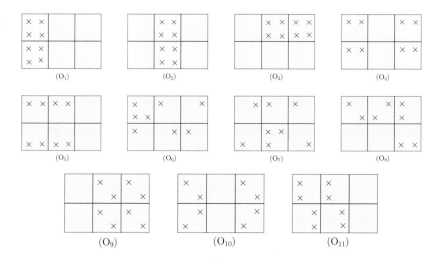

In the case of $Y_1$ the last three points must lie in $\Lambda_1$ and in the case of $Y_2$ they must lie in $\Lambda_2$. For $Y_1$ the only 8-element sets that intersect each of $\{O_1, O_2, \ldots, O_8\}$

## 7.1 Eight Octads Suffice

in 0, 2 or 4 points are

$$(O_a) \quad \text{or} \quad (O_b).$$

The first case $O_a$ is, of course, what happens in the MOG and so in this case we have extended $U = \langle \Omega, O_1, O_2, \ldots, O_8 \rangle$ to a 10-dimensional subspace whose annihilator is an odd line. We have seen in Lemma 7.4 that such spaces extend uniquely to $\mathscr{C}$ and we are done. So we must assume that $Y_1$ extends to $O_b$.

Similarly with $Y_2$ there are just two possibilities:

$$(O_c) \quad \text{or} \quad (O_d).$$

Again $O_c$ is what happens in the MOG and so Lemma 7.4 shows that this configuration extends uniquely to $\mathscr{C}$. So we must assume that $Y_2$ extends to $O_d$. But now we find that

$$(O_b) + (\delta_b \in U) = (\tau_b)$$

and

$$(O_d) + (\delta_d \in U) = (\tau_d).$$

But, as we see, $\tau_b$ and $\tau_d$ have six points in common and so cannot lie in any $S(5, 8, 24)$; thus the extensions to $O_b$ and $O_d$ are impossible and the extensions must occur within $\mathscr{C}$.

So we have shown that $\{O_1, O_2, \ldots, O_8\}$ extends to a unique Steiner system $S(5, 8, 24)$. It remains to prove the last assertion that claims that any set of seven octads of $\mathscr{C}$, $\{O_1, O_2, \ldots, O_7\}$ say, can be extended to more than one copy of the binary Golay code. If

$$U = \langle \Omega, U_1, U_2, \ldots, U_7 \rangle,$$

then $U^o$ must be a 4-dimensional purely sextet subspace of $\mathscr{C}^*$ all of whose 3-dimensional subspaces are copies of Pet$_3$. There is in fact just one class of such subspaces a copy of which can be obtained by extending the Pet$_3$ subspace by the sextet $\chi_8$. The stabilizer of this subspace, which is known as Pet$_4$, is isomorphic to $S_5$ and may be generated by p4a and p4b below

$$\begin{aligned}
\text{p4a} &= (1\ 9\ \infty\ 0\ 16\ 11)(2\ 15\ 8\ 22\ 21\ 7) \\
&\quad (3\ 19\ 6\ 4\ 20\ 13)(5\ 18\ 17\ 14\ 10\ 12);
\end{aligned}$$

$$\begin{aligned}
\text{p4b} &= (1\ 7\ 19)(2\ 9\ 20)(3\ 11\ 12)(4\ 22\ 5) \\
&\quad (6\ 14\ 8)(10\ \infty\ 13)(15\ 16\ 17)(18\ 0\ 21);
\end{aligned}$$

$$\begin{aligned}
\text{cp4} &= (1\ 18\ 13\ 8)(2\ 16\ 12\ 4)(3\ 22\ 9\ 17) \\
&\quad (5\ 20\ 15\ 11)(6\ 7\ 0\ 10)(14\ 19\ 21\ \infty).
\end{aligned}$$

| × | 0 | 1 | 1 | 1 | 1 | 1 | 1 | × | 0 | 1 | 1 |
|---|---|---|---|---|---|---|---|---|---|---|---|
| 0 | × | 2 | 2 | 2 | 2 | 2 | 2 | 0 | × | 2 | 2 |
| 0 | × | 3 | 3 | 3 | 3 | 3 | 3 | 0 | × | 3 | 3 |
| 0 | × | 4 | 4 | 4 | 4 | 4 | 4 | 0 | × | 4 | 4 |

$(\chi_1)$ $\qquad\qquad$ $(\chi_2)$

| × | 2 | 3 | 0 | 0 | 1 | × | 1 | 0 | 2 | × | 4 |
|---|---|---|---|---|---|---|---|---|---|---|---|
| 1 | 3 | 3 | 4 | 4 | 1 | 2 | × | 1 | 0 | 0 | 3 |
| 2 | 1 | 4 | 2 | × | 0 | 2 | 3 | 2 | 4 | 0 | × |
| 3 | × | 0 | 2 | × | 4 | 3 | 1 | 4 | 1 | 3 | 4 |

$(\chi_4)$ $\qquad\qquad$ $(\chi_8)$

Figure 7.7 Pet$_4$, a purely sextet subspace with no even lines

This subgroup of M$_{24}$ acts transitively on the points of $\Omega$ and on the 15 sextets of Pet$_4$. It centralizes a cyclic subgroup of order 4, in S$_{24}$ and outside M$_{24}$, generated by cp4 above. The element cp4 does not fix Pet$_4$; however, its square that is given here as $\rho$ does fix it. This can be verified from the set of generating octads given in Figure 7.8 where for convenience $\rho$ is also displayed in the MOG.

$$\begin{aligned}
\rho := &\ (1\ 13)(2\ 12)(3\ 9)(4\ 16)(5\ 15)(6\ 0) \\
&\ (7\ 10)(8\ 18)(11\ 20)(14\ 21)(17\ 22)(19\ \infty)
\end{aligned}$$

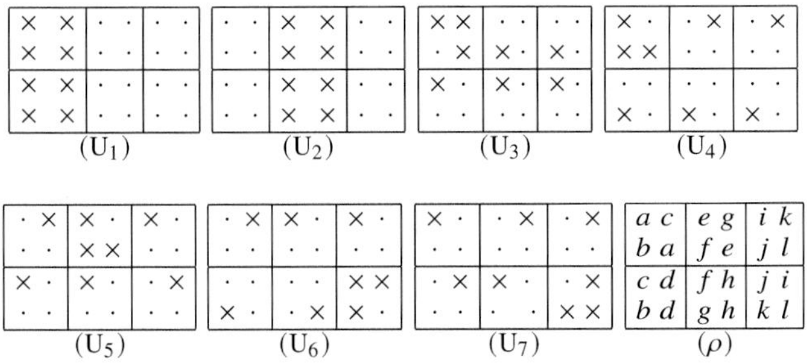

## 7.1 Eight Octads Suffice

Figure 7.8 The Pet$_4$ space octads

The 8-dimensional subspace annihilated by Pet$_4$ is spanned by

$$U = \langle \Omega, U_1, U_2, \ldots, U_7 \rangle,$$

from Figure 7.8 as these seven octads are linearly independent of one another and annihilate the four sextets in Figure 7.7. The element $\rho$ fixes each of $U_i$ for $i$ in $1, \ldots, 5$ and we see that

$$U_6^\rho = U_2 + U_3 + U_4 + U_6 \text{ and } U_7^\rho = U_2 + U_3 + U_4 + U_7 + \Omega,$$

and so $\rho$ fixes the space $U$. However, $\rho$ does not fix $\mathscr{C}$ (consider, for instance, the image of the first and third columns of the MOG) and so we have

$$U \subset \mathscr{C} \cap \mathscr{C}^\rho \subsetneq \mathscr{C}.$$

Since this was the only candidate for an 8-dimensional space that extends uniquely to $\mathscr{C}$ and it fails, we see that our theorem is indeed best possible. □

This gives us the simplest possible test for verifying if a permutation of $\Omega$ lies in our copy of $M_{24}$:

**Corollary** *If $\pi \in S_{24}$ then $\pi \in M_{24}$ if, and only if, the image under $\pi$ of each of $O_1, O_2, \ldots, O_8$ of Figure 7.6 intersects each of $O_1, O_2, \ldots, O_{11}$ evenly.*

*Proof* We know that $\mathscr{C}^\perp = \mathscr{C}$, and so if $X \subset \Omega$ then

$$X \in \mathscr{C} \iff (X, U) = 0 \text{ for all } U \in \mathscr{C}$$
$$\iff (X, U_i) = 0 \text{ for } \{U_1, \ldots, U_{12}\} \text{ a basis for } \mathscr{C}.$$

But our octads $O_1, \ldots, O_{11}$ are chosen so as to make a basis together with $\Omega$, and any even subset is orthogonal to $\Omega$. So even intersection of $O_i^\pi$ with these 11 octads ensures that $O_i^\pi \in \mathscr{C}$. But we have seen that

$$\langle O_1, O_2, \ldots, O_8 \rangle \text{ and } \langle O_1^\pi, O_2^\pi, \ldots, O_8^\pi \rangle$$

extend to unique copies of $\mathscr{C}$, namely $\mathscr{C}$ itself. Thus $\mathscr{C}^\pi = \mathscr{C}$ and $\pi \in M_{24}$. □

**Corollary**  *The projective special linear group* $L_2(23)$ *is a subgroup of* $M_{24}$ *acting on the projective line as labelled in the MOG.*

*Proof*  We may readily use this test to check that

$$\begin{aligned}\alpha &= x \mapsto x+1 &\equiv& \quad (\infty)(0\ 1\ 2\ \ldots 22) \text{ and} \\ \gamma &= x \mapsto -\tfrac{1}{x} &\equiv& \quad (\infty\ 0)(1\ 22)(2\ 11)(3\ 15)(4\ 17)(5\ 9) \\ & & & \quad (6\ 19)(7\ 13)(8\ 20)(10\ 16)(12\ 21)(14\ 18)\end{aligned}$$

lie in $M_{24}$. Together they generate $L_2(23)$. □

### 7.1.1 Some Useful Elements of $M_{24}$

We now have a useful test that enables us to tell whether or not a permutation of $S_{24}$ lies in our copy of $M_{24}$. There are though several classes of elements that may readily be written down.

#### Elements of Cycle Shape $1^8 \cdot 2^8$

The element of order two that fixes every point of $\Lambda_1$ and bodily interchanges $\Lambda_2$ and $\Lambda_3$ visibly preserves the Steiner system and has cycle shape $1^8 \cdot 2^8$. Indeed, from the manner in which the sets of four special tetrads in the 16-ad $\Omega \setminus \Lambda_1 = \Lambda_2 \cup \Lambda_3$ for instance are defined, it is clear that permutations fixing $\Lambda_1$ pointwise and fixing the rows of $\Omega \setminus \Lambda_1$ whilst interchanging its columns in pairs, or fixing its columns whilst interchanging its rows in pairs preserve the set of special tetrads in $\Lambda_2 \cup \Lambda_3$. These generate an elementary abelian group of order 16 consisting of fixed point free permutations on $\Omega \setminus \Lambda_1$ whose elements include every transposition on its 16 points. This means that the set of all octads intersecting $\Lambda_1$ in four points is preserved by these elements, and Lemma 5.2 tells us that this implies that the set of all octads is preserved. Thus, given any octad O and any two points $i$ and $j$ outside O there is an element of $M_{24}$ fixing each point of O and interchanging $i$ and $j$. Examples of how this works in practice are given in Figure 7.9.

#### Elements of Cycle Shape $2^{12}$

Figure 7.10(a) shows how two involutions of cycle shape $1^8 \cdot 2^8$ may be multiplied together to get a fixed point free involution $\pi$. Further scrutiny of this involution shows that its 12 transpositions may be paired in a unique manner so as to form the tetrads of a sextet $\chi(\pi)$, so its centralizer will fix $\chi(\pi)$. Furthermore, this $\pi$ fixes a dodecad $D(\pi)$ that cuts $\chi(\pi)$ as $2^6$ (in fact, there are $2^5$ such dodecads), and $\chi(\pi)$ and $D(\pi)$ determine $\pi$. In Figure 7.10 we give some

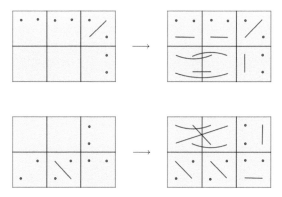

Figure 7.9 Constructing involutions of cycle shape $1^8 . 2^8$

examples of this class of involutions, together with their defining sextets and dodecads.

### Elements of Cycle Shape $1^6 . 3^6$

Our proof of the uniqueness of the Steiner system S(5, 8, 24) showed that the group $M_{24}$ acts sharply transitively on ordered sets $\{x_1, x_2, \ldots, x_7\}$ where $\{x_1, x_2, \ldots, x_6\}$ lie in an octad and $x_7$ is a point outside it. This implies that $M_{24}$ has just two orbits on hexads of points of $\Omega$: *special* hexads of type $S_6$ that lie in an octad, and *umbral* hexads of type $U_6$ that do not. It implies, furthermore, that the stabilizer of either an $S_6$ or a $U_6$ acts as the symmetric group $S_6$ on its six points, and that in the case of a $U_6$ any non-trivial element fixing each of its points must act fixed point freely on the remainder. The order of the group shows that such an element must have order 3 and so cycle shape $1^6 . 3^6$.

Let $\tau$ be such an element and suppose that its fixed point set is the canonical umbral hexad that is the top row of the MOG. Now the octad containing any five points of an umbral hexad must be fixed by $\tau$ and so the six triples that comprise its 3-cycles are determined: They are simply the triples that complete five points of the $U_6$ to an octad. In the case of $\tau$ these triples are the remaining three points of the columns. Indeed, we see that an umbral hexad $U$ is congruent modulo $\mathscr{C}$ to a sextet (by, for instance, adding one of the six octads that contain five points of $U$ to $U$) that has a fixed point and a 3-cycle in each tetrad. We are free to choose the sense of rotation of one of the 3-cycles, then deduce the other senses by consideration of the images of suitable octads. Figure 7.11 illustrates this construction: choose an umbral hexad $U$; identify the triple that complete each five points of $U$ to an octad; choose a sense of cycling one of these triples

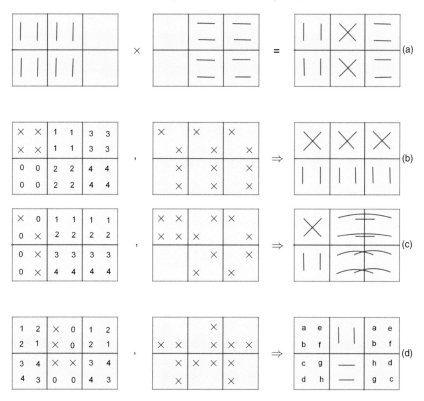

Figure 7.10 Constructing involutions of cycle shape $2^{12}$

and then complete the other senses by following the images of suitable octads. We have seen that the stabilizer of a $U_6$ in $M_{24}$ acts as the symmetric group $S_6$ on its points and that this group must normalize the cyclic subgroup of order 3 that fixes every point of the $U_6$. However, the 2nd and 4th elements in Figure 7.11 generate an extraspecial group of order 27 that is a Sylow 3-subgroup of this normalizing subgroup. This subgroup can contain no copy of the direct product $3 \times A_6$, whose Sylow 3-subgroup is abelian, and so the normalizer of an element of order 3 in this conjugacy class must be $3 \cdot S_6$, the *non-split* triple cover of $S_6$.

### 7.1.2 The Orbits of $M_{24}$ on Subsets of $\Omega$

As earlier, we denote the power set of $\Omega$, the collection of all subsets of $\Omega$, by $P(\Omega)$. Being quintuply transitive on $\Omega$, $M_{24}$ has remarkably few orbits on

## 7.1 Eight Octads Suffice

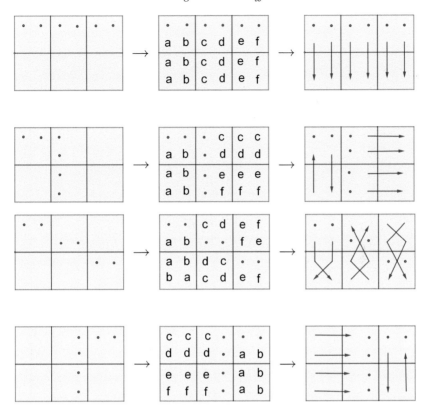

Figure 7.11 Constructing elements of cycle shape $1^6 . 3^6$

$k$-subsets. Todd classified these orbits, finding that there are 49, and Conway exhibited them in a convenient diagram that we reproduce here as Figure 7.12. An orbit on $k$-subsets will be denoted by $R_k$ where $R$ is one of $S, T$ or $U$, standing for Special, Transverse or Umbral. The special octads of $\mathscr{C}$ are in $S_8$ and, for $1 \leq k \leq 12$, $S_k$ consists of $k$-subsets that are contained in a special octad or contain one. The dodecads of the Golay code $\mathscr{C}$ are the *umbral dodecads* in $U_{12}$. The diagram tells us, for each $k$-subset, what orbit of $(k-1)$-subsets we shall arrive in if we remove a point, and what orbit of $(k+1)$-subsets we shall arrive in if we add a point. Thus, given a member of $T_8$, we see that there is a unique point we can remove to leave an $S_7$ and removal of

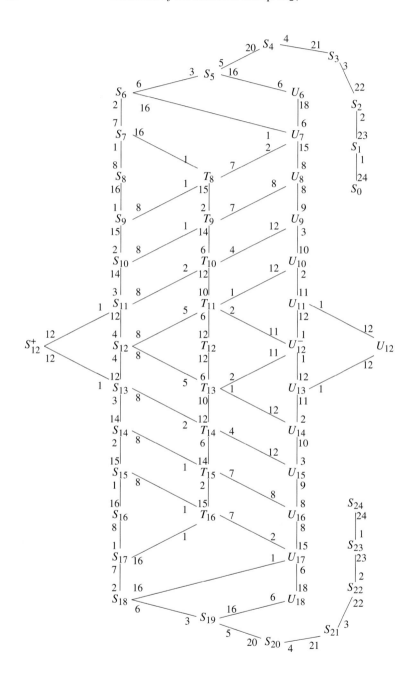

Figure 7.12 The action of $M_{24}$ on $P(\Omega)$

## 7.1 Eight Octads Suffice

any of the other seven leaves a $U_7$; also that there is a unique point whose addition results in an $S_9$ whilst addition of any of the remaining 16 points results in a $T_9$. It is useful in classifying these orbits to ask what a particular $k$-subset is congruent to modulo $\mathscr{C}$. Thus, for instance, if we remove from a $T_8$ labelled T the point that leaves an $S_7$, and then add the point that leads to an octad $O$, then we see that $T + O$ is a duad consisting of the point we have removed and the point we have added.

Subsets of size 12 are unusual in that there are two additional orbits: $S_{12}^+$, the extraspecial dodecads, consist of three tetrads of a sextet and thus contain three octads; and $U_{12}^-$, the *penumbral dodecads* contain all but one point of an umbral dodecad (so they are congruent modulo $\mathscr{C}$ to a duad). Sets of size greater than 12 are described by the same adjective as their complement, and the reader will observe that Figure 7.12 is vertically symmetrical.

### Some Other Interesting Subsets of $\Omega$

As we have seen, the canonical $U_6$ is the top row of the MOG, which is congruent modulo $\mathscr{C}$ to the columns of the MOG. It is the fixed point set of elements of a class of 3-elements, in this case the element that fixes the top row of the MOG and cycles the other three rows while preserving the columns. Its stabilizer is a subgroup of shape $3 \cdot S_6$ that is the normalizer of the aforementioned 3-element.

Two further orbits that are of considerable interest are $U_8$ and $T_{12}$ in that in each of these cases the subgroup fixing such a subset acts transitively on its points. In Figures 7.13 and 7.14 we give canonical examples of each of them together with generators for their stabilizing subgroups.

As can be deduced from Figure 7.12, a $U_8$ consists of four pairs of points ($A, B, C$ and $D$) the removal of any one of which results in an $S_6$; its stabilizing group has shape $2^4 : S_4$. The four pairs that complete these four $S_6$s to octads ($a, b, c$ and $d$) together comprise a disjoint $U_8$, and $aBCD$ and so on are octads. The subset $\Omega \setminus \{A, B, C, D, a, b, c, d\}$ is an octad of $\mathscr{C}$. In Figure 7.13 we exhibit the canonical example of a $U_8$ which is congruent modulo $\mathscr{C}$ to the columns of the MOG, and hence its stabilizing group fixes that sextet. The top-left permutation flips one of the pairs while fixing the other three; those labelled ($A$ $B$) and ($B$ $C$ $D$) generate a copy of $S_4$ acting on the four pairs; and the bottom-right permutation interchanges the two $U_8$s.

Analogously, from Figure 7.12 we see that removal of any point from a $T_{12}$ leads to a $T_{11}$, and that there is a unique point whose further removal leads to a $U_{10}$. There is then a unique pair of points that can be adjoined to this $U_{10}$ to give a $U_{12}$, which is of course a dodecad of the Golay code. Thus modulo $\mathscr{C}$ a $T_{12}$ is congruent to a tetrad, and it consists of six pairs of points whose removal leads

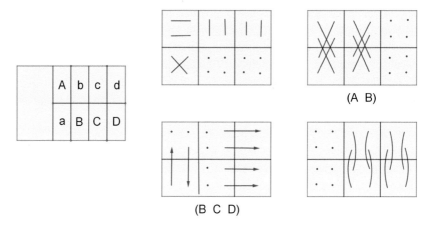

Figure 7.13 The canonical $U_8$ preserved by a subgroup of shape $2^4 : S_4$

to $U_{10}$s. Naturally we choose our canonical $T_{12}$ such that this fixed sextet is the columns of the MOG. We label these six pairs $A, B, C, D, E$ and $F$ and denote by $x$ the pair that must replace $X$ to give a dodecad, for $X = A, B, C, D, E$ or $F$; thus $aBCDEF$ is a dodecad of the Golay code $\mathscr{C}$. The involution of cycle shape $2^{12}$ that fixes each of these pairs lies in $M_{24}$ and, in the notation of Section 8.1.4, for our canonical $T_{12}$ in Figure 7.14(a) it is $(\bar{\omega}\ \omega\ \bar{\omega}\ \omega\ \bar{\omega}\ \omega)$. So the subgroup of $M_{24}$ preserving a $T_{12}$ and its complement must fix the associated sextet and centralize this involution. The involution $(\omega\ 1\ \omega\ 1\ \omega\ 1)$ interchanges our $T_{12}$ and its complement, and if we label the columns of the MOG with the projective line $P_1(5) = \{\infty, 0, 1, 2, 3, 4\}$ then the stabilizer of the two complementary $T_{12}$s has shape

$$(2^2 \times L_2(5)) : 2,$$

where the action on the columns is as indicated in Figure 7.14. Finally note that the subgroup of this group fixing the two $T_{12}$s has shape

$$(2 \times L_2(5))\dot{}2$$

in which the outer half contains no involution but contains elements of order 4 that square to the central element and normalize the linear group.

## 7.2 Using a Computer Package

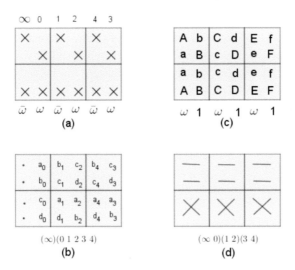

Figure 7.14 The canonical $T_{12}$ and its complement preserved by a subgroup of shape $(2^2 \times L_2(5)) : 2$.

## 7.2 Using a Computer Package

The Miracle Octad Generator was devised for working with the Golay code and the Mathieu group $M_{24}$ in the days before personal computers were widely available. Nowadays computer packages such as MAGMA, see Bosma and Cannon (1994), and GAP, see Neubueser et al. (2006), handle permutation groups of very high degree with immense ease, and for them permutations on 24 letters, or indeed on a few thousand points such as the 1771 sextets, is child's play. Indeed it would be perverse not to make use of these wonderful packages if one has access to them. However, I would maintain that combining the techniques described in this book with computation leads to a deeper understanding of the beautiful structures with which we are dealing, and so in Chapter 8 I shall describe how you can obtain in MAGMA canonical copies of each of the main subgroups dealt with here. The computer code is in general self-explanatory; where it is not we shall provide a subtext.

### 7.2.1 The Group $M_{24}$ Itself

As has been mentioned earlier, the 24 points of $\Omega$ are labelled with the points of the projective line $P_1(23)$ because $M_{24}$ contains copies of the projective special linear group $L_2(23)$. Indeed, the only transitive (and thus doubly transitive) subgroups of $M_{24}$ containing elements of order 23 are copies of this group. But involutions in $L_2(23)$ are fixed point free, and so any element of cycle shape $1^8.2^8$ that moves the fixed point of the 23-cycle will, together with it, generate $M_{24}$. I usually choose the element that interchanges the 1st and 2nd, 3rd and 4th columns of the MOG whilst fixing the rows. Thus

```
> m24:=sub<Sym(24)|
> (1,2,3,4,5,6,7,8,9,10,11,12,13,14,15,16,17,18,19,20,21,22,23),
> (24,23)(3,19)(6,15)(9,5)(11,1)(4,20)(16,14)(13,21)>;
> #m24;
244823040
```

Here 0 is denoted by 23 and $\infty$ by 24.

# 8
# The Maximal Subgroups of $M_{24}$

This chapter is in two parts.[1] The first part describes each of the maximal subgroups of $M_{24}$ and the role it plays in the group. Generators (as permutations on 24 points) are given and MAGMA code is provided that enables the reader immediately to obtain a canonical copy of each of them. The second part provides a combinatorial proof, as appeared in the author's PhD thesis, that this list of maximal subgroups is complete.

Firstly we require generators for $M_{24}$ itself, as described in Section 7.2.1. Thus we take

$\alpha : x \mapsto x+1 \equiv (\infty)(0\ 1\ 2\ 3\ 4\ 5\ 6\ 7\ 8\ 9\ 10\ 11\ 12\ 13\ 14\ 15\ 16\ 17\ 18\ 19\ 20\ 21\ 22)$

and

$\gamma \equiv (\infty\ 0)(11\ 1)(3\ 19)(4\ 20)(6\ 15)(16\ 14)(9\ 5)(13\ 21),$

where $\gamma$ interchanges the 1st and 2nd, 3rd and 4th, columns of the MOG while fixing its rows. When working in MAGMA we must replace 0 by 23 and $\infty$ by 24.

## 8.1 The Maximal Subgroups of $M_{24}$

Any maximal subgroup of a simple group $G$ is either the normalizer in $G$ of an elementary abelian $p$-group, that is to say an abelian group all of whose non-trivial elements have order the prime $p$, or it is the normalizer of a direct product of isomorphic non-abelian simple groups. Subgroups of the first type are said to be *local* and the classfication of them is the *local analysis*; it can be tricky to carry out but is nonetheless systematic. Deciding whether or not $G$ possesses subgroups isomorphic to the simple group $H$ can be very difficult, especially if

---

[1] Ivanov (2018) covers some of the material in this chapter, but from a different point of view.

68                The Maximal Subgroups of $M_{24}$

$H$ is small in comparison to $G$. For instance, whether the smallest Janko group $J_1$ was a subgroup of the Monster group M remained an open question for several years, see Wilson (1986). The group $M_{24}$ is tiny in comparison to the Monster and yet a similar problem arose. Todd (1966) lists some of the maximal subgroups of $M_{24}$. He did not claim that this list was complete but a thesis was produced claiming that it was. Some time later in Caltech, McKay and Wales were performing an early computer calculation and found a transitive copy of $L_2(7)$ in $M_{24}$ that was not contained in any of the maximal subgroups in Todd's list. This small subgroup turned out to be maximal itself and Todd's original eight subgroups together with this addition form a complete list of maximal subgroups of $M_{24}$. Subsequently, the hole in the argument was identified and corrected, see Choi (1972). We discuss each of the maximal subgroups in the following sections.

### 8.1.1 The Stabilizer of a Monad

The stabilizer of a point in $M_{24}$ is the quadruply transitive sporadic simple group $M_{23}$ that preserves the Steiner system $S(4, 7, 23)$. In a similar way as mentioned earlier, the only maximal subgroups of $M_{23}$ that contain elements of order 23 are Frobenius groups of order $23 \times 11$. Thus an element $\alpha$ of order 23 together with a $1^8.2^8$ involution that fixes the fixed point of $\alpha$ will generate a copy of $M_{23}$. We take

$$\delta = (11\ 1)(22\ 2)(4\ 20)(18\ 10)(16\ 14)(8\ 17)(13\ 21)(12\ 7),$$

which fixes the points in the first two columns of the MOG and interchanges the 3rd and 4th, and 5th and 6th columns while preserving the rows. Then

```
> m23:=sub<m24|
> (1,2,3,4,5,6,7,8,9,10,11,12,13,14,15,16,17,18,19,20,21,22,23),
> (11,1)(22,2)(4,20)(18,10)(16,14)(8,17)(13,21)(12,7)>;
> #m23;
10200960
```

### 8.1.2 The Stabilizer of a Duad

The stabilizer of a pair of points is $M_{22}:2$, the automorphism group of the triply transitive Mathieu group $M_{22}$ that preserves the Steiner system $S(3, 6, 22)$. If we take $\infty$ and 0 as the fixed pair, then the group may be generated by

$$\beta = \quad x \mapsto 2x \quad \equiv \quad (1\ 2\ 4\ 8\ 16\ 9\ 18\ 13\ 3\ 6\ 12)$$
$$(5\ 10\ 20\ 17\ 11\ 22\ 21\ 19\ 15\ 7\ 14)$$

### 8.1 The Maximal Subgroups of $M_{24}$

together with $\gamma$ above, where $\beta$ is a linear fractional map of $L_2(23)$ of order 11. This element $\beta$ together with $\delta$ from the previous paragraph generate the simple group $M_{22}$.

```
> autm22:=sub<m24|
> (1,2,4,8,16,9,18,13,3,6,12)
> (5,10,20,17,11,22,21,19,15,7,14),
> (24,23)(11,1)(3,19)(4,20)(6,15)(16,14)(9,5)(13,21)>;
> #autm22;
887040
> m22:=sub<m24|
> (1,2,4,8,16,9,18,13,3,6,12)
> (5,10,20,17,11,22,21,19,15,7,14),
> (11,1)(22,2)(4,20)(18,10)(16,14)(8,17)(13,21)(12,7)>;
> #m22;
443520
```

#### 8.1.3 The Stabilizer of a Triad

The stabilizer of a three points, which is naturally known as $M_{21}$, is isomorphic to the projective special linear group $L_3(4)$, and the stabilizer of the triad they comprise is of shape $L_3(4) : S_3$. Of course, it preserves a Steiner system $S(2, 5, 21)$, which is the 21-point projective plane whose points are the $(4^3 - 1)/(4 - 1) = 21$ 1-dimensional subspaces of a 3-dimensional vector space

|  |  |  | $\xrightarrow{x}$ |  |  |  |  |  |
|---|---|---|---|---|---|---|---|---|
|  |  | 0 | 1 | $\omega$ | $\bar{\omega}$ |  |  |  |
| (100) 0 | (010) $\infty$ | (001) | (101) | ($\omega$01) | ($\bar{\omega}$01) | 0 |  |  |
| (110) 1 | × | (011) | (111) | ($\omega$11) | ($\bar{\omega}$11) | 1 | $\downarrow y$ |  |
| (1$\omega$0) $\omega$ | × | (0$\omega$1) | (1$\omega$1) | ($\omega\omega$1) | ($\bar{\omega}\omega$1) | $\omega$ |  |  |
| (1$\bar{\omega}$0) $\bar{\omega}$ | × | (0$\bar{\omega}$1) | (1$\bar{\omega}$1) | ($\omega\bar{\omega}$1) | ($\bar{\omega}\bar{\omega}$1) | $\bar{\omega}$ |  |  |

Figure 8.1 The 21-point projective plane as it appears in the MOG

over the Galois field $\mathbb{F}_4$, and whose lines are the

$$(4^3 - 1)(4^3 - 4)/(4^2 - 1)(4^2 - 4) = 21$$

2-dimensional subspaces. This configuration can be seen clearly in the MOG as is illustrated in Figure 8.1, where the 1-dimensional subspaces are represented by row vectors $(xyz)$ so that $3 \times 3$ matrices of $L_3(4)$ act on the right in the same way as our permutations do. The three fixed points are those marked with an × in the figure and the 'line at infinity' $z = 0$ consists of the five remaining points of $\Lambda_1$. All other vectors are normalized to have $z = 1$. The symbol below the vector in $\Lambda_1$ denotes the slope $m$ of the four lines that intersect $z = 0$ in that point. With this understanding, all octads containing the three fixed points have linear equations as $y = mx + c$:

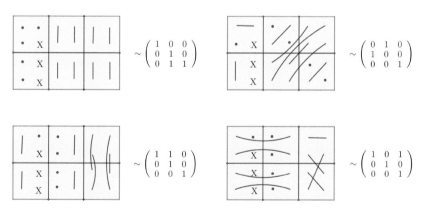

Using Figure 8.1 we may readily write elements of $M_{21}$ fixing the three points $\{5, 15, 19\}$ as $3 \times 3$ matrices over $\mathbb{F}_4$. In Figure 8.2 we give a number of examples of how an element of $M_{21}$ may be represented by a matrix of $SL_3(4)$, where the centre of the matrix group, namely multiplication by $\omega$, has been implicitly factored out.

Figure 8.2 The correspondence between elements of $M_{21}$ and matrices of $SL_3(4)$

As we have noted, $M_{21}$ possesses outer automorphisms that act as $S_3$ on the three points of the fixed triad. They are realized by the field automorphism of

### 8.1 The Maximal Subgroups of $M_{24}$

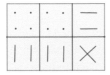
The field automorphism $\omega \leftrightarrow \bar{\omega}$

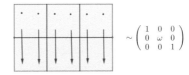
The diagonal automorphism

Figure 8.3 The outer automorphisms of $M_{21}$

$L_3(4)$ that interchanges $\omega$ and $\bar{\omega}$, and the diagonal automorphism. The manner in which these appear in the MOG is shown in Figure 8.3. Note that the duality automorphism of $M_{21}$ that interchanges points and lines cannot occur here. Computationally, it is easier to ask MAGMA for the stabilizer in $M_{24}$ of the three points in question, as indeed we could have done with the monad and duad stabilizers. Thus

```
> autm21:=Stabilizer(m24,{19,15,5});
> #autm21;
120960
> m21:=Stabilizer(m24,[19,15,5]);
> #m21;
20160
```

#### 8.1.4 The Stabilizer of a Sextet

We know that there are

$$\binom{24}{4}\bigg/6 = 1771$$

sextets, so we can work out the order of the stabilizer of a sextet. Moreover, in our discussion of elements of cycle shape $1^6.3^6$ we noted that the stabilizer in $M_{24}$ of a $U_6$, such as the top row of the MOG, acts as the symmetric group $S_6$ on its 6 points and thus on the tetrads of the sextet to which it is congruent modulo $\mathscr{C}$. Thus the kernel of this action on tetrads has order

$$|M_{24}|/(1771 \times 6!) = 3.2^6.$$

The element of order three itself, which we called $\tau$ when the umbral hexad is the top row of the MOG, fixes every tetrad and is thus in this kernel. But we have seen that, fixing any two columns of the MOG pointwise, there are three involutions of cycle shape $1^8.2^8$ that act as Klein fourgroups on the remaining four tetrads, making a total of $\binom{6}{2} \times 3 = 45$ such elements, all of which commute with one another. The elementary abelian group they generate is normalized by $\tau$. But now, if we revert to the language of the hexacode and let 1 stand for a permutation that interchanges the 1st and 2nd, 3rd and 4th elements in a column, $\omega$ stand for interchanging 1st and 3rd, 2nd and 4th, and $\bar{\omega}$ stand for interchanging 1st and 4th, 2nd and 3rd, then these 45 elements correspond to the 45 elements of the hexacode that involve two 0s. They thus generate a normal subgroup of order 64 whose elements are in one to one correspondence with codewords of the hexacode. Generators of this may be seen in Figure 8.4.

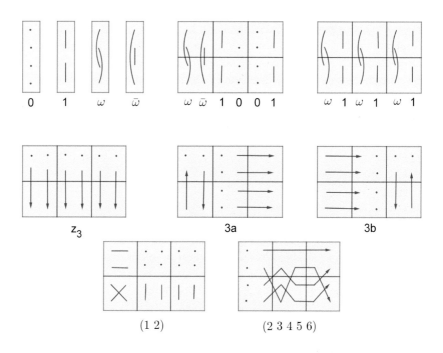

Figure 8.4 Generators of the stabilizer of a sextet

Conjugation of elements of this elementary abelian group of order $2^6$ by $\tau$ (= $z_3$ in Figure 8.4) has the effect of mapping

$$\tau : 1 \mapsto \omega \mapsto \bar{\omega} \mapsto 1.$$

The remaining $(64 - 45 - 1) = 18$ fixed point free involutions in this subgroup fall into six fourgroups the elements of each of which are cycled by $\tau$:

$$\left\{ \begin{array}{c} (1\ 1\ \omega\ \omega\ \bar{\omega}\ \bar{\omega}) \\ (\omega\ \omega\ \bar{\omega}\ \bar{\omega}\ 1\ 1) \\ (\bar{\omega}\ \bar{\omega}\ 1\ 1\ \omega\ \omega) \end{array} \right\} \quad \left\{ \begin{array}{c} (1\ 1\ \bar{\omega}\ \bar{\omega}\ \omega\ \omega) \\ (\omega\ \omega\ 1\ 1\ \bar{\omega}\ \bar{\omega}) \\ (\bar{\omega}\ \bar{\omega}\ \omega\ \omega\ 1\ 1) \end{array} \right\} \quad \left\{ \begin{array}{c} (1\ \omega\ 1\ \omega\ \omega\ 1) \\ (\omega\ \bar{\omega}\ \omega\ \bar{\omega}\ \bar{\omega}\ \omega) \\ (\bar{\omega}\ 1\ \bar{\omega}\ 1\ 1\ \bar{\omega}) \end{array} \right\}$$

$$\left\{ \begin{array}{c} (1\ \omega\ \omega\ 1\ 1\ \omega) \\ (\omega\ \bar{\omega}\ \bar{\omega}\ \omega\ \omega\ \bar{\omega}) \\ (\bar{\omega}\ 1\ 1\ \bar{\omega}\ \bar{\omega}\ 1) \end{array} \right\} \quad \left\{ \begin{array}{c} (1\ \bar{\omega}\ 1\ \bar{\omega}\ 1\ \bar{\omega}) \\ (\omega\ 1\ \omega\ 1\ \omega\ 1) \\ (\bar{\omega}\ \omega\ \bar{\omega}\ \omega\ \bar{\omega}\ \omega) \end{array} \right\} \quad \left\{ \begin{array}{c} (1\ \bar{\omega}\ \bar{\omega}\ 1\ \bar{\omega}\ 1) \\ (\omega\ 1\ 1\ \omega\ 1\ \omega) \\ (\bar{\omega}\ \omega\ \omega\ \bar{\omega}\ \omega\ \bar{\omega}) \end{array} \right\}$$

The symmetric group $S_6$, which is a homomorphic image of the sextet group, acts on the six tetrads (the six columns of the MOG) and these six fourgroups in a non-permutation identical manner. Taking the columns of the MOG as canonical sextet we have in MAGMA:

```
> sext1:={{24,3,6,9},{23,19,15,5},{11,4,16,13},
> {1,20,14,21},{22,18,8,12},{2,10,17,7}};
> sextgp:=Stabilizer(m24,sext1);
> #sextgp;
138240
> #m24/#sextgp;
1771
```

The ways in which the tetrads of two sextets can intersect with one another are given in Lemma 5.1. The following MAGMA code gives the number of sextets in each suborbit, in the order in which they appear in that lemma. In this command: **f** denotes the function from the permutation group on 24 letters to one on 1771 letters; **gp** is the resulting permutation group of degree 1771; and **kk** denotes the kernel of the homomorphism, which in this case is the identity.

```
> f,gp,kk:=CosetAction(m24,sextgp);
> oost1:=Orbits(f(sextgp));
> #oost1;
4
> [#oost1[i]:i in [1..4]];
[ 1, 90, 240, 1440 ]
```

### 8.1.5 The Octad Stabilizer

From the proof of uniqueness of the Steiner system $S(5, 8, 24)$, we know that every even permutation of the eight points of the fixed octad is permissible and that there is an elementary abelian group of order 16 fixing every point of the octad. We also know the order of this group as it has index 759 in $|M_{24}|$, and so may deduce that the stabilizer of an octad has shape

$$2^4 : A_8.$$

We shall choose $\Lambda_1$ as the fixed octad and, fixing the top-left point of the complementary 16-ad $\Lambda_2 \cup \Lambda_3$, we obtain a copy of $A_8 \cong L_4(2)$ acting simultaneously on the eight points of $\Lambda_1$ and the remaining 15 points of $\Lambda_2 \cup \Lambda_3$. In Figure 8.5 we have labelled the 15 non-zero vectors of the underlying 4-dimensional vector space

$$V_c = \{(x, y, z, t) \mid x, y, z, t \in \mathbb{Z}_2\}$$

so that the $xy$-subspace lies along the horizontal axis and the $zt$-subspace lies on the vertical axis as indicated. We use row vectors in $V_c$ as our matrices need to act on the right-hand side in order for multiplication to be consistent with the usual multiplication of permutations. In Figure 8.6(i) we give examples of how permutations fixing the octad $\Lambda_1$ and the top-left point of $\Lambda_2 \cup \Lambda_3$, the zero vector, may be written as $4 \times 4$ matrices over $\mathbb{Z}_2$. However, the translations

$$\tau_a : v \mapsto v + a \text{ for } a \in V_c$$

may be incorporated in a $5 \times 5$ matrix for note that

$$\left(\begin{array}{c|c} A & 0 \\ \hline a & 1 \end{array}\right) \cdot \left(\begin{array}{c|c} B & 0 \\ \hline b & 1 \end{array}\right) = \left(\begin{array}{c|c} AB & 0 \\ \hline aB + b & 1 \end{array}\right),$$

|          | | $e_1$ | $e_2$ | $e_1 + e_2$ |
|---|---|---|---|---|
|          | (0000) | (1000) | (0100) | (1100) |
| $e_3$    | (0010) | (1010) | (0110) | (1110) |
| $e_4$    | (0001) | (1001) | (0101) | (1101) |
| $e_3 + e_4$ | (0011) | (1011) | (0111) | (1111) |

Figure 8.5 The isomorphism $A_8 \cong L_4(2)$

## 8.1 The Maximal Subgroups of $M_{24}$

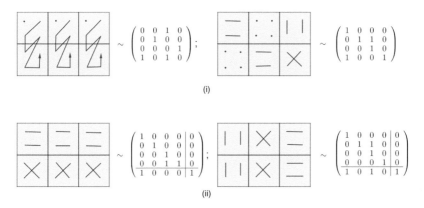

Figure 8.6 The stabilizer of an octad as matrices over $\mathbb{Z}_2$

where $A$ and $B$ are in $L_4(2)$ and $a$ and $b$ are 4-dimensional row vectors. Note that

$$(v \mid 1)\left(\begin{array}{c|c} A & 0 \\ \hline a & 1 \end{array}\right) = (vA + a \mid 1) \text{ and so } (v \mid 1)\left(\begin{array}{c|c} I_4 & 0 \\ \hline a & 1 \end{array}\right) = (v + a \mid 1),$$

the latter corresponding to the translation $\tau_a$. In Figure 8.6(ii) we demonstrate how any element of the subgroup of shape $2^4 : A_8$ stabilizing an octad may be represented by a non-singular $5 \times 5$ matrix over $\mathbb{Z}_2$.

We of course choose $\Lambda_1$, the first brick of the MOG, as our canonical octad and in MAGMA this gives:

```
> oct1:={24,3,6,9,23,19,15,5};
> octgp:=Stabilizer(m24,oct1);
> #octgp;
322560
> #m24/#octgp;
759
```

### The Sylow 2-subgroup of $M_{24}$

There are 759 octads and so, since this number is odd, the octad stabilizing subgroup contains Sylow 2-subgroups of $M_{24}$ or, to put it another way, any Sylow 2-subgroup fixes an octad. Similarly, since there are 1771 sextets and 3795 trios, any Sylow 2-subgroup of $M_{24}$ fixes a sextet and a trio. However, we have seen that the octad stabilizer is isomorphic to a subgroup of the linear group $L_5(2)$ and contains the subgroup consisting of all *unitriangular* matrices, that is to all those matrices with 1s down the main diagonal and 0s above it.

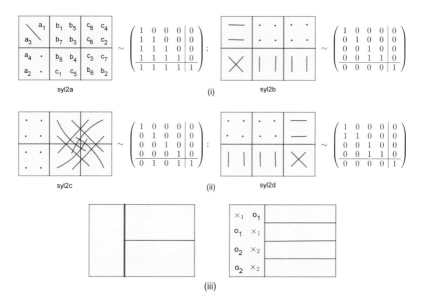

Figure 8.7 Generators for a Sylow 2-subgroup of $M_{24}$ as they appear in $L_5(2)$

There are 10 entries in such a matrix that can be either 0 or 1 and so this group has order $2^{10}$ as required. In Figure 8.7 we give generators for this Sylow subgroup both as permutations in the MOG and as $5 \times 5$ unitriangular matrices in $L_5(2)$. The octad, trio and sextet fixed by this subgroup are as indicated in Figure 8.7(iii). The fixed octad is, of course, $\Lambda_1$, but the sextet involves rows of $\Lambda_2 \cup \Lambda_3$ rather than columns. Note that the stabilizer of this octad, trio and sextet has order three times the order of the desired Sylow subgroup; indeed the element 3a of Figure 8.4 preserves each of them. In order to obtain the given Sylow 2-subgroup, we must also preserve the pairing $(\infty\ 19)(0\ 3)(6\ 9)(15\ 5)$ in the fixed octad, as indicated in the final diagram of Figure 8.7. We verify this in the following MAGMA calculation:

```
> sext2:={{24,19,15,5},{23,3,6,9},{11,1,22,2},
> {4,20,18,10},{16,14,8,17},{13,21,12,7}};
> sextgp2:=Stabilizer(m24,sext2);
> trio2:={{24,3,6,9,23,19,15,5},{11,1,22,2,4,20,18,10},
> {16,14,8,17,13,21,12,7}};
> triogp2:=Stabilizer(m24,trio2);
> e2:=s24!(24,19)(23,3)(6,9)(15,5);
> ce2:=Centralizer(m24,e2);
> #(ce2 meet octgp meet sextgp2 meet triogp2);
```

1024

Of course we may ask directly for a Sylow 2-subgroup in MAGMA

```
> syl2:=SylowSubgroup(m24,2);
> #syl2;
1024
```

However, the above process is informative and gives the version in which the correspondence with the unitriangular matrices is apparent.

### 8.1.6 The Stabilizer of a Dodecad

The most interesting maximal subgroup of $M_{24}$ is the stabilizer of a pair of complementary dodecads, which is known as a *duum*. This is the group

$$\text{Aut } M_{12} \cong M_{12} : 2,$$

where the outer half of the group is interchanging the two dodecads and the simple group $M_{12}$ is acting non-permutation identically on them. The structure of the binary Golay code shows us that there are 2576 dodecads; we first show

**Lemma 8.1** *The Mathieu group $M_{24}$ acts transitively on the 2576 dodecads and the stabilizer of a dodecad acts sharply quintuply transitively on its 12 points.*

*Proof* Dodecads occur in the Golay code as the symmetric difference of two octads that intersect in two points. Transitivity on octads and the action on the eight points of a fixed octad show that we may permute any octad to $\Lambda_1$ and any two points of it to the top two as shown as ×s in

| × | × | *o* | *o* | *o* |
|---|---|---|---|---|
|   |   |   |   | *o* |
|   |   |   |   | *o* |
|   |   |   |   | *o* |

Now Todd's triangle, see Figure 3.1, tells us that there are 16 octads intersecting a given octad, $\Lambda_1$ in this case, in a given two points one of which is shown in the above figure. But no non-trivial element of the $2^4$ of translations, see Figure 2.8, can fix such an octad. [This is visibly the case but also note that fixing such an octad would imply odd permutation action on it, which does not occur in $M_{24}$.] Thus this elementary abelian group of order 16 acts regularly on the 16 octads intersecting $\Lambda_1$ in the two points indicated, and there is just one orbit on dodecads.

Suppose now that D is a fixed dodecad. We know the order of its stabilizer to be $|M_{24}|/2576 = 95\,040$. Given two ordered quintuples of points of D, $[x_1, x_2, \ldots, x_5]$ and $[y_1, y_2, \ldots, y_5]$ then there is certainly an element, $\pi$ say, of $M_{24}$ mapping $x_i$ to $y_i$ for $i = 1, \ldots, 5$. If the octads $X$ and $Y$ defined by these quintuples intersect D in $\{x_1, x_2, \ldots, x_6\}$ and $\{y_1, y_2, \ldots, y_6\}$ respectively then, multiplying if necessary by an element of order 3 that fixes $y_1, \ldots, y_5$ and cycles the remaining three points of $Y$, we may assume that such a permutation maps $x_6$ to $y_6$. But we have seen that the group of $2^4$ translations that fix every point of $Y$ acts transitively on the set of 16 octads intersecting $Y$ in its two remaining points $Y \setminus \{y_1, \ldots, y_6\}$, one of which yields D by symmetric differencing. Multiplying $\pi$ by an appropriate translation if necessary we have obtained a permutation that fixes D and maps $x_i$ to $y_i$ for $i = 1, \ldots, 5$. Thus the order of the stabilizer of a dodecad is divisible by $12 \cdot 11 \cdot 10 \cdot 9 \cdot 8 = 95\,040$, which we know to be its order. So $M_{12}$ acts sharply quintuply transitively on the points of D as asserted. □

We saw in Section 4.2 that a pair of points in a dodecad defines a partition of the complementary dodecad into two special hexads. The subgroup of $M_{12}$ fixing such a partition has order $95\,040/66 = 1440$; it is the automorphism group of the alternating group $A_6$:

$$\text{Aut } A_6 \cong S_6 : 2 \cong \text{P}\Gamma\text{L}_2(9),$$

see Figure 9.10. The three subgroups with index 2 in this group are $S_6 \cong \text{P}\Sigma\text{L}_2(9)$, $\text{PGL}_2(9)$ and $M_{10}$, see Conway et al. (1985, page 5). The subgroups isomorphic to $S_6$ must act non-permutation identically on the two orbits of length 6 as, otherwise, transpositions on both orbits would lead to elements fixing eight points of a dodecad, which we soon see is impossible. This implies that the stabilizer in $M_{12}$ of a subset of five points of the dodecad fixes the remaining point of the special hexad they define and remains transitive on the complementary six points. Thus $M_{12}$ acts transitively on the non-special hexads. The orbits are given by

| | | | |
|---|---|---|---|
| $2 + 10$ | : | $6^2$ | $\text{Aut}A_6 \cong A_6.2^2$ |
| $2 + 10$ | : | $6 + 6$ | $S_6 \cong \text{P}\Sigma\text{L}_2(9)$ |
| $2 + 10$ | : | $6^2$ | $\text{PGL}_2(9)$ |
| $1 + 1 + 10$ | : | $6^2$ | $M_{10}$ |
| $1 + 1 + 10$ | : | $6 + 6$ | $A_6 \cong \text{PSL}_2(9)$ |
| $2 + 10$ | : | $1 + 5 + 6$ | $S_5 \cong \text{PGL}_2(5)$ |

Here $\text{P}\Sigma\text{L}_2(9)$ denotes $\text{PSL}_2(9)$ together with the field automorphism of $\mathbb{F}_9$ and $\text{P}\Gamma\text{L}_2(9)$ denotes $\text{PGL}_2(9)$ together with the field automorphism, see Section

## 8.1 The Maximal Subgroups of $M_{24}$

15.3.1. The group $M_{12}$, and in particular the manner in which it embeds in the MOG, is investigated in more detail in Chapter 9 where we take the dodecad consisting of the symmetric difference of the top row of the MOG and its 4th, 5th and 6th columns in order to emphasize the connection with the 9-point affine plane. However we frequently prefer to take the symmetric difference of the top row of the MOG and the 2nd, 4th and 6th columns as canonical. Thus in MAGMA we have

```
> dod1:={24,19,15,5,11,20,14,21,22,10,17,7};
> dodgp:=Stabilizer(m24,dod1);
> #dodgp;
95040
> duumgp:=Normalizer(m24,dodgp);
> #duumgp;
190080
```

Now any two distinct dodecads must intersect in 0, 4, 6 or 8 points as their sum must be the whole set $\Omega$, a 16-ad, a dodecad or an octad. Thus the intersection matrix for two dua must be one of

$$\begin{pmatrix} 12 & \cdot \\ \cdot & 12 \end{pmatrix} \begin{pmatrix} 4 & 8 \\ 8 & 4 \end{pmatrix} \begin{pmatrix} 6 & 6 \\ 6 & 6 \end{pmatrix}.$$

We show that these are, in fact, orbits of the stabilizer of the initial duum. First note that $T$, a given set of four points in a dodecad $D$, extends to a special hexad of $D$ in four ways, each of which extends to an octad having two points in the complementary dodecad $\Omega \setminus D$. This gives five of the tetrads of the sextet defined by $T$, and so the final tetrad must lie entirely in $\Omega \setminus D$. So every tetrad of $D$ is contained in a unique octad cutting 4.4 across the associated duum. The sum of this octad and $D$ gives a dodecad cutting 8.4 across the duum. There are thus

$$\binom{12}{4} = 495$$

dua cutting the initial duum as in the second matrix. Since $M_{12}$ is 5-transitive, this is a single orbit.

Furthermore, if a dodecad $D'$ intersects $D$ in six points, then those six points cannot be a special hexad of $D$ or the sum of $D'$ with the octad containing that special hexad would be a $\mathscr{C}$-set properly contained in $\Omega \setminus D$. So $D \cap D'$ must be a non-special hexad, and $M_{12}$ acts transitively on these. Again, if two dodecads $D'$ and $D''$ have the same intersection with $D$ then $D'+D''$ is a $\mathscr{C}$-set contained in $\Omega \setminus D$ and so is either $\phi$ or $\Omega \setminus D$. So there is just one orbit on dua cutting the initial duum as in the third matrix, and it contains

$$\binom{12}{5}.6.2/6.2 = 792 \text{ dua},$$

and we have $1 + 495 + 792 = 1288$ as required. The following MAGMA code confirms this result where: **f** denotes the function from a permutation group on 24 letters to a permutation group on 1288 letters; **gp** denotes the image group of degree 1288; and **kk** denotes the kernel of the homomorphism, which in this case is trivial.

```
> f,gp,kk:=CosetAction(m24,duumgp);
> oost1:=Orbits(f(duumgp));
> #oost1;
3
> [#oost1[i]:i in [1..3]];
[ 1, 495, 792 ]
```

### 8.1.7 The Stabilizer of a Trio

A decomposition of the 24 points of $\Omega$ into three disjoint octads, such as the three bricks $\Lambda_1, \Lambda_2$ and $\Lambda_3$, is known as a *trio*. Todd's triangle, see Figure 3.1, tells us that there are 30 octads disjoint from a given octad, which fall into 15 complementary pairs. Thus there are $759 \times 15/3 = 3795$ trios. The stabilizer of an octad acts transitively on these 15 pairs and so $M_{24}$ has a single orbit on trios.

Now the six tetrads of a sextet may be grouped in pairs to form a trio in 15 ways and so we see that a trio may be *refined* to a sextet in

$$(1771 \times 15)/3795 = 7$$

ways, the 7 *refinements* of the trio. These are precisely the 7 sextets defined by the tetrads of the Point space, see Figure 4.1, which we know form a subspace of dimension 3. We also know that there is a copy of the symmetric group $S_3$ bodily permuting the three bricks, and we have seen in Figure 3.4 generators for a copy of the linear group $L_3(2)$ acting on this Point space. The intersections of the kernels of the actions on the 3 bricks and on the 7 refinements thus has order $|M_{24}|/(3795 \times 6 \times 168) = 64$. This group is elementary abelian, generated by those translational symmetries shown in the first three diagrams of Figure 2.8 acting on each of the three 16-ads $\Lambda_i \cup \Lambda_j, i \neq j$. So it consists of one of the permutations shown in Figure 8.8 in each brick so that their product is the identity. Note that each of these permutations on a brick fixes the tetrads of 3 sextets of the Point space whilst complementing those of the other 4; thus it

## 8.1 The Maximal Subgroups of $M_{24}$

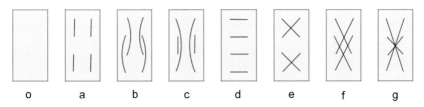

Figure 8.8 The elementary abelian $2^6$ in the trio group

corresponds to a Line. The labelling of the permutations is consistent with this correspondence as in Figure 4.2. So the elementary abelian $2^6$ consists of

$$\{[x, y, z] \mid x, y, z \in \{a, b, \ldots, g\}, \ x.y.z = 1\}.$$

Thus the trio group has shape

$$2^6 : (S_3 \times L_3(2)).$$

There are four ways in which two distinct trios can intersect one another, that is to say the action of $M_{24}$ on trios has rank 5, see Figure 12.5 and Table 12.1. The corresponding intersection matrices are

$$\begin{pmatrix} 8 & \cdot & \cdot \\ \cdot & 8 & \cdot \\ \cdot & \cdot & 8 \end{pmatrix} \begin{pmatrix} 8 & \cdot & \cdot \\ \cdot & 4 & 4 \\ \cdot & 4 & 4 \end{pmatrix} \begin{pmatrix} \cdot & 4 & 4 \\ 4 & \cdot & 4 \\ 4 & 4 & \cdot \end{pmatrix} \begin{pmatrix} 2 & 2 & 4 \\ 2 & 2 & 4 \\ 4 & 4 & \cdot \end{pmatrix} \begin{pmatrix} 4 & 2 & 2 \\ 2 & 4 & 2 \\ 2 & 2 & 4 \end{pmatrix}.$$

The stabilizer of the MOG trio is given in MAGMA by

```
> trio1:={{24,3,6,9,23,19,15,5},{11,4,16,13,1,20,14,21},
> {22,18,8,12,2,10,17,7}};
> triogp:=Stabilizer(m24,trio1);
> #triogp;
64512
> #m24/#triogp;
3795
```

We may now use MAGMA to obtain the permutation action of $M_{24}$ on the 3795 trios. In the following code: **f** is the function mapping a permutation on 24 letters to one on 3795 letters; **gp** denotes the image group of degree 3795; and **kk** is the kernel of the homomorphism, which is trivial in this case. We may then ask for the orbits of the stabilizer of a trio on the remaining trios, and we find that they have lengths

$$[1, 42, 56, 1008, 2688]$$

in the order displayed in the earlier intersection matrices.

```
> f,gp,kk:=CosetAction(m24,triogp);
> oost1:=Orbits(f(triogp));
> #oost1;
5
> [#oost1[i]:i in [1..5]];
[ 1, 42, 56, 1008, 2688 ]
```

### 8.1.8 The Linear Group $L_2(23)$

The 24 points of $\Omega$ were labelled with the projective line $P_1(23)$ as we claim that the linear group $L_2(23)$ preserves the Steiner system and thus is a subgroup of $M_{24}$. The minimal test for membership of $M_{24}$ produced in Section 7.1 may now be applied using Figure 7.6 to show that the linear fractional maps

$$\alpha : x \mapsto x+1 \equiv (\infty)(0\,1\,2\,3\,4\,5\,6\,7\,8\,9\,10\,11\,12\,13\,14\,15\,16\,17\,18\,19\,20\,21\,22)$$

and

$$\beta : x \mapsto -1/x \equiv (\infty\,0)(1\,22)(2\,11)(3\,15)(4\,17)(5\,9) \\ (6\,19)(7\,13)(8\,20)(10\,16)(12\,21)(14\,18)$$

that together generate $L_2(23)$ do indeed satisfy the test.

Moreover, the subgroup

$$\left\langle x \mapsto -\frac{1}{x}, x \mapsto \frac{5x+8}{x-5} \right\rangle \cong D_8$$

preserves $\Lambda_1$ having actions $(\infty\,0)(3\,15)(6\,19)(9\,5)$ and $(\infty\,5)(0\,3)(6\,15)(9\,19)$ respectively on its points. No further elements of $L_2(23)$ can fix $\Lambda_1$ as the stabilizer of a point in $L_2(23)$ is a Frobenius group of shape $23:11$, and thus $\Lambda_1$ has $|L_2(23)|/8 = 759$ images under its action and we may conclude that $L_2(23)$ acts transitively on octads. For convenience we refer to this subgroup as the *projective group*.

### 8.1.9 The Octern Group

Todd's list of maximal subgroups of $M_{24}$ which, as I have mentioned, he did not claim to be complete, in fact omitted one class of subgroups isomorphic to the linear group $L_2(7)$ of order 168. It is known as the *Octern group* $O^n$ as it acts transitively and imprimitively on the 24 points of $\Omega$ with 8 blocks of imprimitivity of size 3 known as *terns*. In fact it is the centralizer within $M_{24}$ of an element $\tau$ of cycle shape $3^8$ *outside* $M_{24}$. Generators of a representative

## 8.2 Completeness of the List of Maximal Subgroups

of this class of subgroups are given in Figure 7.3, where the elements of $\Omega$ are arranged as a $3 \times 8$ array whose columns are the 8 terns. If we denote the tern in the column labelled $i$ by $\underline{i}$ then we find that

$$\infty \cup \underline{1} \cup \underline{2} \cup \underline{4}$$

is a dodecad of the Golay code, as are its 14 images under the action of $O^n$. These 14 dodecads, together with the whole set $\Omega$ and the empty set $\phi$, form a 4-dimensional subspace, $\mathscr{D}$ say, of $\mathscr{C}$ invariant under the action of $O^n$.

In fact $\mathscr{C}$ decomposes under the action of $O^n$ into the sum of this subspace and an 8-dimensional subspace $V_8$ consisting of all $\mathscr{C}$-sets that intersect every tern evenly:

$$\mathscr{C} = \mathscr{D} \oplus V_8.$$

The annihilator of $V_8$ in $\mathscr{C}^*$ is the 4-dimensional space spanned by the 8 terns themselves that contains the 3-dimensional subspace $Oct_3$. The set of 24 duads that lie in a single tern span a complementary subspace of $\mathscr{C}^*$ of dimension 8.

```
> tau:=s24!(24,22,11)(23,20,8)(15,16,18)(5,4,2)(9,1,17)
>(19,14,7)(6,21,12)(3,13,10);
> octerngp:=Centralizer(m24,tau);
> #octerngp;
168
```

It is this small subgroup, which had lain unacknowledged for many years, that gives rise to the constructions of $M_{24}$ described in Chapter 13.

## 8.2 Proof of Completeness of the List of Maximal Subgroups

In this section we produce a short combinatorial argument, which is taken from Curtis (1972, 1977), that this list of nine maximal subgroups is complete.

**Theorem 8.2** *The following is a complete list of maximal subgroups of* $M_{24}$:

(1) *The* Monad stabilizer *of shape* $M_{23}$;
(2) *The* Duad stabilizer *of shape* $M_{22} : 2$;
(3) *The* Triad stabilizer *of shape* $M_{21} : S_3$;
(4) *The* Sextet stabilizer *of shape* $2^6 : 3\cdot S_6$;
(5) *The* Octad stabilizer *of shape* $2^4 : A_8$;
(6) *The* Duum stabilizer *of shape* $M_{12} : 2$;
(7) *The* Trio stabilizer *of shape* $2^6 : (S_3 \times L_2(7))$;

(8) *The* Octern group *of shape* $L_2(7)$;
(9) *The* Projective group *of shape* $L_2(23)$.

*Proof* The method of proof will be to show that any subgroup of $M_{24}$ is contained in a conjugate of at least one of the above, but that there is no containment between them. Let $G$ be a subgroup of $M_{24}$. We shall consider two cases: (a) $G$ acts intransitively on $\Omega$; (b) $G$ acts transitively on $\Omega$.

*Case (a).* Suppose that $G$ is an intransitive subgroup of $M_{24}$ and let $\Lambda \subset \Omega$ be an orbit of $G$. The $\Lambda$ must be congruent modulo $\mathscr{C}$ to one of: the empty set $\phi$, a monad, a duad, a triad or a sextet. Thus $G$ must be contained in one of the octad stabilizer, the dodecad stabilizer (which is contained in the duum stabilizer), or the monad, duad, triad or sextet stabilizer.

*Case (b).* If $G$ is a transitive subgroup of $M_{24}$ it either acts imprimitively with blocks of size 2, 3, 4, 6, 8 or 12, or it acts primitively. We first prove two lemmas that enable us to deduce that such a subgroup must act doubly transitively on the blocks of imprimitivity or, in the primitive case, on the 24 points of $\Omega$.

**Lemma 8.3** *Let $G$ be a permutation group of composite order acting primitively on an even set $\Lambda$. Then $G_a$, the stabilizer in $G$ of a point $a \in \Lambda$, cannot have an orbit of length 1 or 2.*

*Proof* If $G_a$ fixes a further point, then its action on $\Lambda$ must be regular. But a regular group can only be primitive if it is of prime order, which $G$ is not. So let us assume that $G_a$ has an orbit of length 2. Denote a particular 2-orbit of $G_a$ by $T(a)$ and draw a *directed* graph in which $a$ is joined to the two points in $T(a)$. If $b \in T(a) \Leftrightarrow a \in T(b)$ then the graph is undirected of valence 2 and so it consists of a union of polygons. If there is more than one polygon, then they form blocks of imprimitivity. Contradiction. But if there is just one polygon, then it has an even number of vertices and so pairs of opposite vertices form blocks of imprimitivity.

So we may assume that the graph is directed with every vertex having two edges entering it and two edges leaving it. Now draw two paths simultaneously from $a$ where a move consists of going along an edge in a positive direction and the next edge in a negative direction. Note that these paths are well defined, and that they would be swapped by a permutation fixing $a$ and interchanging the two points of $T(a)$. Should these two paths meet at a point other than $a$ then that point would form a block of imprimitivity together with $a$. Thus the paths must eventually meet at $a$ having traversed two polygons of even length. But we may now define a fresh *undirected* graph of valence 2 in which $a$ is joined

## 8.2 Completeness of the List of Maximal Subgroups

to the two points that are opposite $a$ in these two polygons. This returns us to the previously considered case. □

**Lemma 8.4** *Let $G$ be a group of composite order acting primitively on a set $\Lambda$ and suppose that $a \in \Lambda$. If $n_k$ is the largest orbit length of $G_a$ acting on $\Lambda$ and $n_j$ is any other orbit length, $n_j > 1$, then $(n_j, n_k) \neq 1$.*

*Proof* Firstly suppose that $U \leq H$, a permutation group acting on the set $\Lambda$, and let $a \in \Lambda$. Then we claim that if $(|a^H|, |H:U|) = 1$, then $a^U = a^H$. That is to say, if the index of $U$ in $H$ is coprime to the length of the $H$-orbit containing $a$, then $U$ acts transitively on $a^H$. To see this, observe that

$$|a^H| = |H : H_a| \mid |H : H_a|.|H_a : U_a| = |H : U_a| = |H : U|.|U : U_a|,$$

and so $|a^H| \mid |U : U_a| = |a^U| \leq |a^H|$, and equality follows.

Now let $\Delta = \{x \in \Lambda \mid |x^{G_a}| = n_k\}$ and let $b \in \Lambda$ be such that $|b^{G_a}| = n_j$ with $(n_j, n_k) = 1$. Then

$$(|G_a : G_{ab}|, |x^{G_a}|) = (n_j, n_k) = 1 \text{ for all } x \in \Delta.$$

Therefore, from the above result,

$$|x^{G_b}| \geq |x^{G_{ab}}| = |x^{G_a}| = n_k \Rightarrow |x^{G_b}| = n_k \text{ by maximality.}$$

That is, if $\hat{\Delta} = \{x \in \Lambda \mid |x^{G_b}| = n_k\}$ then $\hat{\Delta} \supseteq \Delta$. But similarly $\Delta \supseteq \hat{\Delta}$. Therefore, $\hat{\Delta} = \Delta$. So $\Delta$ is fixed by $\langle G_a, G_b \rangle$. But $G$ is primitive and not regular since it is of composite order. Thus $\langle G_a, G_b \rangle = G$. This implies that $\Delta$ is fixed by $G$, which contradicts the imprimitivity of $G$. □

**Corollary** *Any subgroup of $M_{24}$ acting primitively on 4, 6, 8, 12 or 24 blocks in $\Omega$ acts doubly transitively on those blocks.*

*Proof* Every orbit length (on blocks) of the stabilizer of a block must have length at least 3 and have a factor in common with the largest orbit length. These conditions alone prove the result for 4, 6, 8 and 12. For 24, which is to say primitive action on $\Omega$, there are the following possibilities:

(i)   15   5   3,
(ii)  12   3   [4 4],
(iii) 10   5   [4 4],
(iv)  6    3   [3 3] 4 4,

where the square brackets denote possible fusion. In cases (i) and (ii) for $a \in \Omega$ we denote the 3-orbit of $G_a$ by $T(a)$ and that, since

$$|b^{G_a}| = |G_a : G_{ab}| = |G_b : G_{ba}| = |a^{G_b}|,$$

we have $a \in T(b) \Leftrightarrow b \in T(a)$. So we can draw an undirected graph of valence 3 in which $a$ is joined to $b$ if, and only if, $b \in T(a)$. In these circumstances we ask where the two further points of $T(b) \setminus \{a\}$ lie, noting that if $S(a)$ is another orbit of $G_a$ then

$$\begin{pmatrix} \text{Number of edges from a} \\ \text{point of } T(a) \text{ to points in} \\ S(a) \end{pmatrix} \times |T(a)| = \begin{pmatrix} \text{Number of edges from a} \\ \text{point of } S(a) \text{ to points in} \\ T(a) \end{pmatrix} \times |S(a)|.$$

Clearly $|S(a)| = 4, 5, 8, 12$ or $15$ imples this number must be zero. Thus $T(b) \setminus \{a\} \subset T(a)$ and the graph consists of disjoint complete four graphs that are blocks of imprimitivity. A contradiction.

In case (iii) we simply ask where the three points completing the 5-orbit to an octad can lie.

In case (iv) we ask what type of 10-element subset the 6-orbit together with one of the 4-orbits can be. Consulting Figure 7.12 we see that

- if it is a $U_{10}$ there is a unique pair of points completing it to a $U_{12}$,
- if it is a $T_{10}$ there is a unique pair of points completing it to an $S_{12}^+$,
- if it is an $S_{10}$ it contains a unique pair whose removal leaves an $S_8$.

Observing that there is nowhere for these two points to lie, we conclude that $G$ must act doubly transitively on $\Omega$. □

We now deal with the different block sizes one by one.

*Blocks of size 12.* We have $\Omega = X \cup Y$ with $|X| = |Y| = 12$. Then, modulo $\mathscr{C}$, we have $X \equiv Y \equiv \phi$, a duad or a sextet, and so $G$ is contained in the stabilizer of an octad, a duum, a duad or a sextet.

*Blocks of size 8.* So we have $\Omega = X \cup Y \cup Z$ with $|X| = |Y| = |Z| = 8$. If one of these blocks is an octad, they all are and they form a trio that is visibly fixed by $G$. They cannot be $T_8$s as $G_X$ could not then act transitively on the points of $X$ (we see from Figure 7.12 that a $T_8$ contains a unique point that when removed leaves an $S_7$). So $X, Y$ and $Z$ must be $U_8$s. But, as we saw in Section 7.12, the stabilizer of a $U_8$ fixes a unique octad, and the transitivity of $G$ implies that the three octads associated with our three blocks must form a trio, stabilized by $G$.

*Blocks of size 6.* So $\Omega = X \cup Y \cup Z \cup T$ with $|X| = |Y| = |X| = |T| = 6$. By the corollary $G$ acts doubly transitively on these four blocks, and by transitivity $X$ must be a $U_6$ (if an $S_6$, where do the two points that complete it to an octad lie?). We consider $X \cup Y$ modulo $\mathscr{C}$. If $X \cup Y \equiv \phi$ modulo $\mathscr{C}$ then $X \equiv Y \equiv Z \equiv T$ modulo $\mathscr{C}$, which is a sextet, and so $G$ is in the sextet stabilizer. Otherwise $X \cup Y$ is congruent to a sextet modulo $\mathscr{C}$, and the three pairings of the blocks define

## 8.2 Completeness of the List of Maximal Subgroups

a 2-dimensional subspace of sextets. We know that these are just the even and odd lines that were discussed in Section 7.1 and in each case the stabilizer of such a configuration fixes a trio. Referring to Figure 7.1, in the case of the even line, this is the unique trio for which each of the three sextets is a refinement. For each sextet of an odd line, just two of its tetrads cut the tetrads of the other two as $1^4.0^2$. The union of these two tetrads is, of course, an octad and the three octads so obtained form a trio that is thus stabilized by the subgroup fixing the odd line.

*Blocks of size 4.* By Corollary 8.2 $G$ acts doubly transitively on the six blocks. Now the columns of the MOG may be rearranged, keeping the top row as is, so that the union of any two rows is a dodecad. This arrangement Q allows a copy of $L_2(5)$ acting on the columns and fixing the rows, together with a copy of the alternating group $A_4$ acting on the rows and fixing the columns, extended by an odd element of $PGL_2(5)$ on the columns fixing the rows times an odd permutation of the rows fixing the columns. An example of this last type would be

$$[(1\,2\,4\,3) \times (c\,d)]_Q = (6\,9)(5\,19)(11\,1\,2\,22)(4\,14\,10\,8)(16\,20\,17\,18)(13\,21\,7\,12).$$

So the subgroup corresponding to this arrangement has shape

$$(L_2(5) \times A_4) : 2.$$

It is simply a partition of the 24 points of $\Omega$ into four umbral hexads such that the union of any two is a dodecad: For convenience, we shall call such a partition a *quadum*, by analogy with a sextet which is six tetrads any two of which form an octad. Each hexad is congruent to the same sextet modulo $\mathscr{C}$, namely the columns.

By Corollary 8.2 we may assume that one block is the $\infty$-column in Figure 8.9 and that the element $\rho = [(0\,1\,2\,3\,4) \times I]_Q$ cycles the other five. Centralizing this element we have the subgroup $[I \times A_4]_Q$ and normalizing it we have for instance the element $[(1\,2\,4\,3) \times (c\,d)]_Q$. Each of the other blocks must have a point in each of the four rows of $Q$. Note that if another column of $Q$ is a block,

| Q | ∞ | 0 | 1 | 2 | 3 | 4 |
|---|---|---|---|---|---|---|
| a | ∞ | 0 | 11 | 1 | 22 | 2 |
| b | 3 | 15 | 4 | 14 | 8 | 10 |
| c | 6 | 5 | 16 | 21 | 12 | 17 |
| d | 9 | 19 | 13 | 20 | 18 | 7 |

Figure 8.9 The four umbral hexads 4 × 6 array

then they all are and $G$ fixes the resulting sextet. Moreover, if another block has three points in a column, then it together with the first column form a $T_8$ which is congruent to a duad modulo $\mathscr{C}$; the resulting set of 15 duads must be preserved by $G$ that contradicts the transitivity of $G$. So the other five blocks can have at most two points in any column. Using the normalizer of $\rho$ described earlier, we find that there are just six more possibilities, displayed here in the quadum arrangement:

|   |   |   |   |   |   |
|---|---|---|---|---|---|
| × | o | · | · | · | · |
| × | o | · | · | · | · |
| × | · | o | · | · | · |
| × | · | · | o | · | · |

(i)

|   |   |   |   |   |   |
|---|---|---|---|---|---|
| × | o | · | · | · | · |
| × | o | · | · | · | · |
| × | · | o | · | · | · |
| × | · | · | · | o | · |

(ii)

|   |   |   |   |   |   |
|---|---|---|---|---|---|
| × | o | · | · | · | · |
| × | o | · | · | · | · |
| × | · | o | · | · | · |
| × | · | · | · | · | o |

(iii)

|   |   |   |   |   |   |
|---|---|---|---|---|---|
| × | o | · | · | · | · |
| × | o | · | · | · | · |
| × | · | o | · | · | · |
| × | · | o | · | · | · |

(iv)

|   |   |   |   |   |   |
|---|---|---|---|---|---|
| × | · | o | · | · | · |
| × | · | · | o | · | · |
| × | · | · | · | o | · |
| × | · | · | · | o | · |

(v)

|   |   |   |   |   |   |
|---|---|---|---|---|---|
| × | · | o | · | · | · |
| × | · | · | o | · | · |
| × | · | · | · | o | · |
| × | · | · | · | · | o |

(vi)

Now the union of any two blocks must form a $U_8$ (an $S_8$ would lead to a sextet and a $T_8$ would contradict transitivity), and a $U_8$ falls into four pairs such that the union of any three of them is an $S_6$. For double transitivity to hold, these pairings must split across the blocks in the same way for each of the 15 $U_8$s. In cases (i)–(v) this fails to happen; however, case (vi) does admit a doubly transitive group preserving it. Labelling the blocks $\infty, \underline{0}, \ldots, \underline{4}$ as indicated in the following diagram we can check that this group is isomorphic to $\mathrm{PGL}_2(5)$ generated by:

$$\alpha = (\infty)(\underline{0}\ \underline{1}\ \underline{2}\ \underline{3}\ \underline{4}) \quad \sim \quad (11\ 1\ 22\ 2\ 0)(14\ 8\ 10\ 15\ 4)$$
$$(12\ 17\ 5\ 16\ 21)(7\ 19\ 13\ 20\ 18)$$

and

$$\gamma = (\underline{\infty}\ \underline{0})(\underline{1}\ \underline{2})(\underline{3}\ \underline{4}) \quad \sim \quad (\infty\ 12)(3\ 7)(6\ 11)(9\ 14)(1\ 5)(8\ 22)$$
$$(17\ 13)(19\ 10)(2\ 4)(15\ 18)(16\ 0)(20\ 21).$$

We may verify that this group fixes the duum

$$\{\infty, 0, 1, 2, 7, 9, 11, 13, 18, 19, 20, 22\}/\{3, 4, 5, 6, 8, 10, 12, 14, 15, 16, 17, 21\},$$

and so may be ruled out of our current investigation. However, this partition of $\Omega$ into six blocks of size 4 is sufficiently interesting to merit a combinatorial construction of the two dodecads in this invariant duum. In the following diagram, which refers now to the MOG arrangement, the blocks are denoted

## 8.2 Completeness of the List of Maximal Subgroups

by ×s and the other five tetrads of each of the associated sextets are denoted by $0, \ldots, 4$.

| × | 0 | 1 | 2 | 3 | 4 |
|---|---|---|---|---|---|
| × | 0 | 1 | 2 | 3 | 4 |
| × | 0 | 1 | 2 | 3 | 4 |
| × | 0 | 1 | 2 | 3 | 4 |

$\underline{\infty} \sim \{\infty, 3, 6, 9\}$

| 3 | 4 | × | 0 | 3 | 4 |
|---|---|---|---|---|---|
| 2 | 2 | 2 | 1 | 0 | 0 |
| 4 | 3 | 0 | × | 3 | 4 |
| 1 | 1 | 2 | 1 | × | × |

$\underline{0} \sim \{7, 11, 12, 14\}$

| 0 | 1 | 1 | × | 3 | 0 |
|---|---|---|---|---|---|
| 2 | × | 0 | 2 | 4 | 0 |
| 1 | 3 | 4 | 2 | × | × |
| 4 | 1 | 2 | 3 | 4 | 3 |

$\underline{1} \sim \{1, 8, 17, 19\}$

| 1 | 1 | 2 | 2 | × | 4 |
|---|---|---|---|---|---|
| 3 | 4 | 4 | 3 | 3 | × |
| 2 | 1 | 2 | 1 | 3 | 0 |
| 0 | × | × | 0 | 4 | 0 |

$\underline{2} \sim \{5, 10, 13, 22\}$

| 3 | 2 | 3 | 4 | 4 | × |
|---|---|---|---|---|---|
| 0 | 1 | 3 | × | 0 | 0 |
| 4 | × | × | 2 | 3 | 2 |
| 1 | 1 | 2 | 4 | 1 | 0 |

$\underline{3} \sim \{2, 15, 16, 20\}$

| 2 | × | 4 | 2 | 1 | 1 |
|---|---|---|---|---|---|
| 0 | 0 | × | 4 | × | 2 |
| 1 | 0 | 3 | 2 | 4 | 3 |
| 3 | 4 | 3 | × | 1 | 0 |

$\underline{4} \sim \{0, 4, 18, 21\}$

We draw a graph on the 24 points of $\Omega$ in which two points are joined if, and only if, they occur together in just one non-block tetrad of these six sextets. This graph is regular of degree 5 and disconnected with two connected components of size 12. These components are the two dodecads forming the invariant duum, and the graph defined on each of them is that defined on the 12 faces of a regular dodecahedron with two faces joined if they share an edge. This configuration is displayed in Figure 8.10.

*Blocks of size 3.* By Corollary 8.2 we may assume that $G$ acts doubly transitively on the eight blocks. Now the normalizer in $M_{24}$ of an element of cycle shape $3^8$ is a subgroup of shape $S_3 \times L_2(7)$ in which the linear group acts

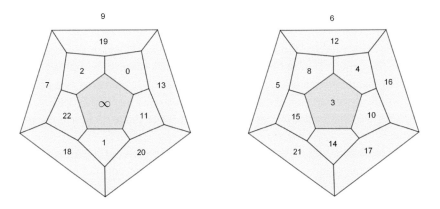

Figure 8.10 The dodecahedra defined on a duum by the blocks of type (vi)

| $B$ | $\infty$ | 0 | 1 | 2 | 3 | 4 | 5 | 6 |
|---|---|---|---|---|---|---|---|---|
| $a$ | $\infty$ | 0 | 3 | 6 | 19 | 9 | 5 | 15 |
| $b$ | 11 | 1 | 4 | 16 | 20 | 13 | 21 | 14 |
| $c$ | 22 | 2 | 18 | 8 | 10 | 12 | 7 | 17 |

Figure 8.11 The 3 × 8 brick arrangement

identically on three octads of a trio, and the $S_3$ permutes the three octads. Generators for this subgroup may be seen in Figure 8.5 where two elements act in an identical manner on the three bricks of the MOG; the $S_3$ bodily permutes the three bricks. For convenience we exhibit the MOG trio as a 3 × 8 array in Figure 8.11 where the copy of $L_2(7)$ permutes the columns whilst preserving the rows, and the $S_3$ permutes the rows whilst preserving the columns. An element of order 7 fixing one block whilst cycling the other seven may taken as $[(\infty)(0\ 1\ 2\ 3\ 4\ 5\ 6) \times I]_B$; commuting with this element we have $[I \times S_3]_B$ and normalizing it we have $[(\infty)(0)(1\ 2\ 4)(3\ 6\ 5)]_B$. Under the action of this subgroup there are the following possibilities:

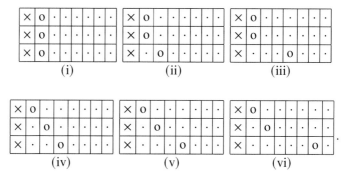

Case (i) visibly has the full Brick subgroup preserving it, and this is contained in the trio group. That the blocks admit no further symmetries follows from the fact that the sextets that are congruent modulo $\mathscr{C}$ to the union of a pair of blocks are all refinements of the MOG trio and no other trio. This, incidentally, explains why the centralizer of an element of cycle shape $3^8$ lies in the trio group.

Double transitivity on the blocks means that the union of a pair of blocks cannot be congruent to an $S_6$ as the stabilizing subgroup could not preserve the 28 pairs of points completing these $S_6$s to octads. Both Cases (ii) and (iii) fail to satisfy this condition and so may be eliminated.

Case (iv) cannot have a doubly transitive group acting on it as the sextets congruent to pairs of blocks do not intersect the blocks in the same manner.

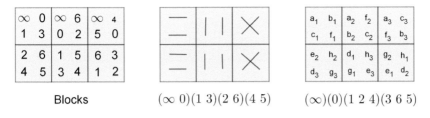

Figure 8.12 The blocks of size 3 in Case (v)

Case (vi) has a copy of the Octern group, isomorphic to $L_2(7)$ acting on it, see Figure 7.3.

This leaves Case (v). In Figure 8.12 we exhibit the eight blocks in the MOG, together with two permutations of $M_{24}$ that preserve the block system, in each case giving their action on the blocks. The element of order 7 that we have used in our analysis of the various possibilities together with these two permutations generate a soluble group of shape $2^3 : 7 : 3$. Using this soluble subgroup we find that the only octads that have a point in each of the eight blocks of this system are the bricks of the MOG, and thus this group is contained in the trio group. It is, nonetheless, an interesting partition of the 24 points affording as it does a transitive stabilizing subgroup, and so we display a $3 \times 8$ array whose columns are the blocks and whose rows are the fixed trio, namely the bricks of the MOG:

|   | $\infty$ | 0 | 1 | 2 | 3 | 4 | 5 | 6 |
|---|---|---|---|---|---|---|---|---|
| $a$ | $\infty$ | 0 | 3 | 6 | 19 | 9 | 5 | 15 |
| $b$ | 11 | 4 | 16 | 20 | 13 | 21 | 14 | 1 |
| $c$ | 22 | 10 | 12 | 7 | 17 | 2 | 18 | 8 |

With respect to this array, the generators of the stabilizing subgroup are

$$\langle [(\infty)(0\ 1\ 2\ 3\ 4\ 5\ 6) \times I],\ [(\infty)(0)(1\ 2\ 4)(3\ 6\ 5) \times (a\ b\ c)],$$
$$[(\infty\ 0)(1\ 3)(2\ 6)(4\ 5) \times I] \rangle \cong 2^3 : 7 : 3$$

As with the octern group, this subgroup is the centralizer in $M_{24}$ of an element of order 3 outside $M_{24}$, and in both cases the stabilizing subgroup has order 168. However, in the octern group case that group is simple and maximal; in this case it is soluble and contained in the trio group.

*Blocks of size 2.* By Corollary 8.2 $G$ acts doubly transitively on the 12 blocks, and so we may assume that the element $\alpha : x \mapsto 2x$ of $L_2(23)$ lies in $G$, and so our first block will be $\{\infty, 0\}$. The element $\alpha$ is normalized within $M_{24}$ by a cyclic group of order 10 that we present here as the product of an element $\beta$

| $L$ | $\infty$ | 0 | 1 | 2 | 3 | 4 | 5 | 6 | 7 | 8 | 9 | 10 |
|---|---|---|---|---|---|---|---|---|---|---|---|---|
| $a$ | 0 | 1 | 2 | 4 | 8 | 16 | 9 | 18 | 13 | 3 | 6 | 12 |
| $b$ | $\infty$ | 5 | 10 | 20 | 17 | 11 | 22 | 21 | 19 | 15 | 7 | 14 |

Figure 8.13 The pairing preserved by a subgroup isomorphic to $\mathrm{PGL}_2(11)$

of order 5 cubing $\alpha$ and an element $\gamma$ of order 2 inverting it. This group can be seen acting on a $2 \times 12$ array with respect to which we have

$$\alpha = [(\infty)(0\ 1\ 2\ 3\ 4\ 5\ 6\ 7\ 8\ 9\ 10) \times I],$$
$$\beta = [(\infty)(0)(1\ 3\ 9\ 5\ 4)(2\ 6\ 7\ 10\ 8) \times I],$$
$$\gamma = [(\infty)(0)(1\ 10)(2\ 9)(3\ 8)(4\ 7)(5\ 6) \times (a\ b)].$$

Under the action of this normalizing subgroup, there are just three orbits on pairs that are contenders for the second block: (i) $\{1, 5\}$; (ii) $\{1, 10\}$; and (iii) $\{1, 20\}$. For a given block system to admit a doubly transitive group, every pair of blocks must be contained in the same number of $S_6$s that are unions of blocks. This does not hold in Cases (ii) and (iii), so we must now consider Case (i) in which the blocks are the columns in Figure 8.13. Now the array in Figure 8.13 is preserved by a subgroup of $M_{24}$ isomorphic to $\mathrm{PGL}_2(11)$ in which the even elements fix the two rows and the odd elements interchange them. These two rows are dodecads of the Golay code and so this subgroup fixes a duum, and is contained in a copy of $M_{12} : 2$. We must show that the pairing cannot be preserved by any larger subgroup. But there are just two dodecads that contain one point from each column, namely the two rows of Figure 8.13. This is readily seen as, if there was another such dodecad, then its symmetric difference with the top row would be an octad consisting of a union of blocks. There is no such octad and so the duum of Figure 8.13 is uniquely defined.

We note in passing that, analogously to the two sets of imprimitivity with blocks of size 3, we see that this copy of $\mathrm{PGL}_2(11)$ is the centralizer in $M_{24}$ of an element of order 2 outside $M_{24}$.

*G acts primitively on $\Omega$.* By Corollary 8.2 $G$ acts doubly transitively on $\Omega$ and so we may assume that $\alpha : x \mapsto x + 1$ of $L_2(23)$ lies in $G$; it is normalized by $\beta : x \mapsto 2x$ to generate a Frobenius group of shape $23 : 11$. By Sylow's theorem this must also be the normalizer of $\alpha$ in $M_{24}$ as it has index congruent to 1 modulo 23. Now if $N_G(\alpha) = \langle \alpha \rangle$ then by Burnside's theorem $\langle \alpha \rangle$ has a normal complement; that is to say, $G$ has a subgroup $H$ such that $G = H : \langle \alpha \rangle$ with $H \cap \langle \alpha \rangle = \langle 1 \rangle$. Since $23 \nmid |H|$, $\alpha$ must normalize a Sylow 2-subgroup, and hence normalize its centre. But $\alpha$ cannot commute with an involution and so this centre must contain more than $2^{11} - 1$ elements, which is impossible.

## 8.2 Completeness of the List of Maximal Subgroups

So
$$N_G(\alpha) = \langle \alpha, \beta \rangle \cong 23 : 11.$$

Now $N_{M_{24}}(\beta) \cong 11 : 10$ as in the previous section (again by Sylow's theorem the index of a subgroup of shape $11 : 10$ in $M_{24}$ is congruent to 1 modulo 11, and the normalizer of such an element in the symmetric group $S_{24}$ has order $2^2.11^2.10$ so no larger normalizer is possible). If $N_G(\beta) = \langle \beta \rangle$ then, appealing to Burnside again, $G$ has a normal subgroup $K$ such that $G = K : \langle \beta \rangle$ and $K \cap \langle \beta \rangle = \langle 1 \rangle$. But $K$ must act doubly transitively on $\Omega$ and the above arguments apply to it, and so $11 \mid |K|$ which is a contradiction. So we have

$$C_{11} \cong \langle \beta \rangle \subsetneq N_G(\beta) \subseteq N_{M_{24}}(\beta) \cong 11 : 10.$$

There are then three possibilities:

(1) $N_G(\beta) \cong 11 : 2$ when $\langle N_G(\beta), \alpha \rangle \cong L_2(23)$;
(2) $N_G(\beta) \cong 11 : 5$ when $\langle N_G(\beta), \alpha \rangle \cong M_{23}$; and
(3) $N_G(\beta) \cong 11 : 10$ when $\langle N_G(\beta), \alpha \rangle \cong M_{24}$.

Suppose there is a subgroup $G$ satisfying

$$L_2(23) \subsetneq G \subsetneq M_{24}.$$

Then we have $N_G(\alpha) \cong 23 : 11$ and $N_G(\beta) \cong 11 : 2$ and so we require

$$
\begin{aligned}
|G : L_2(23)| &\equiv 1 \bmod 11, \text{ and}\\
|G : L_2(23)| &\equiv 1 \bmod 23, \text{ and}\\
|G : L_2(23)| &\text{ divides } 2^7.3^2.5.7.
\end{aligned}
$$

A consideration of this index modulo 253 shows that there is no solution.

*Possible containment between these nine subgroups.* We see that there is no possible containment between these nine subgroups, either by consideration of the combinatorial object they preserve or by their orders, except that we must confirm that the Octern group is not contained in any of the other eight. But being transitive, $O^n$ clearly does not fix a monad, a duad, a triad or an octad, and its order does not divide either the sextet group, the duum group or the projective group $L_2(23)$. This leaves the trio group. But from Figure 7.3 we see that the only possible trio fixed by $O^n$ is the one fixed by the element of order 7, and this is not preserved by the other generator. Thus we conclude that every subgroup of $M_{24}$ is contained in a conjugate of one of the nine groups listed, and that there is no containment between them. □

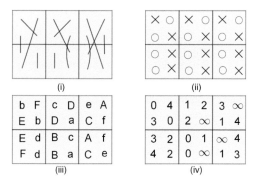

Figure 8.14 An involution $\rho$ in $S_{24}$ but outside $M_{24}$ pairs of whose transpositions generate the Pet$_5$ subspace of $C^*$.

## 8.3 Other Interesting Subgroups

### 8.3.1 The Pet$_5$ Subgroup

Pet$_5$ is a 5-dimensional purely sextet subspace of $C^*$, the dual space of the binary Golay code $C$. It was mentioned briefly in our proof that *Eight Octads Suffice* to define $C$, see Section 7.1, as it contains a 4-dimensional subspace Pet$_4$ with the property that all of its 2-dimensional subspaces are odd lines. If it were possible for 7 octads to be sufficient to define a unique copy of $C$ then, together with the whole set $\Omega$, they would have to span the annihilator of Pet$_4$, since it is the only 4-dimensional subspace of $C^*$ with this property. Thus we had to show that the 8-dimensional subspace of $C$ obtained in this way extends to more than one copy of $C$.

The subgroup of $M_{24}$ preserving a copy of Pet$_5$ is the centralizer in $M_{24}$ of a fixed point free involution in $S_{24}$ that lies outside $M_{24}$. The canonical such involution $\rho$ is shown in Figure 8.14, where

$\rho =$

$(\infty\, 19)(15\, 5)(0\, 9)(3\, 6)(11\, 14)(20\, 21)(1\, 4)(16\, 13)(22\, 7)(10\, 17)(2\, 8)(18\, 12).$

The 31 sextets in the 5-dimensional subspace Pet$_5$ are those defined by pairs of transpositions of $\rho$. Six of these are special in that they lie in no even line of Pet$_5$ and tetrads defining these six sextets are displayed in Figure 8.14(iv); they are the *blocks* of the arrangement. Clearly the sum of these six blocks is 0 as their union is $\Omega$. There are 32 dodecads fixed by $\rho$ of which 30 contain the union of two blocks; the remaining two form a duum, which is thus fixed by the subgroup of $M_{24}$ stabilizing Pet$_5$. This subgroup, which thus lies in a copy of

### 8.3 Other Interesting Subgroups

$M_{12} : 2$, is isomorphic to $PGL_2(5)$ and it may be generated by $\alpha$ and $\gamma$ given by

|  | Element in $M_{24}$ | Element in $PGL_2(5)$ |
|---|---|---|
| $\alpha :=$ | (13 18 1 3 0)(16 12 4 6 9) | $\sim (\infty)(0\ 1\ 2\ 3\ 4)$ |
|  | ($\infty$ 11 5 7 17)(19 14 15 22 10) |  |
| $\gamma :=$ | (2 19)(8 $\infty$)(20 13)(21 16)(11 6)(12 22) | $\sim (\infty\ 0)(1\ 3)(2\ 4)$ |
|  | (14 3)(18 7)(1 10)(4 17)(5 0)(15 9) |  |

We have naturally chosen our copy of $Pet_5$ so that the fixed duum is canonical, namely the symmetric difference of the top row of the MOG and its 2nd, 4th and 6th columns, together with its complement, as displayed in Figure 8.14(ii). The 15 sums of two blocks, which are thus labelled with unordered pairs from $P_1(5)$, form an orbit and together with the empty set they form the aforementioned 4-dimensional subspace known as $Pet_4$; it is thus stabilized by the full $Pet_5$ subgroup. The sums of three blocks form the final orbit of sextets, giving

$$1 + 6 + \binom{6}{2} + \binom{6}{3}\bigg/2 = 1 + 6 + 15 + 10 = 32.$$

This space contains 15 even lines, such as

$$\{\infty 0, \infty 14 \sim 023, \infty 23 \sim 014\},$$

which form the lines of a Petersen graph, hence the name. The edges are thus labelled with the 15 unordered pairs of blocks, and the edges with the partitions into 2 threes.

#### 8.3.2 The $T_{12}$ Subgroup

As we saw in Section 7.1.2 the stabilizer of two complementary $T_{12}$s has shape

$$(2^2 \times L_2(5)) : 2.$$

It cannot be maximal in $M_{24}$ as its stabilizer preserves the sextet to which the $T_{12}$s are congruent modulo $\mathscr{C}$; it is, however, of interest as it acts transitively on $\Omega$ preserving a partition into 12 pairs that define an involution of $M_{24}$ of cycle shape $2^{12}$. In Figure 8.15(c) we exhibit this pairing for the canonical $T_{12}$, which is shown in capital letters; generators for the subgroup of $M_{24}$ preserving the pair of complementary $T_{12}$s it defines are given in Figure 7.14. Note that this $T_{12}$ corresponds to the hexacodeword $[\bar{\omega}\ \omega\ \bar{\omega}\ \omega\ \bar{\omega}\ \omega]$, however it does not lie in $\mathscr{C}$ as it has evenly many entries in each column but does not intersect the top row of the MOG evenly. Complementing on any column produces a $\mathscr{C}$-set. The normal Klein 4-group in the stabilizer of these two complementary

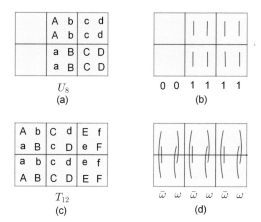

Figure 8.15 The canonical $T_{12}$ and its stabilizing subgroup

$T_{12}$s is generated by the element shown here as $[\omega\ 1\ \omega\ 1\ \omega\ 1]$ together with the central involution $[\bar\omega\ \omega\ \bar\omega\ \omega\ \bar\omega\ \omega]$. Figure 8.15 shows how pairings defined by a $U_8$, (a) and (b), and a $T_{12}$, (c) and (d), extend uniquely to involutions of shape $1^8.2^8$ and $2^{12}$ respectively. A $1^8.2^8$ involution arises in this way from $(\binom{8}{3} \times 4)/4 = 56$ different $U_8$s, whilst a $2^{12}$ involution arises from $2^5$ different $T_{12}$s. Thus we may confirm the orders of the stabilizers of a $U_8$ and a $T_{12}$ given on pages 63 and 64.

# 9

# The Mathieu Group $M_{12}$

## 9.1 The Steiner System $S(5, 6, 12)$

The Steiner system $S(5, 6, 12)$, which is unique up to relabelling of the 12 points, consists of

$$\binom{12}{5} \bigg/ \binom{6}{5} = 132$$

6-element subsets known as *special hexads* or simply *hexads* of a 12-element set with the property that any 5 of the 12 points lie together in precisely one of them. The group of permutations of the 12 points that preserve this system of hexads is the famous Mathieu group $M_{12}$, which acts sharply quintuply transitively on the 12 points and thus has order $12.11.10.9.8 = 95\,040$. The group $M_{12}$ contains subgroups isomorphic to the linear group $L_2(11)$ acting transitively in its natural action as linear fractional maps on the 12 points of the projective line $P_1(11)$, and so we usually label the 12 points $\{\infty, 0, 1, \ldots, X\}$ where X denotes 10. Fixing a point, $\infty$ say, we obtain the Mathieu group $M_{11}$ that thus preserves a Steiner system $S(4, 5, 11)$, namely the

$$\binom{11}{4} \bigg/ \binom{5}{4} = 66$$

pentads obtained by removing $\infty$ from those hexads that contain it. Unlike $M_{12}$ and $M_{11}$, which are simple groups, the stabilizer of two points, $\infty$ and 0 say, is not simple and we have

$$M_{10} \cong A_6\dot{}2,$$

containing the alternating group of order 360 with index 2 and preserving a Steiner system $S(3, 4, 10)$. The 'upper dot' indicates the fact that this extension is *non-split*; that is to say there are no elements of order 2 outside the $A_6$.

If we allow permutations that interchange $\infty$ and 0, then we obtain the full automorphism group of $A_6$:

$$\mathrm{Aut}(A_6) \cong A_6^{\cdot} 2^2 \cong \mathrm{P\Gamma L}_2(9),$$

the projective general linear group $\mathrm{PGL}_2(9)$ extended by the field automorphism that cubes every element of the field, thus interchanging $i$ and $-i$. The 10 points are, of course, labelled with the projective line

$$P_1(9) = \{\infty, 0, \pm 1, \pm i, \pm 1 \pm i \ : \ 1 + 1 + 1 = 0, i^2 = -1\}.$$

The Steiner system $S(3, 4, 10)$ may then be taken as the 30 images under the action of $\mathrm{P\Gamma L}_2(9)$ of $\{\infty, 0, 1, -1\}$, a tetrad that is fixed by a subgroup isomorphic to $2 \times \mathrm{PGL}_2(3) \cong 2 \times S_4$ of order 48.

## 9.2 Π, the Affine Plane of Order 3

Fixing three points, $\infty$, 0 and 1 say, we obtain a Steiner system $S(2, 3, 9)$, namely the triples that extend $\{\infty, 0, 1\}$ to a hexad; this is a copy of the affine plane of order 3 whose points are the nine vectors in a 2-dimensional vector space over $\mathbb{Z}_3$ and whose triples are the cosets of its 1-dimensional subspaces, as described in Chapter 2, see Figure 9.1. In these $3 \times 3$ arrays, the triples appear as the lines of a *generalized noughts and crosses* or *tic-tac-toe* board. They fall into four sets of three parallel lines with slopes $\infty, 0, 1$ and $-1$. For instance, the line $y = -x + 1$ consists of the three points

$$\left\{ \begin{pmatrix} 0 \\ 1 \end{pmatrix}, \begin{pmatrix} 1 \\ 0 \end{pmatrix}, \begin{pmatrix} -1 \\ -1 \end{pmatrix} \right\},$$

which is shown as the points labelled 0 in Figure 9.1(iv).

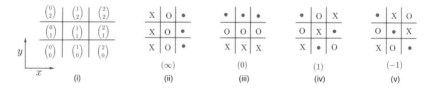

Figure 9.1 The 9-point affine plane

The 3-point stabilizer is a group of order 72 and shape $3^2 : Q_8$, where $Q_8$ denotes the quaternion group of order 8, and the subgroup stabilizing a triple, $\{\infty, 0, 1\}$ say, is the *Hessian group* H of order 432 and shape $3^2 : \mathrm{GL}_2(3) \cong 3^2 : 2S_4$, where the $S_4$ is acting on the four sets of parallel lines. It has a normal subgroup of order 9 consisting of all translations $v \mapsto v + a$, extended by the general linear group of all non-singular $2 \times 2$ matrices over $\mathbb{Z}_3$, the integers modulo 3. Elements generating this Hessian group are shown diagrammatically in Figure 9.2.

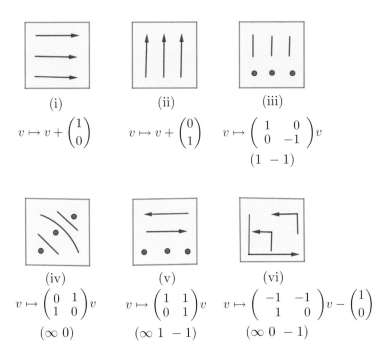

Figure 9.2 Elements generating the Hessian group $H \cong 3^2 : 2S_4$

## 9.3 Action of H on Subsets of Points of Π

From Figure 9.2, which exhibits elements as affine transformations and also indicates how they act on lines of various slopes ($\infty, 0, 1$ and $-1$), we see that H acts doubly transitively on the points, and hence transitively on the 12 lines. Fixing two points is a subgroup isomorphic to $S_3$, which fixes the third point of the line containing the two fixed points and acts transitively on the remaining six points. Thus H acts transitively on the $(9 \times 8 \times 6)/(3 \times 2) = 72$ triples of points that do not constitute a line. We refer to the labelling of the nine points

100                    The Mathieu Group $M_{12}$

|  | ∞ |  |  | 0 |  |  | 1 |  |
|---|---|---|---|---|---|---|---|---|
| 5 | 2 | 6 | 8 | X | 2 | 3 | 6 | X |
| 9 | 7 | X | 4 | 9 | 6 | 7 | 4 | 2 |
| 8 | 4 | 3 | 3 | 7 | 5 | 5 | 9 | 8 |

$\quad\quad\{\infty,0\}\{1,-1\}\quad\quad\quad\{\infty,1\}\{0,-1\}\quad\quad\quad\{\infty,-1\}\{0,1\}$

$\quad\quad\quad\quad$(i)$\quad\quad\quad\quad\quad\quad\quad\quad$(ii)$\quad\quad\quad\quad\quad\quad\quad\quad$(iii)

Figure 9.3 Three arrangements of the 9-point affine plane corresponding to ∞, 0 and 1 in the Kitten triangle

as shown in Figure 9.3(i) as the ∞–*arrangement* and observe that in it $\{2,6,X\}$ is such a *triangle*. It is fixed by the copy of $S_3$ generated by the affine elements:

$$r \equiv v \mapsto \begin{pmatrix} -1 & -1 \\ 1 & 0 \end{pmatrix} v - \begin{pmatrix} 1 \\ 0 \end{pmatrix} = (2\ X\ 6)(3\ 5\ 8)(4\ 7\ 9),$$

$$f_\infty \equiv v \mapsto \begin{pmatrix} 0 & 1 \\ 1 & 0 \end{pmatrix} v = (2\ X)(3\ 5)(4\ 9).$$

The element $r$ demonstrates that H acts transitively on sets of four points no three of which lie on a line; these are known as *quadrangles*. The three

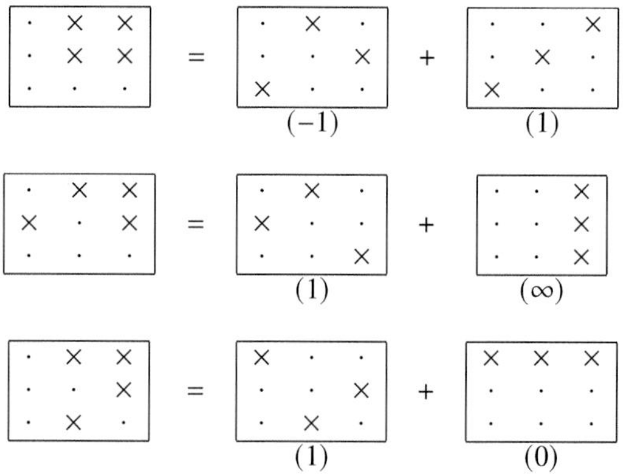

Figure 9.4 Quadrangles containing $\{2,6,X\}$

quadrangles containing $\{2,6,X\}$ are $\{2,6,X,7\}$, $\{2,6,X,9\}$ and $\{2,6,X,4\}$; each of these may be expressed uniquely as the symmetric difference of two lines as shown in Figure 9.4, but just one of these quadrangles is the sum of two *perpendicular* lines in the ∞–arrangement, that is, slopes ∞ and 0 or 1 and −1. Thus $\{2,6,X,7\}$ is the sum of a line of slope 1 and a line of slope −1 in the

∞-arrangement; whereas, for instance, {2, 6, X, 9} is the sum of a line of slope 1 and a line of slope ∞. The sets of four points that are the sums of two lines that are perpendicular to one another, for which all possibilities are shown in Figure 9.5(ii), are called *squares*. Their complements are sets of five points that are uniquely the union of two perpendicular lines; they are called *crosses*; all possible crosses are shown in Figure 9.5(i). Now the permutation $r$ permutes the slopes of lines in the ∞-arrangement as $(\infty\ 0\ -1)(1)$; that is to say it takes lines with slope ∞ to lines with slope 0, lines with slope 0 to lines with slope $-1$ and lines with slope $-1$ to lines with slope ∞. It simply cycles the three lines with slope 1. Figure 9.3 shows the three images of the ∞-arrangement under the action of $r$: the ∞-, 0- and 1-arrangements. We see that the two lines that sum to the quadrangle {2, 6, X, 9} are perpendicular in the 0-arrangement, and so it is a cross in this arrangement. Similarly the quadrangle {2, 6, X, 4} is a cross in the 1-arrangement. So we see that every set of four points of Π that does not contain a line forms a square in precisely one of the three arrangements of Figure 9.3.

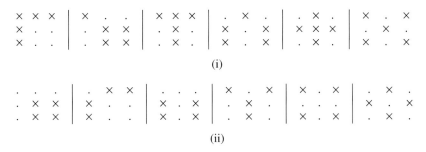

Figure 9.5 Crosses and squares: unions of two perpendicular lines in the 9-point plane and their complements

## 9.4 The Kitten

Having examined the 9-point affine plane in some detail, we may now produce a device for recognizing and reading off the hexads of the Steiner system S(5, 6, 12). We do this by gluing together three arrangements of the plane which are the images of the ∞-arrangement under the action of the element $r$. It is known affectionately as the Kitten as it does for $M_{12}$ what the MOG does for $M_{24}$. Thus in Figure 9.6 the ∞-, 0- and 1-arrangements are as shown in Figure 9.3, which also indicates which sets of lines (as labelled in the ∞-arrangement) are perpendicular to one another. Thus lines of slopes ∞ and 1 in the ∞-arrangement become perpendicular in the 0-arrangement. The element $r$

rotates the Kitten through 120° and cycles the three arrangements; the name of the arrangement ($\infty$, 0 or 1) is known as its *vertex*. The three planes inevitably appear as rhombi in Figure 9.6 as part of the equilateral triangle; however, in Figure 9.7 we have distorted the triangle so that for each of the examples, the relevant plane is shown as a square.

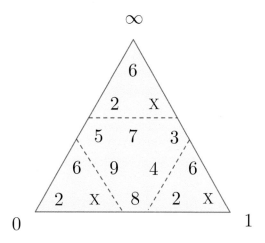

Figure 9.6 The Kitten – a device for recognizing the hexads of the Steiner system $S(5, 6, 12)$

The hexads are now readily described:

(i) $\{\infty, 0, 1\}$ together with a line of $\Pi$ (12 such);
(ii) the union of two parallel lines of $\Pi$ (12 such);
(iii) the union of two intersecting lines of $\Pi$ together with the vertex of the arrangement in which they are perpendicular ($12 \times 9/2 = 54$ such);
(iv) the symmetric difference of two intersecting lines together with the two vertices of the two arrangements in which they are not perpendicular (54 such).

## 9.5 Recognizing Hexads in the Kitten

Certainly the points $\infty$, 0 and 1 together with one of the 12 lines of the affine plane form a hexad of the system. Thus $\{\infty, 0, 1, 6, 2, 5\}$ and $\{\infty, 0, 1, 6, 9, 4\}$ are hexads. If we fix, say, the $\infty$-arrangement then for each of the 9 points of the affine plane there are two pairs of perpendicular lines through it. The union of two perpendicular lines is a set of 5 points that together with $\infty$ forms a hexad

## 9.7 Finding the Hexad Containing a Given 5 Points

of the system. The complement of such a hexad consists of the remaining two fixed points, in this case 0 and 1, together with the symmetric difference of two perpendicular lines in the ∞-arrangement. Thus for each of the 9 points we obtain two hexads for each arrangement giving a total of $9 \times 2 \times 3 = 54$ hexads. The complement of each of these hexads is also a hexad of the system, giving a total of

$$2 \times (12 + 54) = 132.$$

In Figure 9.5 we show the readily recognizable configurations that appear as the union of perpendicular lines in the noughts and crosses board. Thus $\{\infty, 2, 7, 4, 3, 8\}$, $\{0, 1, 6, X, 5, 9\}$, $\{0, 5, 3, 9, 2, 8\}$ and $\{\infty, 1, 7, 6, 4, X\}$ are hexads of the system.

## 9.6 Mnemonic for Reconstructing the Kitten

As has been mentioned, the Mathieu group $M_{12}$ contains copies of the linear group $L_2(11)$; thus, with the labelling chosen, all even linear fractional maps preserve the Steiner system. We have:

$$r : x \mapsto \frac{1}{1-x} \cong (\infty\ 0\ 1)(2\ X\ 6)(3\ 5\ 8)(4\ 7\ 9).$$

To construct the Kitten we take an equilateral triangle and label its vertices ∞, 0 and 1, anticlockwise say, with ∞ at the apex. We then write 6, 7, 8 on the vertical altitude below ∞ and fill in the rest of the labels so that $r$ corresponds to a rotation through $120^o$ anticlockwise.

## 9.7 Finding the Hexad Containing a Given 5 Points

As explained earlier, we refer to ∞, 0 and 1 as the *vertices*; the union of two perpendicular lines is called a *cross* and the symmetric difference of two perpendicular lines is called a *square*. Recall that the group of symmetries of the plane acts transitively on the sets of three points that do not lie on a line, and that such a set is contained in a unique square in each of the three arrangements, a different one for each of the three arrangements. Given a set of five points $U$, we seek $x$ so that $U \cup \{x\}$ is a hexad. If $U$ contains all three vertices, then we simply complete the remaining two points to a line in the affine plane. If $U$ contains two vertices, 0 and 1 say, then we first check if the remaining 3 points form a line of the affine plane; if they do then we simply adjoin ∞. If not, we look in the ∞-arrangement and complete the remaining 3 points to a square. If

$U$ contains just one vertex, $\infty$ say, then we check in the $\infty$-arrangement to see if the remaining 4 points can be extended to a cross. If not, then these four points must form a square in either the 0-arrangement or the 1-arrangement, and the required hexad is obtained by adjoining 1 or 0 respectively.

**Example 9.1** Complete $\{\infty, 2, 3, 4, 5\}$ to a hexad.

In the 1-arrangement we see that $\{2, 3, 4, 5\}$ is a square; thus $\{\infty, 0, 2, 3, 4, 5\}$ is the required hexad.

**Example 9.2** Complete $\{0, 2, 3, 5, 9\}$ to a hexad.

We see that $\{2, 3, 5, 9\}$ can be extended to a cross in the 0-arrangement by adding 8, and so the required hexad is $\{0, 2, 3, 5, 8, 9\}$.

**Example 9.3** Complete $\{1, 3, 4, 5, 9\}$ to a hexad.

We see that $\{3, 4, 5, 9\}$ does not complete to a cross in the 1-arrangement. However, $\{3, 4, 5, 9\}$ is a square in the $\infty$-arrangement, and so the required hexad is $\{0, 1, 3, 4, 5, 9\}$.

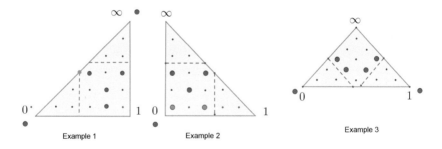

Figure 9.7 The hexads of the three examples as they appear in the Kitten

## 9.8 Useful Generators Using the Kitten

Given the construction of the Kitten and the description of the hexads, it is clear that the reflections in the three altitudes are in $M_{12}$. These are

$$\begin{aligned} f_\infty &\equiv (0\ 1)(2\ X)(5\ 3)(9\ 4), \\ f_0 &\equiv (\infty\ 1)(6\ X)(5\ 8)(4\ 7), \\ f_1 &\equiv (\infty\ 0)(2\ 6)(3\ 8)(7\ 9). \end{aligned}$$

If we let

$$a : x \mapsto x + 1 \cong (0\ 1\ 2\ 3\ 4\ 5\ 6\ 7\ 8\ 9\ X),$$

then we have

$$\langle a, f_0 \rangle \cong M_{12}, \quad \langle a, f_\infty \rangle \cong M_{11}, \quad \langle a, f_1 \rangle \cong M_{11},$$

where in the first occurrence, $M_{11}$ is the stabilizer of $\infty$ and in the second, it acts transitively on the 12 points.

## 9.9 A Class of Involutions in $M_{12}$ Using the Kitten

The group $M_{12}$ contains a class of involutions of cycle shape $1^4 \cdot 2^4$. Clearly, if $\sigma$ is such an involution, then the hexad containing three of the fixed points of $\sigma$ and one of its four transpositions must be fixed by $\sigma$ and so also contain the fourth fixed point. Since $M_{12}$ acts quintuply transitively on the 12 points, and the stabilizer of four points is a quaternion group $Q_8$ we may choose the four fixed points at random and there will be a unique involution fixing these four points and transposing the pairs of points that complete these four to a hexad.

**Example 9.4** Find the involution of $M_{12}$ that fixes 3, 5, 7 and 9.

Now $\{3, 5, 7, 9\}$ can be completed to a cross in the 0-arrangement by adding X, and so (0 X) is one of the four transpositions. Also $\{3, 5, 7, 9, 4, 6\}$ is the union of two parallel lines in the affine plane and so (4 6) is another transposition. In the $\infty$-arrangement $\{2, 3, 5, 7, 9\}$ is a cross, and in the 1-arrangement $\{3, 5, 7, 9, 8\}$ is a cross. Thus ($\infty$ 2) and (1 8) are also transpositions and the required involution is

$$(3)(5)(7)(9)(0\ X)(4\ 6)(\infty\ 2)(1\ 8).$$

## 9.10 $M_{12}$ as a Subgroup of $M_{24}$

As we have seen, see Section 8.1.6, the stabilizer of a dodecad $D$ in $M_{24}$ is a subgroup isomorphic to $M_{12}$, and the outer automorphism of this copy of $M_{12}$ interchanges $D$ with its complement $\Omega \setminus D$. Thus the stabilizer of the duum $\{D : \Omega \setminus D\}$ has shape $M_{12} : 2$. The hexads of the Steiner system $S(5, 6, 12)$ defined on $D$ that is preserved by such a copy of $M_{12}$ are just those subsets of size 6 that lie in octads, the other two points of which are in the complementary dodecad. Every pair of points in $D$ determine a unique partition of $\Omega \setminus D$ into two special hexads. In this chapter we favour a slightly different canonical dodecad from that used in Section 8.3.1 and take the symmetric difference of the top row of the MOG and the 4th, 5th and 6th columns. Thus the $3 \times 3$ array in the bottom right of the MOG will correspond to the familiar 9-point affine plane. In Figure 9.8 we exhibit this dodecad, showing how it is obtained as the symmetric difference of two octads.

Figure 9.8 The canonical dodecad as the symmetric difference of two octads

In Figure 9.9 we exhibit these partitions when one of the hexads is the union of two parallel lines in the plane $\Pi$. We thus obtain $\binom{12}{2} \times 2 = 132$ hexads in the complementary dodecad and these are, of course, the hexads of the Steiner system $S(5, 6, 12)$ that the subgroup $M_{12}$ preserves. The subgroup of $M_{12}$ fixing such a partition into two hexads is isomorphic to $S_6$ acting non-permutation identically on the two hexads. In Figure 9.10 we exhibit an element of order 2 acting with cycle shape $2^3$ on one hexad and cycle shape $1^4.2$ on the other; and an element of order 3 with cycle shapes $1^3.3$ and $3^2$ on the two hexads.

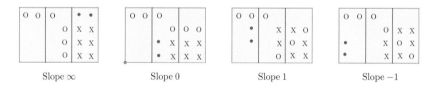

Figure 9.9 Some octads having six points in the canonical dodecad

## 9.11 Embedding the Kitten in the MOG

We now label the highlighted dodecad with the points of the 12-point projective line as it appears in the Kitten, Figure 9.6. The vertices $\infty, 0$ and $1$ label the top

## 9.11 Embedding the Kitten in the MOG

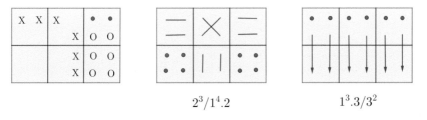

$2^3/1^4.2$      $1^3.3/3^2$

Figure 9.10 The non-permutation identical action of $S_6$ on two sets of size 6

| $\infty$ | 0 | 1 | $\infty_1$ | $\infty_2$ | $\infty_3$ |
|---|---|---|---|---|---|
| $-1_1$ | $1_1$ | $0_1$ | 6 | 2 | 5 |
| $-1_2$ | $1_2$ | $0_2$ | 7 | 9 | X |
| $-1_3$ | $1_3$ | $0_3$ | 8 | 3 | 4 |

Figure 9.11 Embedding the Kitten in the MOG

points of the first three columns and the 9-point plane appears in the remaining $3 \times 3$ block, see Figure 9.11.

In Figure 9.12 we show how the elements $f_\infty$, $f_0$ and $f_1$ of $M_{12}$ extend to elements of $M_{24}$. The element $d$ fuses the three vertices with the 9-point plane within $M_{12}$. The element $\alpha$ that consists of a reflection of the MOG in its vertical bisector followed by bodily interchanging the bottom two rows (note that neither of these two constituents lies within $M_{24}$) interchanges the canonical dodecad with its complement and so may be used to generate $M_{12} : 2$.

Recall that the lines of $\Pi$ have slope $\infty$, 0, 1 or $-1$. In Figure 9.11 two lines of slope $\infty$ complete to an octad with two points from the complementary dodecad labelled $\infty_i$; two lines of slope 0 complete to octads with two points labelled $0_i$, and so on. In Figure 9.9 we exhibit four octads that have 6 points in $\Pi$, one corresponding to slopes of each type. So in Figure 9.9 corresponding to slope $m$ the two dots form an octad together with either $\{\infty, 0, 1\}$ and one line of slope $m$, or with the other two lines of slope $m$.

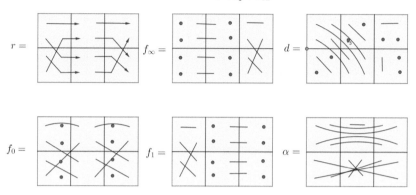

Figure 9.12 Useful elements of $M_{12}$ in the MOG

## 9.12 The Outer Automorphism of $S_6$

In general the automorphism group of the alternating group $A_n$ is the symmetric group $S_n$. However, the case $n = 6$ is exceptional in that $S_6$ acts on sets of 6 letters in two distinct ways, and an outer automorphism interchanges the two sets. In order to see this we introduce some classical terminology due to J. J. Sylvester. Suppose that the first 6 letters on which $S_6$ acts is $\mathcal{P} = \{\infty, 0, 1, 2, 3, 4\}$, then the 15 partitions of $\mathcal{P}$ into three pairs are referred to as *synthemes*. There are, of course, 15 pairs of points in $\mathcal{P}$ and so numerically it may be possible to partition the 15 pairs into five distinct synthemes. Indeed the reader will have no difficulty in convincing himself or herself that there are precisely six ways of doing this, as shown in Figure 9.13. These sets of five synthemes are known as *synthematic totals* or simply *totals*. [The point being that if $ab.cd.ef$ is in a total, then the syntheme in that total that includes the pair $ac$ must be $ac.be.df$ or $ac.bf.de$; but once either of these is chosen, the total is determined. So every syntheme will occur in precisely two totals and we must have $15 \times 2/5 = 6$ totals.]

| ∞0.14.23 | ∞0.14.23 | ∞1.20.34 | ∞2.31.40 | ∞3.42.01 | ∞4.03.12 |
|---|---|---|---|---|---|
| ∞1.20.34 | ∞1.30.42 | ∞2.41.03 | ∞3.02.14 | ∞4.13.20 | ∞0.24.31 |
| ∞2.31.40 | ∞3.21.04 | ∞4.32.10 | ∞0.43.21 | ∞1.04.32 | ∞2.10.43 |
| ∞3.42.01 | ∞2.43.10 | ∞3.04.21 | ∞4.10.32 | ∞0.21.43 | ∞1.32.04 |
| ∞4.03.12 | ∞4.02.31 | ∞0.13.42 | ∞1.24.03 | ∞2.30.14 | ∞3.41.20 |
| $T_1$ | $T_2$ | $T_3$ | $T_4$ | $T_5$ | $T_6$ |
| ∞/01234 | ∞/01324 | ∞/01243 | ∞/01342 | ∞/01432 | ∞/01423 |

Figure 9.13 The six synthematic totals

## 9.12 The Outer Automorphism of $S_6$

It is clear from the manner in which the totals were obtained that any permutation of the points of $\mathcal{P}$ must induce a permutation of $\mathcal{T}$, the set of totals. Indeed, we may readily check that the permutation ($\infty$ 0) induces the permutation $(T_1\ T_2)(T_3\ T_6)(T_4\ T_5)$ on the totals; so a transposition of cycle shape $1^4.2$ induces a permutation of cycle shape $2^3$ on the totals.

These totals are extremely cumbersome objects and so we introduce an abbreviation. First note that $T_1$ is preserved by

$$\langle (0\ 1\ 2\ 3\ 4), (\infty\ 0)(1\ 4), (1\ 2\ 4\ 3) \rangle \cong \mathrm{PGL}_2(5)$$

of order 120, confirming that it has just $6!/120 = 6$ images under the action of $S_6$, and so the notation will need to be preserved by such a subgroup. The symbol

$$a/bcdef$$

will stand for the total

$$\begin{array}{l} ab.cf.de \\ ac.db.ef \\ ad.ec.fb \\ ae.fd.bc \\ af.be.cd \end{array}.$$

So the portion to the right of the forward slash is interpreted as a 5-cycle $(b\ c\ d\ e\ f)$ the syntheme containing $ab$ will be $ab.cf.de$, and the syntheme containing $ad$ will be $ad.ce.bf$. A given total has many names of this form but the triple transitivity of $\mathrm{PGL}_2(5)$ means that we can always choose the one of form $\infty/01xyz$ and the six permutations of $\{2, 3, 4\}$ give all possible totals as shown in Figure 9.13.

As has been mentioned earlier, fixing two points in one of the two dodecads of a duum determines a partition of the complementary dodecad into two sets of size 6 each of which makes an octad of the $S(5, 8, 24)$ with the two fixed points. For us the two fixed points will be the top point in each of the first two columns of the MOG, which lie in the canonical dodecad of Figure 9.8. The complementary dodecad is then partitioned into two hexads as shown in Figure 9.14. The first of these two hexads is labelled with points of $\mathcal{P}$ and the second with those of $\mathcal{T}$. We have already observed that a transposition on $\mathcal{P}$ acts as a permutation of shape $2^3$ on $\mathcal{T}$, and this is demonstrated by transposition $(2\ 3)$ in Figure 9.15. The stabilizer of two points in $M_{12}$, denoted by $M_{10}$, has order $10.9.8 = 720$; it consists of the alternating group $A_6$ acting on the two hexads, extended by a coset interchanging them.

If we further allow interchanging of the two fixed points, then we obtain the

| | | $\infty 23 \over 014$ | $\infty/ \over 01234$ | $\infty/ \over 01243$ | $\infty/ \over 01423$ |
|---|---|---|---|---|---|
| $\infty$ | 0 | $\infty/ \over 01324$ | $\infty 14 \over 023$ | $\infty 13 \over 024$ | $\infty 24 \over 013$ |
| 3 | 1 | $\infty/ \over 01342$ | $\infty 12 \over 034$ | $\infty 01 \over 234$ | $\infty 03 \over 124$ |
| 2 | 4 | $\infty/ \over 01432$ | $\infty 34 \over 012$ | $\infty 02 \over 134$ | $\infty 04 \over 123$ |

| $\infty$ | • | • | • |
|---|---|---|---|
| • | $-1+i$ | $i$ | $1+i$ |
| • | $-1$ | $0$ | $1$ |
| • | $-1-i$ | $-i$ | $1-i$ |

Figure 9.14 The isomorphism between $S_6$ and $P\Sigma L_2(9)$

full automorphism group of $A_6$. This group acts on the remaining 10 points of the first dodecad as the linear group $P\Gamma L_2(9)$, that is to say the linear group $PGL_2(9)$ extended by the field automorphism interchanging $i$ and $-i$. Thus

$$\text{Aut}(A_6) \cong M_{10} : 2 \cong P\Gamma L_2(9).$$

This group thus contains three subgroups of index 2, namely

$$S_6 \cong P\Sigma L_2(9) \ [hF]; \quad M_{10} \ [Hf] \text{ and } PGL_2(9) \ [HF],$$

where $P\Sigma L_2(9)$ denotes $PSL_2(9)$ extended by the field automorphism $\alpha$ shown in Figure 9.15 as acting as $(\infty \, 4)(1 \, 3)(0 \, 2)$ on $\mathcal{P}$. The symbol in square brackets refers to the action of the outer coset in each case: $h$ means it fixes the two hexads, $H$ means that it interchanges them; $f$ means that it does not interchange the two fixed points, $F$ means that it does, see also page 78.

The 10 remaining points of the first dodecad are, of course, labelled with the projective line $P_1(9) = \{\infty\} \cup \mathbb{F}_9$. Figure 9.15 gives useful permutations preserving this $2 + 10 + (6 + 6)$ partition of the 24 points of $\Omega$ and thus exhibits the simultaneous actions on $6 + 6$ and 10 points. The final permutation of Figure 9.15 interchanges the two hexads and the two fixed points, and thus lies in $PGL_2(9)$.

### An Irresistible Digression: The Hoffman–Singleton Graph

As we have seen, the stabilizer of an octad in $M_{24}$ is a subgroup of shape $2^4 : A_8$ acting on the 8 points of the octad and the 16 points of the complementary 16-ad. Fixing a point in the 16-ad is a copy of $A_8$ acting as $L_4(2)$ on the remaining 15 points of the 16-ad, and if we now fix a point in the octad we are left with a copy of $A_7$ acting on the remaining *heptad* of 7 points of the octad and transitively

## 9.12 The Outer Automorphism of $S_6$

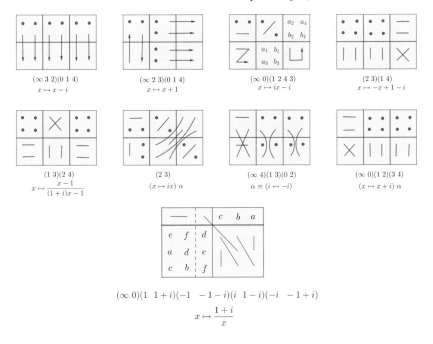

Figure 9.15 Permutations in $P\Gamma L_2(9)$ and Aut $A_6$

on the 15 points of the 16-ad. We may use these simultaneous actions to define one of the most symmetrical and fascinating graphs that exist.

We take as vertices the $\binom{7}{3} = 35$ triples of points in the heptad, together with the remaining 15 points of the 16-ad. Now join two triples if they are disjoint from one another; thus every triple is joined to four other triples.

The unique octad containing a given triple together with the two fixed points must have a further 3 points in the 16-ad, as it already has 4 points in the fixed octad. Join the triple to each of those 3 points. We have

$$35 \times 3/15 = 7,$$

and so every point of the 15-orbit must be joined to seven triples. Thus we have constructed a regular graph $\Gamma$ of valence 7 on 50 vertices as illustrated in Figure 9.16(i). It is the celebrated Hoffman–Singleton graph; it has *diameter* (the maximum distance between two points) 2 and *girth* (the length of shortest cycles) 5, and is thus a Moore graph.

$\Gamma$ is remarkably symmetrical: not only is its automorphism group transitive on its 50 vertices, but the stabilizer of a vertex acts as the full symmetric

112    *The Mathieu Group* $M_{12}$

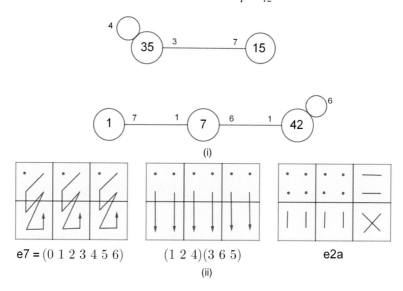

Figure 9.16 Diagrams of the Hoffman–Singleton graph and elements of $M_{24}$ preserving it

| ★ | 0 | ★ | (0, 1/23645) [42] | ∞ [50] | (0, 1/23654) [49] |
|---|---|---|---|---|---|
| 1 | 3 | (1, 0/23456) [36] | (3, 0/12645) [38] | (1, 0/23465) [43] | (3, 0/12465) [45] |
| 2 | 6 | (2, 0/13546) [37] | (6, 0/12534) [41] | (2, 0/13654) [44] | (6, 0/12543) [48] |
| 4 | 5 | (4, 0/12356) [39] | (5, 0/12436) [40] | (4, 0/12653) [46] | (5, 0/12364) [47] |

Figure 9.17 Vertices of the Hoffman–Singleton graph in the MOG

group $S_7$ on its 7 joins, and acts transitively on the remaining 42 vertices. Figure 9.16(i) illustrates these suborbits. If we label the fixed vertex ∞ and it joins $\{0, 1, \ldots, 6\}$ then the points in the 42-orbit may be labelled $(i, T)$, where $i \in \{0, 1, \ldots, 6\}$ and $T$ is a synthematic total on the remaining 6 points. We now have

$$i \text{ joins } (i, T) \text{ and}$$

## 9.12 The Outer Automorphism of $S_6$

$(i, T)$ joins $(j, T^{(i\ j)})$ for $i \neq j$.

To see how this corresponds to the definition of $\Gamma$ using the MOG, we fix the octad $\Lambda_1$, its top left point and the top left point of $\Lambda_2 \cup \Lambda_3$ to obtain our $A_7$. We now choose $\infty$ to be the top left point of $\Lambda_3$, as shown in Figure 9.17. This means that the familiar 7-element and 3-element shown in Figure 9.16 fix $\infty$ and preserve the graph. We label the remaining points of $\Lambda_1$ so that this element acts as (0 1 2 3 4 5 6); then the 7 triples that are joined to $\infty$ correspond to the 7 special tetrads of the 16-ad that contain $\star$ (top left) and $\infty$. Thus the vertex labelled 0 corresponds the triple 365, and the other joins of $\infty$ follow by applying the 7-element.

|    | triple | graph         |    | triple | graph          |    | graph         |
|----|--------|---------------|----|--------|----------------|----|---------------|
| 1  | 406    | 1             | 22 | 125    | $(1,0/23645)$  | 43 | $(1,0/23465)$ |
| 2  | 510    | 2             | 23 | 236    | $(2,0/13456)$  | 44 | $(2,0/13654)$ |
| 3  | 621    | 3             | 24 | 340    | $(3,0/12546)$  | 45 | $(3,0/12465)$ |
| 4  | 032    | 4             | 25 | 451    | $(4,0/12635)$  | 46 | $(4,0/12653)$ |
| 5  | 143    | 5             | 26 | 562    | $(5,0/12346)$  | 47 | $(5,0/12364)$ |
| 6  | 254    | 6             | 27 | 603    | $(6,0/12435)$  | 48 | $(6,0/12543)$ |
| 7  | 365    | 0             | 28 | 014    | $(0,1/23546)$  | 49 | $(0,1/23654)$ |
| 8  | 235    | $(1,0/23546)$ | 29 | 135    | $(1,0/23564)$  | 50 | $\infty$      |
| 9  | 346    | $(2,0/13645)$ | 30 | 246    | $(2,0/13564)$  |    |               |
| 10 | 450    | $(3,0/12456)$ | 31 | 350    | $(3,0/12564)$  |    |               |
| 11 | 561    | $(4,0/12536)$ | 32 | 461    | $(4,0/12563)$  |    |               |
| 12 | 602    | $(5,0/12643)$ | 33 | 502    | $(5,0/12463)$  |    |               |
| 13 | 013    | $(6,0/12345)$ | 34 | 613    | $(6,0/12453)$  |    |               |
| 14 | 124    | $(0,1/23456)$ | 35 | 024    | $(0,1/23564)$  |    |               |
| 15 | 123    | $(1,0/23654)$ | 36 |        | $(1,0/23456)$  |    |               |
| 16 | 234    | $(2,0/13465)$ | 37 |        | $(2,0/13546)$  |    |               |
| 17 | 345    | $(3,0/12654)$ | 38 |        | $(3,0/12645)$  |    |               |
| 18 | 456    | $(4,0/12365)$ | 39 |        | $(4,0/12356)$  |    |               |
| 19 | 560    | $(5,0/12643)$ | 40 |        | $(5,0/12436)$  |    |               |
| 20 | 601    | $(6,0/12354)$ | 41 |        | $(6,0/12534)$  |    |               |
| 21 | 012    | $(0,1/23465)$ | 42 |        | $(0,1/23645)$  |    |               |

Figure 9.18 The labelling of the vertices of the Hoffman–Singleton graph

In Figure 9.18 we show explicitly how elements $(i, T)$ of the 42-orbit correspond to triples and points of the 16-ad, and for computational purposes label them with the numbers 1 to 50. We may now readily write down the action of our 7-element on the 50 vertices, see e7 below; and similarly the action of e2a, which moves $\infty$, lies in $M_{24}$ and so we can immediately write down its action. It remains to produce an element outside $M_{24}$ that fixes $\infty$ and we choose the transposition (5 6) as our element e2b below.

```
> s50:=Sym(50);
> e7:=s50!(1,2,3,4,5,6,7)(8,9,10,11,12,13,14)(15,16,17,18,19,20,21)
> (22,23,24,25,26,27,28)(29,30,31,32,33,34,35)(36,37,38,39,40,41,42)
> (43,44,45,46,47,48,49);
> e2a:=s50!(1,33)(2,20)(3,25)(4,24)(5,15)(6,30)(8,9)(10,12)(17,23)
> (18,26)(21,28)(22,32)(27,31)(29,34)(37,39)
> (40,41)(43,45)(44,47)(46,48)(49,50);
> e2b:=s50!(5,6)(8,22)(9,37)(10,45)(11,25)(12,41)(13,26)(14,21)
> (15, 29)(16,23)(17,31)(18,39)(19,48)(20,47)(24,38)
> (27,40)(28,42)(30,44)(32,46)(33,34)(35,49)(36,43);
> #sub<s50|e7,e2a,e2b>;
252000
```

As we saw in the previous section, the 6 totals on $\{1\ldots 6\}$ can all be written as $1/23xyz$ for some permutation of $x, y, z$. In Figures 9.17 and 9.16 we have chosen the leading three entries as small as possible, so applying (5 6) is particularly easy.

As we see from the MAGMA code earlier, the group so generated has order 252 000. It is in fact the unitary group $U_3(5)$ extended by an outer automorphism of order 2, see the ATLAS Conway et al. (1985, page 34). The element e2a is in the simple group but e2b is an outer automorphism.

## 9.13 More on the Outer Automorphism of $M_{12}$

As we have seen, the stabilizer of a dodecad, which is isomorphic to $M_{12}$, acts non-permutation identically on the 12 points of that dodecad and on the 12 points of the complementary dodecad. Indeed, the copy of $M_{11}$ fixing a point in one dodecad acts transitively on the complement. However, unlike the situation when $S_6$ acts on two sets of size 6, elements of order 2 and 3 have the same cycle shape on both dodecads, namely $2^6$, $1^4.2^4$, $3^4$ or $1^3.3^3$. In Figure 9.19 though we exhibit an element of order 8 that has cycle shape $8.2.1^2$ on one dodecad and $8.4$ on the other; so its square has cycle shapes $4^2.1^4$ and $4^2.2^2$ respectively in the two actions.

We also exhibit the extension of the element $a$ of order 11 that corresponds to the linear fractional map $x \mapsto x+1$ of $L_2(11)$ to $M_{24}$ where it has cycle shape $1^2.11^2$ and acts as $(0\ 1\ \cdots X)(a_1\ a_2\ \cdots a_X)$. Note that the extension of $f_\infty$ to $M_{24}$ shown in Figure 9.12 together with $a$ generates the copy of $M_{11}$ fixing $\infty$ on the canonical dodecad, but a transitive copy of $M_{11}$ on the complementary dodecad.

As has been mentioned earlier, $M_{12}$ contains copies of the linear group $L_2(11)$ and so elements of order five in $L_2(11)$ must fix at least two of the 132

## 9.13 More on the Outer Automorphism of $M_{12}$

| X | X | X | O | O | O |
|---|---|---|---|---|---|
| O | O | O | X | X | X |
| O | O | O | X | X | X |
| O | O | O | X | X | X |

The canonical duum

| — | • | $e_4$ | $d_1$ | $d_5$ |  |
|---|---|---|---|---|---|
| $e_1$ | $e_3$ | $e_2$ | • | $c_1$ | $c_5$ |
| $d_6$ | $d_4$ | $d_3$ | $c_3$ | $c_8$ | $c_2$ |
| $d_2$ | $d_8$ | $d_7$ | $c_7$ | $c_6$ | $c_4$ |

$1^2.2.8/4.8$

| • | 0 | 1 | $a_8$ | $a_4$ | $a_0$ |
|---|---|---|---|---|---|
| $a_7$ | $a_3$ | $a_6$ | 6 | 2 | 5 |
| $a_X$ | $a_2$ | $a_5$ | 7 | 9 | X |
| $a_1$ | • | $a_9$ | 8 | 3 | 4 |

$a$

Figure 9.19 The canonical duum and the non-permutation identical actions of $M_{12}$ on two sets of size 12

hexads of the Steiner system S(5, 6, 12). The element

$$b := x \mapsto 3x \equiv (1\ 3\ 9\ 5\ 4)(2\ 6\ 7\ X\ 8)(0)(\infty)$$

must fix the hexad containing $\{1, 3, 9, 5, 4\}$ and so the 6th point of this hexad must be $\infty$ or $0$. The only subgroups of $L_2(11)$ strictly containing $\langle b \rangle$ contain copies of the dihedral group $D_{10}$ and elements of order 2 normalizing $\langle b \rangle$ interchange $\infty$ and 0; thus no element outside $\langle b \rangle$ fixes either of these hexads and so each of them has $660/5 = 132$ images, namely the whole of the Steiner system. So we see that $L_2(11)$ preserves two distinct Steiner systems and can thus be extended to a copy of $M_{12}$ in two ways. In one of them $\{1, 3, 9, 5, 4, 0\}$ is a hexad and in the other $\{1, 3, 9, 5, 4, \infty\}$ is; we have chosen the former case, although some authors prefer the other version of $M_{12}$. Note that odd elements of $PGL_2(11)$ such as

$$c := x \mapsto -x \equiv (1\ X)(2\ 9)(3\ 8)(4\ 7)(5\ 6)$$

interchange the two Steiner systems. In Figure 9.20 we give an alternative labelling of the points in which the two dodecads of the duum are each labelled

| $\infty$ | 0 | 1 | 4 | 5 | $\infty$ |
|---|---|---|---|---|---|
| 8 | 6 | 3 | 2 | X | 8 |
| 0 | 2 | 9 | 4 | 6 | 9 |
| 7 | X | 1 | 7 | 3 | 5 |

Figure 9.20 The transitive copy of $L_2(11)$ in $M_{12}$

with the 12-point projective line, one with large numerals and the other with small ones. The elements of $L_2(11)$ act in identical fashion as linear fractional maps on large and small numerals and

$$b := x \mapsto -\frac{1}{x} \equiv (\infty\ 0)(1\ X)(2\ 5)(3\ 7)(4\ 8)(6\ 9)$$

together with

$$a := x \mapsto x + 1 \equiv (0\ 1\ 2\ 3\ 4\ 5\ 6\ 7\ 8\ 9\ X)$$

on both sets of numerals generate the group. However note that in the canonical dodecad, denoted by large numerals $\{1, 3, 9, 5, 4, 0\}$ is a hexad, whilst in its complement $\{1, 3, 9, 5, 4, \infty\}$ is a hexad, as is illustrated in Figure 9.21. Note that the element $\alpha$ in Figure 9.12 interchanges the two dodecads, inverts the element $a$ above and acts as an odd permutation (namely $x \mapsto 5 - x$) on the pairs of corresponding points of the two projective lines.

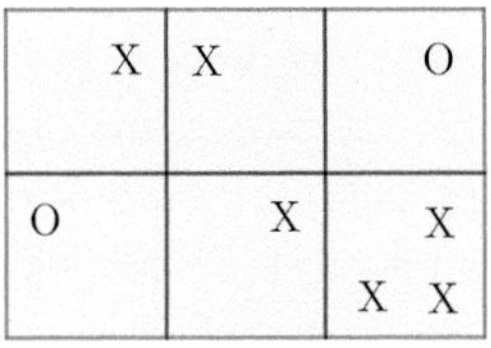

| X | X |   | O |
|---|---|---|---|
| O |   | X | X X |

$\{1, 3, 9, 5, 4, 0, \infty, 0\}$

| O O |   | X | X X |
|---|---|---|---|
|   | X |   |   |
|   | X |   |   |
|   | X |   |   |

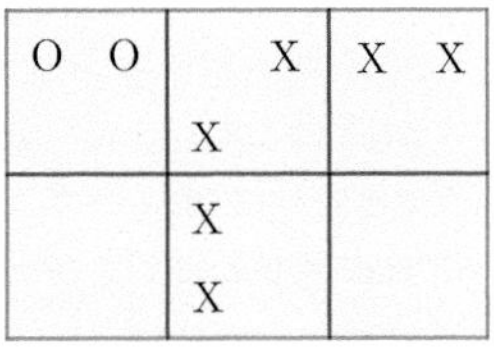

$\{1, 3, 9, 5, 4, \infty, \infty, 0\}$

Figure 9.21 A hexad in each dodecad of a duum, exhibiting the two Steiner systems $S(5, 6, 12)$ preserved by $L_2(11)$

# 10

# The Leech Lattice

## 10.1 Sphere-Packing and Lattices

The vector space of all $n$-tuples of real numbers is denoted by $\mathbb{R}^n$. Thus

$$\mathbb{R}^n = \{(x_1, x_2, \ldots x_n) \mid x_i \in \mathbb{R}\}$$

and if $\mathbf{x} = (x_1, x_2, \ldots, x_n)$ and $\mathbf{y} = (y_1, y_2, \ldots, y_n)$ are two vectors in $\mathbb{R}^n$, then the distance between them is defined by

$$d(\mathbf{x}, \mathbf{y}) = \sqrt{(x_1 - y_1)^2 + (x_2 - y_2)^2 + \cdots + (x_n - y_n)^2},$$

which is simply the extension to $n$-dimensions of the familiar distance formulae in $\mathbb{R}^2$ and $\mathbb{R}^3$ given by Pythagoras' theorem. We may then define the $n$-dimensional sphere in $\mathbb{R}^n$ centred on $\mathbf{a}$ with radius $r$ to be

$$B(\mathbf{a}, r) = \{\mathbf{x} \in \mathbb{R}^n \mid d(\mathbf{x}, \mathbf{a}) = r\}.$$

A *lattice* in $\mathbb{R}^n$, the $n$-dimensional space over the real numbers, is the set of all integral combinations of a set of $n$ linearly independent vectors in $\mathbb{R}^n$. Thus if $\{v_1, v_2, \ldots, v_n\}$ is a linearly independent set of vectors in $\mathbb{R}^n$, then

$$\Gamma = \{m_1 v_1 + m_2 v_2 + \cdots + m_n v_n \mid m_i \in \mathbb{Z}\}$$

is the associated lattice. So, if $n = 2$, and we took $v_1 = (1, 0)$ and $v_2 = (0, 1)$ then

$$\Gamma = \{r v_1 + s v_2 \mid r, s \in \mathbb{Z}\} = \{(r, s) \mid r, s \in \mathbb{Z}\}$$

the set of all points in the plane with integral coefficients. In other words the familiar *square lattice*.

The minimal distance between lattice points equals the minimal distance of lattice points from the origin O, which is to say the length of the shortest vectors in the lattice. Call this distance $\rho$; then spheres with centres placed

# The Leech Lattice

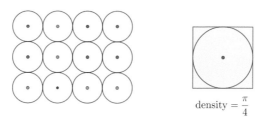

Figure 10.1 The square lattice in $\mathbb{R}^2$ illustrating the density of the associated sphere-packing

at the lattice points and radius $\rho/2$ might touch one another but they will not intersect otherwise. It is natural to ask what proportion of $n$-dimensional space will be occupied by such a set of spheres and, moreover, by a judicious choice of lattice, what is the maximal proportion of $\mathbb{R}^n$ that can be occupied by such a *lattice packing*.[1] In Figure 10.1 we show the sphere-packing based on the square lattice. In order that the circles have unit radius, we choose as lattice points those with *even* integer coefficients. Then the plane is tiled by $2 \times 2$ squares each containing a circle of radius 1; so the proportion of the plane covered by circles, which is to say the *density* $\delta$ of the packing, is given by

$$\delta = \frac{\text{area of circle}}{\text{area of square}} = \frac{\pi}{4}.$$

However it is visually apparent from Figure 10.1 that we can fill a higher proportion of the plane with our circles if we move each row slightly so that it slots into the gaps in the row below. This is accomplished if, instead of taking $v_1 = (2, 0)$ and $v_2 = (0, 2)$ as our generating vectors, we take $v_1 = (2, 0)$ and $v_2 = (1, \sqrt{3})$, see Figure 10.2, thus obtaining the *hexagonal lattice*. The plane is now tiled by equilateral triangles with edge length 2 as shown shaded in Figure 10.2, containing three $60^o$ sectors. So the density is given by

$$\delta = \frac{\text{area of unit semicircle}}{\text{area of equilateral triangle}} = \frac{\pi/2}{\sqrt{3}} = \frac{\pi}{2\sqrt{3}}.$$

Note moreover that this is the tightest possible packing since every circle is surrounded by six other circles that themselves are touching one another. This leads to a second sphere-packing problem: What is the greatest number of unit $n$-dimensional spheres that can touch a given unit sphere without overlapping

---

[1] The more general question of what proportion of $\mathbb{R}^n$ can be occupied by non-overlapping spheres of the same size if we do not restrict their centres to lying at lattice points is remarkably difficult to resolve even in $\mathbb{R}^3$. A vast amount of information about this fascinating topic, including best possible results at the time of its publication, is given in Conway and Sloane (1988).

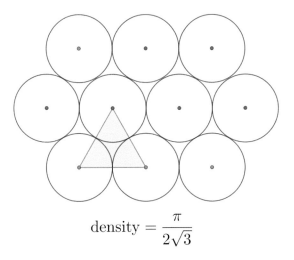

Figure 10.2 The hexagonal lattice in $\mathbb{R}^2$ illustrating the density of the associated sphere-packing

one another? This number is known as the *kissing number*. For $n = 2$ it is clearly 6, but even $n = 3$ is extraordinarily difficult and led to a disagreement between Newton and Gregory in 1694. It is clear that 12 can be accomplished by taking spheres centred at the 12 vertices of a regular icosahedron with radius one half their distance from the centre. But these spheres do not touch one another; indeed consideration of solid angle suggests that one could shuffle these 12 spheres around to make room for a thirteenth. Newton insisted that 12 was the maximum but Gregory felt that 13 could be achieved. Newton was correct but it is surprisingly difficult to prove, see Leech (1956).

## 10.2 The Leech Lattice

In 1965 John Leech, a Cambridge mathematician who had moved into computing, produced a 24-dimensional lattice $\Lambda$ that afforded a remarkably efficient packing, see Leech (1964). It was preserved by $M_{24}$ permuting the 24 vectors in the standard basis of 24-dimensional space, together with sign changes on the $\mathscr{C}$-sets of the 12-dimensional binary Golay code, see Mathieu (1873), Todd (1966) and Curtis (1976). Following the remarkable work of Viazovska, see Cohn et al. (2022), it is now proved that the lattice packing provided by $\Lambda$ is, in fact, the *universal* optimal sphere packing in 24-dimensional space.

There are many different ways of defining $\Lambda$ but the one which we present here follows that given in Conway's *Three lectures on exceptional groups* (1971), and Conway and Sloane (1988, Chapter 10). For convenience, and since it is mainly the symmetries of the lattice $\Lambda$ that concern us, we shall choose a scaling in which all vectors have integral coordinates. As usual we take $\Omega = \{\infty, 0, 1, \ldots, 22\}$ arranged as in the standard MOG:

| ∞ | 0  | 11 | 1  | 22 | 2  |
|---|----|----|----|----|----|
| 3 | 19 | 4  | 20 | 18 | 10 |
| 6 | 15 | 16 | 14 | 8  | 17 |
| 9 | 5  | 13 | 21 | 12 | 7  |

We then take

$$\{v_i \mid i \in \Omega\}$$

as the standard orthonormal basis for $V \cong \mathbb{R}^{24}$; thus every vector of $V$ can be displayed as a $4 \times 6$ array in the MOG diagram. An element $\pi \in M_{24}$ acts on $V$ by permuting the basis elements by mapping $\pi : v_i \mapsto v_{\pi(i)}$ and a $\mathscr{C}$-set $U$ in the Golay code acts by negating on the positions of the codeword $U$. Thus

$$U : \begin{array}{l} v_i \mapsto -v_i \text{ if } i \in U, \\ v_i \mapsto v_i \text{ if } i \notin U. \end{array}$$

For $X \subset \Omega$ we then write

$$v_X = \sum_{i \in X} v_i,$$

thus if $U \in \mathscr{C}_8$ is an octad of the Steiner system, then $2v_U$ denotes the vector having 2 in the positions of the octad $U$ and 0s elsewhere. Now define

$$\Lambda_0 = \langle 2v_U \mid U \in \mathscr{C}_8 \rangle,$$

the lattice consisting of all integral combinations of the 759 vectors defined earlier. Note that

[diagram of MOG arrays showing: array with 2s in first two columns + array with 2s in columns 1 and 3 − array with 2s in columns 2 and 3 = array with 4s in first column]

Thus by the quintuple transitivity of $M_{24}$, we see that $4v_T \in \Lambda_0$ for $T$ any tetrad in $\Omega$. This in turn implies, by subtracting two such vectors corresponding to tetrads that intersect in 3 points, that all vectors of the form $4v_i - 4v_j \in \Lambda_0$ for all $i, j \in \Omega$. This leads to Theorem 24 of Conway (1971):

**Theorem 10.1** *A vector belongs to $\Lambda_0$ if, and only if, (i) all its coordinates are even, (ii) those which are not multiples of 4 lie in the positions of a $\mathscr{C}$-set and (iii) the coordinate sum is a multiple of 16.*

## 10.2 The Leech Lattice

*Proof* The set of vectors $4v_T$ and $4v_i - 4v_j$ spans the lattice of all vectors having all entries a multiple of 4 and coordinate sum a multiple of 16. Moreover the octads span the Golay code $\mathscr{C}$, so the entries that are congruent to 2 modulo 4 lie on a $\mathscr{C}$-set. □

We now define the *Leech lattice*[2] $\Lambda$ to be $\Lambda_0$ extended by the vector

$$-4v_\infty + v_\Omega = -3v_\infty + v_0 + v_1 + \cdots + v_{22} = \begin{array}{|cccccc|} \hline -3 & 1 & 1 & 1 & 1 & 1 \\ 1 & 1 & 1 & 1 & 1 & 1 \\ 1 & 1 & 1 & 1 & 1 & 1 \\ 1 & 1 & 1 & 1 & 1 & 1 \\ \hline \end{array}.$$

Note that

$$\begin{array}{|cccccc|} \hline -3 & 1 & 1 & 1 & 1 & 1 \\ 1 & 1 & 1 & 1 & 1 & 1 \\ 1 & 1 & 1 & 1 & 1 & 1 \\ 1 & 1 & 1 & 1 & 1 & 1 \\ \hline \end{array} - \begin{array}{|cccccc|} \hline -4 & 4 & 0 & 0 & 0 & 0 \\ 0 & 0 & 0 & 0 & 0 & 0 \\ 0 & 0 & 0 & 0 & 0 & 0 \\ 0 & 0 & 0 & 0 & 0 & 0 \\ \hline \end{array} = \begin{array}{|cccccc|} \hline 1 & -3 & 1 & 1 & 1 & 1 \\ 1 & 1 & 1 & 1 & 1 & 1 \\ 1 & 1 & 1 & 1 & 1 & 1 \\ 1 & 1 & 1 & 1 & 1 & 1 \\ \hline \end{array},$$

and so all vectors of the form $v_\Omega - 4v_i$ are in $\Lambda$. Thus $M_{24}$, which clearly fixes $\Lambda_0$, also fixes $\Lambda$ itself. Furthermore sign changes on a $\mathscr{C}$-set clearly preserve the conditions (i), (ii) and (iii) of Theorem 10.1 and so preserve $\Lambda_0$; and acting on $v_\Omega - 4v_\infty$ by negating signs on a $\mathscr{C}$-set (without loss of generality not containing $\infty$) is equivalent to subtracting a vector of $\Lambda_0$ of shape $2v_U$ for $U$ a $\mathscr{C}$-set. So $\Lambda$ is preserved by the group

$$N = 2^{12} : M_{24}.$$

We now have a convenient description of the vectors of $\Lambda$ as given in Theorem 25 of Conway (1971).

**Theorem 10.2** *The vector $(x_\infty, x_0, \ldots, x_{22})$ is in $\Lambda$ if, and only if,*

(i) *all the coordinates $x_i$ are even integers, or all are odd integers;*
(ii) *the coordinate positions on which the $x_i$ are congruent to one another modulo 4 is a $\mathscr{C}$-set of the binary Golay code;*
(iii) *the coordinate sum is congruent to 0 modulo 8 if the entries are even, and to 4 modulo 8 if the entries are odd.*

*Moreover, for $x, y \in \Lambda$ we have the scalar product $x \cdot y$ is a multiple of 8 and $x \cdot x$ is a multiple of 16.*

---

[2] To obtain the unimodular form of $\Lambda$, that is to say a lattice in which a matrix of 24 linearly independent generating vectors has determinant $\pm 1$, we must divide these vectors by $\sqrt{8}$.

*Proof* The three conditions and the statement about inner products hold for the generators and so follow for all vectors by linearity. If a vector satisfies conditions (i) to (iii), then we may subtract a suitable multiple of $v_\Omega - 4v_\infty$ to get a vector whose entries are even and whose coordinate sum is a multiple of 16. Theorem 10.1 then applies. □

Thus all vectors in $\Lambda$ (with this scaling) have norm of form $16m$ for some integer $m \geq 2$. The integer $m$ is called the vector's *type*, and we denote the set of $m$-vectors by $\Lambda_m$. It is a straightforward matter to enumerate all vectors of a particular type under the action of $2^{12} : M_{24}$, see Figure 10.3 where we display all orbits of vectors in $\Lambda_2, \Lambda_3$ and $\Lambda_4$. Each orbit is determined by its shape as vectors in $\mathbb{R}^{24}$ except for those of shape $2^{16}.0^8$, when there are two orbits of the same shape. Congruence of the coordinate sum to 0 modulo 8 implies that such a vector must have an even number of negative coordinates. However, since we can always change sign on a special tetrad of the 16-ad in which the vector lives, and since the special tetrads form a Steiner system $S(3, 4, 16)$, any such vector lies in an orbit either with the vector that has all its coordinates +2, or one that has precisely two $-2$s. We shall see shortly that these behave quite differently from one another in the lattice. In Chapter 11 we shall see that these three types $\Lambda_2, \Lambda_3$ and $\Lambda_4$ form single orbits under the action of the full automorphism group of the lattice. In the meantime it is worth noting how certain simple vectors arise in $\Lambda$ as we have defined it. Thus, if $O_1, O_2$ and $O_3$ are three mutually disjoint octads (such as the three bricks of the MOG) then $2v_{O_1} + 2v_{O_2} + 2v_{O_3} = 2v_\Omega \in \Lambda$, and so

$$(4v_i - v_\Omega) + (4v_j - v_\Omega) + 2v_\Omega = 4v_i + 4v_j \in \Lambda \quad \text{for all } i, j.$$

Then

$$(4v_i + 4v_j) + (4v_i - 4v_j) = 8v_i \in \Lambda \quad \text{for all } i.$$

### 10.2.1 The Factor Space $\Lambda/2\Lambda$

The factor space $\Lambda/2\Lambda$ may be regarded as a 24-dimensional vector space over the field $\mathbb{Z}_2$. Suppose that $u, v \in \Lambda_2 \cup \Lambda_3 \cup \Lambda_4$ and that $u \equiv v$ modulo $2\Lambda$. Then, if $u - v = 2w$ for some $w \in \Lambda$, we have

$$(u - v) \cdot (u - v) = u \cdot u + v \cdot v - 2u \cdot v = 4w \cdot w \geq 8.16.$$

But, since $u \cdot u$ and $v \cdot v$ are of the form $16k$ for $k = 2, 3$ or 4, this can only happen if $u \cdot u = v \cdot v = 4.16$, $w \cdot w = 2.16$ and $u \cdot v = 0$. In other words $u$ and $v$ must be orthogonal 4-vectors and their difference must be twice a 2-vector.

## 10.2 The Leech Lattice

### $\Lambda_2$

| Shape | Number | Description |
|---|---|---|
| $(4^2, 0^{22})$ | $\binom{24}{2} \times 2^2 = 1104$ | $\pm 4$ in two positions; 0s elsewhere |
| $(2^8, 0^{16})$ | $759 \times 2^7 = 97152$ | $\pm 2$ on an octad with evenly many +2s |
| $(-3, 1^{23})$ | $24 \times 2^{12} = 98304$ | 24 choices for the $\pm 3$; $2^{12}$ of sign changes |
| Total | 196 560 | |

### $\Lambda_3$

| Shape | Number | Description |
|---|---|---|
| $(2^{12}, 0^{12})$ | $2576 \times 2^{11} = 5\,275\,648$ | dodecad and sign changes |
| $(3^3, 1^{21})$ | $\binom{24}{3} \times 2^{12} = 8\,290\,304$ | 3 points and sign changes |
| $(4, 2^8, 0^{15})$ | $759 \times 16 \times 2^8 = 3\,108\,864$ | octad, point outside and sign changes |
| $(5, 1^{23})$ | $24 \times 2^{12} = 98\,304$ | position of 5 and sign changes |
| Total | 16 773 120 | |

### $\Lambda_4$

| Shape | Number | Description |
|---|---|---|
| $(2^{16}, 0^8)$ | $759 \times 2^{11} = 1\,554\,432$ | octad, sign changes on $\mathscr{C}$-set |
| $(2^{16}, 0^8)$ | $759 \times 2^{11} \times 15 = 23\,316\,480$ | octad, sign changes on non-$\mathscr{C}$-set |
| $(3^5, 1^{19})$ | $\binom{24}{5} \times 2^{12} = 174\,096\,384$ | 5 points and sign changes |
| $(4^4, 0^{20})$ | $\binom{24}{4} \times 2^4 = 170\,016$ | tetrad and sign changes |
| $(4^2, 2^8, 0^{14})$ | $759 \times \binom{16}{2} \times 2^9 = 46\,632\,960$ | octad, 2 points outside and sign changes |
| $(4, 2^{12}, 0^{11})$ | $2576 \times 12 \times 2^{12} = 126\,615\,552$ | dodecad, point outside and sign changes |
| $(5, 3^2, 1^{21})$ | $24 \times \binom{23}{2} \times 2^{12} = 24\,870\,912$ | a point, a pair and sign changes |
| $(6, 2^7, 0^{16})$ | $759 \times 8 \times 2^7 = 777\,216$ | octad, a point in it and sign changes |
| $(8, 0^{23})$ | $24 \times 2 = 48$ | point and sign change |
| Total | 398 034 000 | |

Figure 10.3 Orbits of vectors of types 2, 3 and 4 in $\Lambda$ under the action of $N$

Moreover $u \cdot (u - v) = u \cdot u = 4.16 = 2u \cdot w$ and so

$$(u - w) \cdot (u - w) = u \cdot u + w \cdot w - 2u \cdot w = 4.16 + 2.16 - 4.16 = 2.16.$$

Thus

$$u = (u - w) + w \text{ and } v = (u - w) - w \text{ where } (u - w), w \in \Lambda_2.$$

Now the only way in which the 4-vector $8v_\infty$ can be expressed as the sum of two 2-vectors is as $8v_\infty = (4v_\infty + 4v_i) + (4v_\infty - 4v_i)$, when it is congruent modulo $2\Lambda$ to the 4-vector $(4v_\infty + 4v_i) - (4v_\infty - 4v_i) = 8v_i$. So $\{\pm 8v_i \mid i \in \Omega\}$ is a set of 48 4-vectors any two of which are congruent to one another modulo $2\Lambda$. As we shall see once we have demonstrated transitivity on vectors in $\Lambda_4$, the set of 4-vectors falls into equivalence classes of size 48 modulo $2\Lambda$. But now

$$1 + |\Lambda_2|/2 + |\Lambda_3|/2 + |\Lambda_4|/48 = 1 + 98\,280 + 8\,386\,560 + 8\,292\,375$$
$$= 16\,777\,216 = 2^{24},$$

and so there are representatives for every congruence class modulo $2\Lambda$ among $\Lambda_2 \cup \Lambda_3 \cup \Lambda_4$; thus every vector in $\Lambda$ is congruent modulo $2\Lambda$ to a unique 2-vector, a unique 3-vector or to 24 mutually orthogonal 4-vectors and their negatives. Such a set of 48 vectors is known as a *frame of reference* or more simply as a *cross*, and $X = \{\pm 8v_i \mid i \in \Omega\}$ is known as the *standard cross*; it is clearly fixed by our group $N = 2^{12} : M_{24}$. In fact we have, see Conway (1971, Thm 26):

**Theorem 10.3** *If $\lambda$ is an automorphism of $\Lambda$ that fixes the standard cross $X$, then $\lambda \in N \cong 2^{12} : M_{24}$.*

*Proof* We must have $\lambda = \pi \epsilon_C$ for some $\pi \in S_{24}$ and $\epsilon_C$ being negation on a set $C \subset \Omega$. We wish to show that $\pi \in M_{24}$ and $C$ is a $\mathscr{C}$-set. We have seen that $2v_\Omega \in \Lambda$, so

$$(2v_\Omega)\lambda = \left(\sum_{i \in \Omega} 2v_i \pi\right)\epsilon_C = \left(\sum_{i \in \Omega} 2v_{i\pi}\right)\epsilon_C = (2v_\Omega)\epsilon_C,$$

which is in $\Lambda$ if, and only if $C$ is a $\mathscr{C}$-set. It remains to show that $\pi \epsilon_C . \epsilon_C = \pi \in M_{24}$. But, if $U \in \mathscr{C}_8$ then

$$(2v_U)\pi = 2v_{U^\pi},$$

which is in $\Lambda$ if, and only if, $U^\pi \in \mathscr{C}_8$ for all $U \in \mathscr{C}_8$. In other words if, and only if, $\pi \in M_{24}$. □

As we shall see in Chapter 11, this fact enables us to work out the order of the group of all automorphisms of $\Lambda$.

## 10.2 The Leech Lattice

Table 10.1 The orbits of $2^{12}{:}M_{24}$ on crosses

$$\begin{pmatrix} 8 & \cdot & \cdot & \cdots & \cdot \\ \cdot & 8 & \cdot & \cdots & \cdot \\ \cdot & \cdot & 8 & \cdots & \cdot \\ \cdot & \cdot & \cdot & \cdots & \cdot \\ \cdot & \cdot & \cdot & \cdots & 8 \end{pmatrix}$$

(i)
1
standard

$$\begin{pmatrix} 4_{4\times 4} & 0 & 0 & 0 & 0 & 0 \\ 0 & 4_{4\times 4} & 0 & 0 & 0 & 0 \\ 0 & 0 & 4_{4\times 4} & 0 & 0 & 0 \\ 0 & 0 & 0 & 4_{4\times 4} & 0 & 0 \\ 0 & 0 & 0 & 0 & 4_{4\times 4} & 0 \\ 0 & 0 & 0 & 0 & 0 & 4_{4\times 4} \end{pmatrix}$$

(ii)
$1\,771.2 = 3\,542$
sextet

$$\left(\begin{array}{cccc|cc} 6 & 2 & \cdots & 2 & & \\ 2 & 6 & \cdots & 2 & & \\ & & \cdots & & 0_{8\times 16} & \\ & & \cdots & & & \\ 2 & 2 & \cdots & 6 & & \\ \hline & 0_{16\times 8} & & & 2_{16\times 16} & \end{array}\right)$$

(iii)
$759.2^6 = 48\,576$
octad

$$\left(\begin{array}{ccc|c} 5 & 3 & 3 & \\ 3 & 5 & 3 & 1_{3\times 21} \\ 3 & 3 & 5 & \\ \hline & 1_{21\times 3} & & T \end{array}\right)$$

(iv)
$\binom{24}{3}.2^{11} = 4\,145\,152$
triad

$$\left(\begin{array}{cc|cccccc} & & 4 & 4 & \cdot & \cdot & \cdot & \cdot \\ & & -4 & -4 & \cdot & \cdot & \cdot & \cdot \\ & & \cdot & \cdot & 4 & 4 & \cdot & \cdot \\ 2_{16\times 8} & & \cdot & \cdot & -4 & -4 & \cdot & \cdot \\ & & \cdot & \cdot & \cdot & \cdot & \cdot & \cdot \\ & & \cdot & \cdot & \cdot & \cdot & 4 & 4 \\ & & \cdot & \cdot & \cdot & \cdot & -4 & -4 \\ \hline 0_{8\times 8} & & & & 2_{8\times 16} & & & \end{array}\right)$$

(v)
$759.15.2^7 = 1\,457\,280$
involution

$$\left(\begin{array}{cc|cc} 4 & & & \\ & 4 & & 2_{12\times 12} \\ & & 4 & \\ \hline & & & 4 \\ 2_{12\times 12} & & & 4 \\ & & & & 4 \end{array}\right)$$

(vi)
$1\,288.2^{11} = 2\,637\,824$
duum

### 10.2.2 The Crosses

Under the action of $N \cong 2^{12} : M_{24}$ there are six orbits on crosses, as shown in Table 10.1 where an entry $d_{m\times n}$ denotes an $m \times n$ submatrix all of whose entries are $\pm d$. Orbit (i) is the standard cross whose stabilizer is our subgroup $N \cong 2^{12} : M_{24}$; the other orbits correspond conveniently to combinatorial objects associated with $M_{24}$. In order to find the other vectors in $\chi(v)$, the cross defined by the 4-vector $v$, we write $v$ as the sum of two 2-vectors in all possible ways, thus $v = u_1 + u_2$ say, when $u_1 - u_2 \in \chi(v)$. Thus

(i) The *standard* cross $X = \{\pm 8v_i \mid i \in \Omega\}$.
(ii) *Sextet* type crosses. To each sextet $S = \{T_0, T_1, \ldots, T_5\}$ there correspond two such crosses, $S^e$ and $S^o$, both consist of 4-vectors that have $\pm 4$ on a tetrad $T_i$ of the sextet $S$ and 0s elsewhere. In the former the number of + signs is even, and in the latter it is odd. Note for instance that if $v$ has 4s in the first column of the MOG then,

| 4 | · | · | · | · | · |
|---|---|---|---|---|---|
| 4 | · | · | · | · | · |
| 4 | · | · | · | · | · |
| 4 | · | · | · | · | · |

$=$

| 2 | 2 | · | · | · | · |
|---|---|---|---|---|---|
| 2 | 2 | · | · | · | · |
| 2 | 2 | · | · | · | · |
| 2 | 2 | · | · | · | · |

$+$

| 2 | −2 | · | · | · | · |
|---|----|---|---|---|---|
| 2 | −2 | · | · | · | · |
| 2 | −2 | · | · | · | · |
| 2 | −2 | · | · | · | · |

and thus

| 2 | 2 | · | · | · | · |
|---|---|---|---|---|---|
| 2 | 2 | · | · | · | · |
| 2 | 2 | · | · | · | · |
| 2 | 2 | · | · | · | · |

$-$

| 2 | −2 | · | · | · | · |
|---|----|---|---|---|---|
| 2 | −2 | · | · | · | · |
| 2 | −2 | · | · | · | · |
| 2 | −2 | · | · | · | · |

$=$

| · | 4 | · | · | · | · |
|---|---|---|---|---|---|
| · | 4 | · | · | · | · |
| · | 4 | · | · | · | · |
| · | 4 | · | · | · | · |

is in $\chi(v)$. For brevity we denote this calculation by

$$(4^4, 0^{20}) = (2^4, 2^4, 0^{16}) + (2^4, (-2)^4, 0^{16})$$
$$\Rightarrow (0^4, 4^4, 0^{16}).$$

If the initial 4-vector has non-zero entries on tetrad $T_0$, then there are five choices for the second tetrad, and $2^3$ choices for the signs on it; thus giving $(5 \times 8)/2 = 20$ further 4-vectors and their negatives. We also have

$$(4^4, 0^{20}) = (4^2, 0^2, 0^{20}) + (0^2, 4^2, 0^{20})$$
$$\Rightarrow (4^2, (-4)^2, 0^{20}),$$

giving a further 3 4-vectors (together with their negatives). There are $1771 \times 2 = 3542$ crosses of this type.

(iii) *Octad* type crosses. In order to have coordinate sum congruent to 0 modulo 8, 4-vectors of shape $6.2^7.0^{16}$ must have an odd number of negative coefficients. As canonical octad type cross we take the one containing $v = (-6, 2^7; 0^{16})$, the others then follow by sign changes on a $\mathscr{C}$-set. Thus

$$(-6, 2^7; 0^{16}) = (-3, 1^7; 1^{16}) + (-3, 1^7; (-1)^{16})$$
$$\Rightarrow (0^8; 2^{16}).$$

## 10.2 The Leech Lattice

Sign changes on octads in the complementary 16-ad lead to there being 16 such 4-vectors (and their negatives).

$$(-6, 2^7; 0^{16}) = (-2, -2, 2^6; 0^{16}) + (-4, 4, 0^6; 0^{16}) \Rightarrow (2, -6, 2^6; 0^{16}),$$

giving a further seven 4-vectors in the cross $\chi(v)$. Note that vectors of shape $(\pm 2^{16}.0^8)$ whose negative coordinates lie in the positions where a $\mathscr{C}$-set intersects the 16-ad lie in this type of cross. There are $759 \times 2^6 = 48\,576$ crosses of this type.

(iv) *Triad* type crosses. As canonical triad type cross, we take the one containing the vector $v = (5, -3^2; 1^{21})$ when we see

$$(5, -3, -3; 1^{21}) = (1, 1, -3; 1^{21}) + (4, -4, 0; 0^{21})$$
$$\Rightarrow (-3, 5, -3; 1^{21})$$

of which there are just 2, and

$$(5, -3^2; 1^{21}) = (3, -1^2; -1^5, 1^{16}) + (2, -2^2; 2^5, 0^{16}) \Rightarrow (1, 1^2; -3^5, 1^{16}),$$

giving a further 21 4-vectors in $\chi(v)$, corresponding to the 21 octads containing a given triad, see Figure 3.1. With sign changes we see that there are $\binom{24}{3} \times 2^{11} = 4\,145\,142$ crosses of this type.

(v) *Involution* type crosses. The 4-vector $v = (2^8; 4^2, 0^{14})$ is visibly the sum of two 2-vectors, thus

$$(2^8.4^2.0^{14}) = (2^8.0^2.0^{14}) + (0^8.4^2.0^{14})$$
$$\Rightarrow (2^8. - 4^2.0^{14}).$$

Furthermore,

$$(2^8; 4^2, 0^{14}) = (2^4, 0^4; 2^2, 2^2, 0^{12}) + (0^4, 2^4; 2^2, -2^2, 0^{12})$$
$$\Rightarrow (2^4, -2^4; 0^2, 4^2, 0^{12})$$

of which there are 14 corresponding to the seven ways in which a pair of points in the complementary 16-ad can be completed to a special tetrad, times 2 sign changes on those pairs of points; and

$$(2^8; 4^2, 0^{14}) = (1^8; 3, 1, -1^7, 1^7) + (1^8; 1, 3, 1^7, -1^7)$$
$$\Rightarrow (0^8; 2, -2, -2^7, 2^7).$$

There are eight of these corresponding to the octads in the complementary 16-ad that contain one of the two points with coordinate 4, but not the other. These correspond to involutions in $M_{24}$ of shape $1^8.2^8$ where the coefficients $\pm 2$ occur on the fixed octad, and the $\pm 4$s occur in the positions of the eight transpositions.

$$
\begin{array}{c}
\underbrace{\begin{array}{|cc|cc|cc|}\hline \cdot & \cdot & -2 & 2 & 2 & 2 \\ \cdot & \cdot & -2 & 2 & 2 & 2 \\ \cdot & \cdot & 2 & 2 & 2 & 2 \\ \cdot & \cdot & 2 & 2 & 2 & 2 \\ \hline\end{array}}_{(v)} = \underbrace{\begin{array}{|cc|cc|cc|}\hline 1 & -1 & -1 & 1 & 1 & 1 \\ 1 & -1 & -1 & 1 & 1 & 1 \\ 1 & -1 & 3 & 1 & 1 & 1 \\ 1 & -1 & -1 & 1 & 1 & 1 \\ \hline\end{array}}_{(u_1)} + \underbrace{\begin{array}{|cc|cc|cc|}\hline -1 & 1 & -1 & 1 & 1 & 1 \\ -1 & 1 & -1 & 1 & 1 & 1 \\ -1 & 1 & -1 & 1 & 1 & 1 \\ -1 & 1 & 3 & 1 & 1 & 1 \\ \hline\end{array}}_{(u_2)}
\end{array}
$$

and thus

$$
\underbrace{\begin{array}{|cc|cc|cc|}\hline 1 & -1 & -1 & 1 & 1 & 1 \\ 1 & -1 & -1 & 1 & 1 & 1 \\ 1 & -1 & 3 & 1 & 1 & 1 \\ 1 & -1 & -1 & 1 & 1 & 1 \\ \hline\end{array}}_{(u_1)} - \underbrace{\begin{array}{|cc|cc|cc|}\hline -1 & 1 & -1 & 1 & 1 & 1 \\ -1 & 1 & -1 & 1 & 1 & 1 \\ -1 & 1 & -1 & 1 & 1 & 1 \\ -1 & 1 & 3 & 1 & 1 & 1 \\ \hline\end{array}}_{(u_2)} = \underbrace{\begin{array}{|cc|cc|cc|}\hline 2 & -2 & \cdot & \cdot & \cdot & \cdot \\ 2 & -2 & \cdot & \cdot & \cdot & \cdot \\ 2 & -2 & 4 & \cdot & \cdot & \cdot \\ 2 & -2 & -4 & \cdot & \cdot & \cdot \\ \hline\end{array}}_{(u_1 - u_2)}
$$

Figure 10.4 Calculating 4-vectors in involution type crosses

In order to clarify the correspondence with a class of involutions in $M_{24}$, and to demonstrate the way that 4-vectors of shape $(-2^2, 2^{14}, 0^8)$ behave differently from $(2^{16}, 0^8)$, we repeat this calculation for the vector $v$ in Figure 10.4. Note that

$$v = (-2^2, 2^{14}, 0^8) = (-3, 1, 1^{14}, 1^8) + (1, -3, 1^{14}, -1^8),$$

leading to

$$(-3, 1, 1^{14}, 1^8) - (1, -3, 1^{14}, -1^8) = (-4, 4, 0^{14}, 2^8) \in \chi(v).$$

The two coordinates $\pm 4$ occur on the eight transpositions of a $1^8.2^8$ involution of $M_{24}$. There are $759 \times 15 \times 2^7 = 1\,457\,280$ crosses of this type.

(vi) *Duum* type crosses. Recall that a *duum* is a pair of complementary dodecads in $\mathscr{C}$. We take $v = (4, 0^{11}; -2, 2^{11})$, noting that the congruence condition means we have to have an odd number of $-2$s on the dodecad. We then have:

$$(4, 0^{11}; -2, 2^{11}) = (3, -1^{11}; 1, 1^{11}) + (1, 1^{11}; -3, 1^{11})$$
$$\Rightarrow (2, -2^{11}; 4, 0^{11}),$$

of which there is just 1;

$$(4, 0^{11}; -2, 2^{11}) = (3, -1^5, 1^6; -1, -1, 1^{10}) + (1, 1^5, -1^6; -1, 3, 1^1)$$
$$\Rightarrow (-2, 2^5, -2^6; 0, 4, 0^{10}),$$

of which there are 11 corresponding to the positions in which the 3 in the second 2-vector can appear; and finally

$$(4, 0^{11}; -2.2^{11}) = (2, 2, 0^{10}; 0, 0^5, 2^6) + (2, -2, 0^{10}; -2, 2^5, 0^6)$$
$$\Rightarrow (0, 4, 0^{10}; 2, -2^5, 2^6),$$

of which there are another 11, corresponding to the 11 positions in which the second 2 in the first vector can appear. There are $1288 \times 2^{11} = 2\,637\,824$ crosses of this type.

**Remark** Note that any automorphism of $\Lambda$ maps the standard cross to another cross. Since the standard cross is a scaled copy of the standard basis of $\mathbb{R}^{24}$, the rows of such an automorphism are 24 mutually orthogonal 4-vectors of a cross, suitably scaled. Thus we may refer to sextet type, octad type and so on, automorphisms of $\Lambda$.

### 10.2.3 The Lorentzian Construction of $\Lambda$

A remarkably concise definition of $\Lambda$ can be obtained by extending 24-dimensional real space to 26-dimensional Lorentz space, see Conway et al. (1982) and Conway and Sloane (1988, page 522). Explicitly, we let $\mathbb{R}^{n,1}$ denote the set of vectors $x = (x_0, x_1, \ldots, x_{n-1} \mid x_n)$ with inner product and norm defined by

$$x \cdot y = x_0 y_0 + \cdots + x_{n-1} y_{n-1} - x_n y_n, \quad N(x) = x \cdot x.$$

The $x_0, \ldots, x_{n-1}$ are known as the *space-like* coordinates, and $x_n$ is the *time-like* coordinate. A lattice $\Lambda$ in $\mathbb{R}^{n,1}$ is said to be *integral* if $x \cdot y \in \mathbb{Z}$ for all $x, y \in \Lambda$ and *unimodular* if it possesses a basis with respect to which the Gram matrix has determinant $\pm 1$. An integral lattice is said to be *even* and of type II if $x \cdot x \in 2\mathbb{Z}$ for all $x \in \Lambda$, and *odd* or type I otherwise. It turns out, see for instance Neumaier and Seidel (1983), that there is a unique odd unimodular lattice for all $n$, and when $n$ is congruent to 1 modulo 8 there is a unique even unimodular lattice. Thus we may take

$$I_{n,1} = \{(x_0, x_1, \ldots, x_{n-1} \mid x_n) \mid x_i \in \mathbb{Z} \text{ for all } i\}, \text{ and}$$

$$II_{n,1} = \left\{ (x_0, x_1, \ldots, x_{n-1} \mid x_n) \;\middle|\; \begin{array}{l} \text{either } x_i \in \mathbb{Z} \text{ for all } i \\ \text{or } x_i \in \mathbb{Z} + \tfrac{1}{2} \text{ for all } i \\ \text{and } x_0 + \cdots + x_{n-1} - x_n \in 2\mathbb{Z} \end{array} \right\}.$$

It will be $II_{25,1}$ that is of particular interest to us. As usual we shall denote the orthogonal complement of the vector $u$ by $u^\perp$; thus, if $u \in I_{25,1}$ or $u \in II_{25,1}$, then dim $u^\perp = 25$. But, if $u$ is an *isotropic* vector, that is to say $u \cdot u = 0$, then $u \in u^\perp$ and so we may factor $u^\perp$ by the 1-dimensional space $\langle u \rangle$ to obtain a 24-dimensional space.

Several years ago Conway and the author produced an isotropic vector, namely,

$$u = (1, 3, 5, \ldots, 45, 47, 51 \mid 145) \in I_{25,1},$$

which had the property that

$$\frac{u^\perp \cap I_{25,1}}{\langle u \rangle} \cong \Lambda, \text{ the Leech lattice.}$$

More recently, as an offshoot of their remarkable classification of the 'deep holes' in $\Lambda$, Conway et al. (1982) found the delightful vector

$$u = (0, 1, 2, \ldots, 24 \mid 70) \in II_{25,1},$$

which has the same property, see Conway and Sloane (1988, Chapter 26). They also give examples of how the orthogonal complements of other isotropic vectors in $II_{25,1}$ give rise to Niemeier lattices in the same manner, see Niemeier (1973).[3]

---

[3] It is worth noting that this is the only non-trivial occasion when the sum of the first $n$ squares is itself a perfect square, see for instance Mordell (1969, page 258).

# 11

# The Conway Group $\cdot$O

The group $N$ of shape $2^{12} : M_{24}$ used to construct $\Lambda$ has three orbits on the shortest vectors in the lattice, but the geometry of the lattice suggested to Leech that its full group of symmetries was much larger. It certainly appeared that the three orbits on 2-vectors should fuse into a single orbit under the action of this putative larger group. For instance, the number of 2-vectors having a certain inner product with a fixed 2-vector was independent of which of the three orbits that fixed vector lay in. As an example, if we work out how many 2-vectors have inner product 16 with a vector of shape $4^2.0^{22}$ and with a vector of shape $2^8.0^{16}$, we find that there are exactly 4600 in each case, as is shown in Table 11.1.

John McKay, who was at the time a PhD student in Edinburgh and who had met Conway at the International Congress of Mathematicians in Moscow in 1966, thought, like Leech, that the full group of symmetries was much larger than $N$; he set about persuading a leading mathematician to investigate. He approached Conway and Conway was hooked! In a beautifully elegant piece of work, Conway was able to produce a further element that preserved the lattice, and so work out the order of its group of automorphisms that was indeed some $10^7$ times larger than the group that had been used in its construction, see Conway (1969a,b).

It should be mentioned that, at a conference two years after Conway had constructed the group, Ernst Witt approached him and informed him that he himself had in fact discovered 'your so-called group' many years earlier. Conway had to take this seriously as Witt had a deep knowledge of the $n$-dimensional lattices involved. However, when Witt went on to say that he considered the order of the group to be far too large to be interesting, Conway doubted his claim as many of the other 24-dimensional lattices have much larger groups of symmetries than $\Lambda$.

The Conway group $\cdot$O is not itself simple as it has a centre of order 2 which, in the 24-dimensional representation described here, is generated by $-I_{24}$. Factoring out the centre $\langle -I_{24} \rangle$ we obtain the Conway simple group $Co_1$.

## 11.1 The Conway Element $\xi_T$

Since the subgroup of symmetries of $\Lambda$ that stabilizes the standard cross X is known to be $N \cong 2^{12} : M_{24}$, Conway sought a new element that would map X to a different cross. The next simplest type of cross is the sextet type $S^e$ or $S^o$, and so he aimed to produce an element that would map X to $S^e$. This was the motivation behind the definition of his $\xi_T$. So let $T$ be a tetrad, such as the first column of the MOG, and let the sextet defined by $T$ be $\{T = T_0, T_1, \ldots, T_5\}$, in our case the columns of the MOG; then the Conway element $\xi_T$ is defined as follows

$$\xi_T : \sum_{i \in \Omega} \lambda_i v_i \mapsto \sum_{i \in \Omega} \eta_i v_i,$$

where

$$\eta_j = \lambda_j - \tfrac{1}{2} \sum_{k \in T_i} \lambda_k \quad \text{for } j \in T_i, i \neq 0,$$
$$\eta_j = -\lambda_j + \tfrac{1}{2} \sum_{k \in T_0} \lambda_k \quad \text{for } j \in T_0.$$

In other words: We take half the sum of the entries in each tetrad from the entries in that tetrad, and then negate on the initial tetrad $T$. We then see that

$$\xi_T : \begin{array}{|cc|cc|cc|} \hline 4 & 4 & \cdot & \cdot & \cdot & \cdot \\ \cdot & \cdot & \cdot & \cdot & \cdot & \cdot \\ \cdot & \cdot & \cdot & \cdot & \cdot & \cdot \\ \cdot & \cdot & \cdot & \cdot & \cdot & \cdot \\ \hline \end{array} \mapsto \begin{array}{|cc|cc|cc|} \hline -2 & 2 & \cdot & \cdot & \cdot & \cdot \\ 2 & -2 & \cdot & \cdot & \cdot & \cdot \\ 2 & -2 & \cdot & \cdot & \cdot & \cdot \\ 2 & -2 & \cdot & \cdot & \cdot & \cdot \\ \hline \end{array}$$

$$\xi_T : \begin{array}{|cccccc|} \hline \cdot & 2 & 2 & 2 & 2 & 2 \\ 2 & \cdot & \cdot & \cdot & \cdot & \cdot \\ 2 & \cdot & \cdot & \cdot & \cdot & \cdot \\ 2 & \cdot & \cdot & \cdot & \cdot & \cdot \\ \hline \end{array} \mapsto \begin{array}{|cccccc|} \hline 3 & 1 & 1 & 1 & 1 & 1 \\ 1 & -1 & -1 & -1 & -1 & -1 \\ 1 & -1 & -1 & -1 & -1 & -1 \\ 1 & -1 & -1 & -1 & -1 & -1 \\ \hline \end{array},$$

thus fusing the three orbits. The next step is to show that $\xi_T$ preserves $\Lambda$. To do this we show that every vector in our chosen generating set, which consists of $\{2v_U : U \in \mathcal{C}_8\}$ together with $v_\Omega - 4v_\infty$, is mapped by $\xi_T$ to a vector in $\Lambda$. An octad $U$ cuts the tetrads of the sextet $S$, that is, the columns of the MOG, as $4^2.0^4$, $2^4.0^2$ or $3.1^5$. The last case was dealt with above in our demonstration that $\xi_T$ fuses the three orbits on 2-vectors. For the others, we see

$$\xi_T : \begin{array}{|cc|cc|cc|} \hline 2 & 2 & \cdot & \cdot & \cdot & \cdot \\ 2 & 2 & \cdot & \cdot & \cdot & \cdot \\ 2 & 2 & \cdot & \cdot & \cdot & \cdot \\ 2 & 2 & \cdot & \cdot & \cdot & \cdot \\ \hline \end{array} \mapsto \begin{array}{|cc|cc|cc|} \hline 2 & -2 & \cdot & \cdot & \cdot & \cdot \\ 2 & -2 & \cdot & \cdot & \cdot & \cdot \\ 2 & -2 & \cdot & \cdot & \cdot & \cdot \\ 2 & -2 & \cdot & \cdot & \cdot & \cdot \\ \hline \end{array}$$

and

$\xi_T:$ 
$\begin{array}{|cccc|cc|}\hline 2 & 2 & 2 & 2 & \cdot & \cdot \\ 2 & 2 & 2 & 2 & \cdot & \cdot \\ \cdot & \cdot & \cdot & \cdot & \cdot & \cdot \\ \cdot & \cdot & \cdot & \cdot & \cdot & \cdot \\ \hline\end{array}$ $\mapsto$ $\begin{array}{|cccc|cc|}\hline \cdot & \cdot & \cdot & \cdot & \cdot & \cdot \\ \cdot & \cdot & \cdot & \cdot & \cdot & \cdot \\ 2 & -2 & -2 & -2 & \cdot & \cdot \\ 2 & -2 & -2 & -2 & \cdot & \cdot \\ \hline\end{array}$,

and finally

$\xi_T:$ $\begin{array}{|cc|cc|cc|}\hline -3 & 1 & 1 & 1 & 1 & 1 \\ 1 & 1 & 1 & 1 & 1 & 1 \\ 1 & 1 & 1 & 1 & 1 & 1 \\ 1 & 1 & 1 & 1 & 1 & 1 \\ \hline\end{array}$ $\mapsto$ $\begin{array}{|cc|cc|cc|}\hline 3 & -1 & -1 & -1 & -1 & -1 \\ -1 & -1 & -1 & -1 & -1 & -1 \\ -1 & -1 & -1 & -1 & -1 & -1 \\ -1 & -1 & -1 & -1 & -1 & -1 \\ \hline\end{array}$.

Note that sign changes on an even number of columns of the MOG lie in $N$ and so the above are the only cases we need to consider.

The resulting group $\langle N, \xi_T \rangle$ that Conway called $\cdot$O or 'dotto', when factored by its centre of order 2, was a new simple group.

## 11.2 Transitivity on Short Vectors

We have already seen that the Conway element $\xi_T$ demonstrates the transitivity of $\cdot$O on vectors in $\Lambda_2$. Transitivity on vectors in $\Lambda_3$, whose orbits under $N$ are exhibited in Figure 10.3, is equally straightforward.

**Theorem 11.1** *The Conway group $\cdot$O acts transitively on 3-vectors.*

*Proof* We simply observe the action of $\xi_T$ on a vector from each of the four orbits of $N$:

$\xi_T:$ $\begin{array}{|cc|cc|cc|}\hline 5 & 1 & 1 & 1 & 1 & 1 \\ 1 & 1 & 1 & 1 & 1 & 1 \\ 1 & 1 & 1 & 1 & 1 & 1 \\ 1 & 1 & 1 & 1 & 1 & 1 \\ \hline\end{array}$ $\leftrightarrow$ $\begin{array}{|cc|cc|cc|}\hline -1 & -1 & -1 & -1 & -1 & -1 \\ 3 & -1 & -1 & -1 & -1 & -1 \\ 3 & -1 & -1 & -1 & -1 & -1 \\ 3 & -1 & -1 & -1 & -1 & -1 \\ \hline\end{array}$;

$\begin{array}{|cc|cc|cc|}\hline 4 & -2 & 2 & 2 & 2 & 2 \\ 2 & \cdot & \cdot & \cdot & \cdot & \cdot \\ 2 & \cdot & \cdot & \cdot & \cdot & \cdot \\ 2 & \cdot & \cdot & \cdot & \cdot & \cdot \\ \hline\end{array}$ $\leftrightarrow$ $\begin{array}{|cc|cc|cc|}\hline 1 & -1 & 1 & 1 & 1 & 1 \\ 3 & 1 & -1 & -1 & -1 & -1 \\ 3 & 1 & -1 & -1 & -1 & -1 \\ 3 & 1 & -1 & -1 & -1 & -1 \\ \hline\end{array}$;

$\begin{array}{|cc|cc|cc|}\hline \cdot & 2 & \cdot & 2 & \cdot & 2 \\ 2 & \cdot & 2 & \cdot & 2 & \cdot \\ 2 & \cdot & 2 & \cdot & 2 & \cdot \\ 2 & \cdot & 2 & \cdot & 2 & \cdot \\ \hline\end{array}$ $\leftrightarrow$ $\begin{array}{|cc|cc|cc|}\hline 3 & 1 & -3 & 1 & -3 & 1 \\ 1 & -1 & -1 & -1 & -1 & -1 \\ 1 & -1 & -1 & -1 & -1 & -1 \\ 1 & -1 & -1 & -1 & -1 & -1 \\ \hline\end{array}$.

134  *The Conway Group* ·O

This demonstrates the fusion of the four orbits.  □

We could prove transitivity on $\Lambda_4$ in a similar manner but, since $N$ has so many orbits on 4-vectors as shown in Figure 10.3, we prefer to prove:

**Theorem 11.2**  *The Conway group* ·O *acts transitively on crosses.*

*Proof*  Let us denote the set of crosses by $\bar\Lambda_4$. We shall show that the vector $8v_\infty$ of the standard cross can be mapped to vectors in each of the other five types of cross using $\xi_T$ and elements of $N$. Thus $\xi_T$ :

| 8 | · | · | · | · | · |
|---|---|---|---|---|---|
| · | · | · | · | · | · |
| · | · | · | · | · | · |
| · | · | · | · | · | · |

(i)  ↔

| −4 | · | · | · | · | · |
|---|---|---|---|---|---|
| 4 | · | · | · | · | · |
| 4 | · | · | · | · | · |
| 4 | · | · | · | · | · |

(ii)

| · | 4 | · | · | · | · |
|---|---|---|---|---|---|
| 4 | · | · | · | · | · |
| 4 | · | · | · | · | · |
| 4 | · | · | · | · | · |

(ii)  ↔

| 6 | 2 | · | · | · | · |
|---|---|---|---|---|---|
| 2 | −2 | · | · | · | · |
| 2 | −2 | · | · | · | · |
| 2 | −2 | · | · | · | · |

(iii)

| −6 | · | 2 | 2 | 2 | 2 |
|---|---|---|---|---|---|
| · | 2 | · | · | · | · |
| · | 2 | · | · | · | · |
| · | 2 | · | · | · | · |

(iii)  ↔

| 3 | −3 | 1 | 1 | 1 | 1 |
|---|---|---|---|---|---|
| −3 | −1 | −1 | −1 | −1 | −1 |
| −3 | −1 | −1 | −1 | −1 | −1 |
| −3 | −1 | −1 | −1 | −1 | −1 |

(iv) ,

| 4 | 2 | 2 | 2 | 2 | 2 |
|---|---|---|---|---|---|
| 2 | 4 | · | · | · | · |
| 2 | · | · | · | · | · |
| 2 | · | · | · | · | · |

(v)  ↔

| 1 | −1 | 1 | 1 | 1 | 1 |
|---|---|---|---|---|---|
| 3 | 1 | −1 | −1 | −1 | −1 |
| 3 | −3 | −1 | −1 | −1 | −1 |
| 3 | −3 | −1 | −1 | −1 | −1 |

(iv)

| 4 | −2 | · | 2 | · | 2 |
|---|---|---|---|---|---|
| 2 | · | 2 | · | 2 | · |
| 2 | · | 2 | · | 2 | · |
| 2 | · | 2 | · | 2 | · |

(vi)  ↔

| 1 | −1 | −3 | 1 | −3 | 1 |
|---|---|---|---|---|---|
| 3 | 1 | −1 | −1 | −1 | −1 |
| 3 | 1 | −1 | −1 | −1 | −1 |
| 3 | 1 | −1 | −1 | −1 | −1 |

(iv) ,

which demonstrates transitivity on the six orbits of $N$ on crosses as shown in Table 10.1. □

**Corollary**  *The Conway group ·O acts transitively on 4-vectors.*

*Proof*  The stabilizer of the standard cross X is $N \cong 2^{12} : M_{24}$ acting transitively on its 48 4-vectors. Since ·O acts transitively on crosses, the stabilizer of any cross acts in a similar manner on its 48 vectors. □

## 11.3 The Order of ·O

We are now in a position to work out the order of ·O. We know the order of the stabilizer of a cross and the total number of crosses is given by

$$1 + 3542 + 48\,576 + 4\,145\,152 + 1\,457\,280 + 2\,637\,824 = 8\,292\,375.$$

Since we have now seen that ·O acts transitively on crosses, we are able to say that

$$|\cdot O| = 2^{12} \times |M_{24}| \times |\bar{\Lambda}_4| = 2^{22}.3^3.5.7.11.23 \times 3^6.5^3.7.13$$
$$= 2^{22}.3^9.5^4.7^2.11.13.23 = 8\,315\,553\,613\,086\,720\,000.$$

## 11.4 Constructing ·O Using MAGMA

The immensely useful MAGMA library contains generators for a wealth of groups including the Conway groups, and we could simply download generators from this source; however, in this section we shall obtain generators ourselves following the construction described earlier, by way of the Conway element $\xi_T$. We shall construct the group ·O as $24 \times 24$ matrices, and to facilitate the computation we shall work over $\mathbb{Z}_3$, the field of integers modulo 3, rather than the rationals. This simplification may be justified by verifying that the 196 560 2-vectors remain distinct when read modulo 3, and both the monomial subgroup $N$ and the Conway element $\xi_T$ act on them in the same manner when written over $\mathbb{Z}_3$ as they do over the rationals. We must first construct $N \cong 2^{12} : M_{24}$ in which the elementary abelian group of order $2^{12}$ is represented by diagonal matrices each with $-1$ in the positions of a $\mathscr{C}$-set, and $M_{24}$ is represented by permutation matrices. That is to say the permutation $\pi$ is represented by a matrix of all 0s except that there is a 1 in the $(i, i^\pi)$ position for $i = 1, \ldots, 24$.

```
> s24:=Sym(24);
> m24:=sub<s24|
```

```
> (1,2,3,4,5,6,7,8,9,10,11,12,13,14,15,16,17,18,19,20,21,22,23),
> (24,23)(3,19)(6,15)(9,5)(11,1)(4,20)(16,14)(13,21)>;
> #m24;
244823040
> z3:=GaloisField(3);
> as:=[i^m24.1:i in [1..24]];
> am:=PermutationMatrix(z3,as);
> bs:=[i^m24.2:i in [1..24]];
> bm:=PermutationMatrix(z3,bs);
> oct1:=[z3!1:i in [1..24]];
> for i in {24,3,6,9,23,19,15,5} do
for> oct1[i] := z3!-1;
for> end for;
> oct1m:=DiagonalMatrix(z3,24,oct1);
> gl24:=GeneralLinearGroup(24,z3);
> Nm:=sub<gl24|am,bm,oct1m>;
> #Nm;
1002795171840
```

In the above we have taken our usual two generators for $M_{24}$: $\alpha$ of order 23 and $\beta$ of order 2, interchanging the 1st and 2nd, 3rd and 4th columns of the MOG, whilst fixing the rows. They are written as sequences, *as* and *bs* respectively, of length 24 with the image of $i$ in the $i$th position. The command **PermutationMatrix** then requires us to stipulate the field over which we are working, namely $\mathbb{Z}_3$, followed by this sequence. It returns matrices *am* and *bm* representing $\alpha$ and $\beta$.

We now adjoin a diagonal matrix that negates on the positions corresponding to the first brick of the MOG. As above the command **DiagonalMatrix** requires the field over which we are working, the dimension $n$ and a sequence of length $n$. As we know, $M_{24}$ acts transitively on the octads, and negation on the 759 octads generates the binary Golay code. Finally our group $N$ is defined as a subgroup of the general linear group of dimension 24 over $\mathbb{Z}_3$. MAGMA confirms that this group does indeed have order $2^{12} \times |M_{24}|$.

We must now construct the Conway element $\xi_T$ over the field $\mathbb{Z}_3$. Recall that this element consists of six $4 \times 4$ matrices on the positions of the tetrads of a sextet. Note that the fraction $\frac{1}{2}$ becomes $-1$ or 2 in $\mathbb{Z}_3$ and so if the sextet were $\lambda = \{\{1,2,3,4\}, \{5,6,7,8\}, \ldots, \{21,22,23,24\}\}$ (which certainly is *not* a sextet of the Steiner system with our labelling of the points of $\Omega$), then $\xi_T$ would have the form:

$$C = \begin{pmatrix} 2 & 1 & 1 & 1 \\ 1 & 2 & 1 & 1 \\ 1 & 1 & 2 & 1 \\ 1 & 1 & 1 & 2 \end{pmatrix}, \text{ and } \xi_T = \begin{pmatrix} -C & 0 & 0 & 0 & 0 & 0 \\ 0 & C & 0 & 0 & 0 & 0 \\ 0 & 0 & C & 0 & 0 & 0 \\ 0 & 0 & 0 & C & 0 & 0 \\ 0 & 0 & 0 & 0 & C & 0 \\ 0 & 0 & 0 & 0 & 0 & C \end{pmatrix}.$$

## 11.4 Constructing ·O Using MAGMA

We shall construct $\xi_T$ as if the above partition was a sextet and then apply a permutation $\pi$ to map this partition into the columns of the MOG.

```
> gl4:=GeneralLinearGroup(4,z3);
> mat4z3:=MatrixRing(z3,4);
> cct:=mat4z3![z3!1:i in [1..16]];
> cc:=mat4z3!cct+Identity(gl4);
> cc;
[2 1 1 1]
[1 2 1 1]
[1 1 2 1]
[1 1 1 2]
> con:=DirectSum(DirectSum(DirectSum(-cc,cc),
DirectSum(cc,cc)),DirectSum(cc,cc));
> pis:=
>[24,3,6,9,23,19,15,5,11,4,16,13,1,20,14,21,22,18,8,12,2,10,17,7];
> pim:=PermutationMatrix(z3,pis);
> xim:=pim^-1*con*pim;
> dotto:=sub<gl24|am,bm,oct1m,xim>;
> #dotto;
8315553613086720000
> Factorisation(#dotto);
[ <2, 22>, <3, 9>, <5, 4>, <7, 2>, <11, 1>, <13, 1>, <23, 1> ]
```

Note that the command **DirectSum(xx,yy)** of an $n \times n$ matrix $xx$ with an $m \times m$ matrix $yy$ results in an $(n+m) \times (n+m)$ matrix with $xx$ and $yy$ down the diagonal and zeros elsewhere. Thus $con$ is a $24 \times 24$ matrix with

$$[-cc, cc, cc, cc, cc, cc]$$

down the main diagonal. We introduce the **MatrixRing** $mat4z3$ so that we can define singular matrices; thus $cct$ is the $4 \times 4$ matrix all of whose entries are 1. Then $C$ is obtained by adding the identity matrix to $cct$. The permutation $\pi$ maps the partition $\lambda$ into the columns of the MOG, which *is* a sextet of the Steiner system, and so we obtain $\xi_T$ by conjugating $con$ by $\pi$. Adjoining this element to $N$, as Conway originally did, we obtain 'dotto' or ·O. MAGMA verifies that its order is as we have calculated it, and then factorizes it for us as

$$2^{22}.3^9.5^4.7^2.11.13.23.$$

**Remark** The standard cross, see Figure 10.1, is simply a scaled copy of the standard basis of $\mathbb{R}^{24}$ and any element of ·O must map it to another cross. Thus the rows of any element of ·O must correspond to 24 mutually orthogonal 4-vectors in a cross, as displayed in Figure 10.1. Thus we may talk about *sextet, octad, triad, involutory* or *duum* type elements of ·O. Elements corresponding to the standard cross are *monomial*. The Conway element $\xi_T$ is, of course, a sextet type element.

Table 11.1 Calculation of 2-vectors having i.p. 16 with a fixed 2-vector

| Fixed vector | Shape | Description | Calculation | Subtotal |
|---|---|---|---|---|
| $4^2.0^{22}$ | $4.0.\pm 4.0^{21}$ | 22 choices for $\pm 4$ | $2\times 22\times 2$ | 88 |
| | $2^2.(\pm 2)^6.0^{16}$ | octads through 2 points and evenly many $-2$s | $77.2^5$ | 2 464 |
| | $3.1.(-1)^7.1^{15}$ | 2 choices for 3 times $2^{10}$ sign changes | $2\times 2^{10}$ | 2 048 |
| | | | Total | 4 600 |
| $2^8.0^{16}$ | $4^2.0^6.0^{16}$ | 2 points of octad | $\binom{8}{2}$ | 28 |
| | $2^6.(-2)^2.0^{16}$ | 2 points of octad | $\binom{8}{2}$ | 28 |
| | $2^4.0^4.(\pm 2)^4.0^{12}$ | 4 points of octad evenly many $-2$s | $\binom{8}{4}\times 4\times 2^3$ | 2 240 |
| | $1^8.-3.1^{15}$ | 16 choices for 3 $2^5$ of sign changes | $16\times 2^5$ | 512 |
| | $3.-1.1^6.(-1)^6.1^{10}$ | 8.7 choices for $-3$ and 1 $2^5$ of sign changes | $8\times 7\times 2^5$ | 1 792 |
| | | | Total | 4 600 |

## 11.5 The Conway Group $Co_2$

We have seen that $\cdot O$ acts transitively on $\Lambda_2$ and so the order of $Co_2$, the stabilizer of a 2-vector in $\cdot O$, is given by

$$|Co_2| = |\cdot O|/|\Lambda_2| = 2^{22}.3^9.5^4.7^2.11.13.23/2^4.3^3.5.7.13$$
$$= 2^{18}.3^6.5^3.7.11.23$$
$$= 42\,305\,421\,312\,000.$$

In Table 11.1 we listed the 2-vectors having inner product 16 with a given fixed 2-vector, for two different fixed vectors. Note that if $v$ and $u_1$ are two 2-vectors with inner product 16 then, writing $u_2 = v - u_1$, we have

$$u_2\cdot u_2 = (v-u_1)\cdot(v-u_1) = v\cdot v + u_1\cdot u_1 - 2u_1\cdot v = 32+32-32 = 32,$$
and $\quad v\cdot u_2 = v\cdot(v-u_1) = v\cdot v - v\cdot u_1 = 32-16 = 16,$

and so $u_2$ is also a 2-vector having inner product 16 with $v = u_1 + u_2$. So the vectors listed in Table 11.1 fall into 2300 pairs as shown in Table 11.2. Each of these pairs spans a triangular lattice as shown in Figure 11.1 and these 2300 2-dimensional sublattices containing $v$ are permuted by the group $Co_2$. Now the orbits of $Co_2$ on these sublattices must be unions of the orbits indicated in Table 11.2 in both of the cases considered. That this can only be accomplished if they all fuse into one orbit can perhaps be most easily seen by considering

## 11.5 The Conway Group $Co_2$

Table 11.2 The 2300 sublattices permuted by $Co_2$

| Fixed 2-vector | Sublattice generators | Number | Modulo 11 |
|---|---|---|---|
| $4^2.0^{22}$ | $\left.\begin{array}{l} 4.0.4.0^{21} \\ 0.4.-4.0^{21} \end{array}\right\}$ | 44 | 0 |
| | $\left.\begin{array}{l} 2^2.2^6.0^{16} \\ 2^2.-2^6.0^{16} \end{array}\right\}$ | 1 232 | 0 |
| | $\left.\begin{array}{l} 3.1.-1^7.1^{15} \\ 1.3.1^7.-1^{15} \end{array}\right\}$ | 1 024 | 1 |
| | Total | 2 300 | 1 |
| $2^8.0^{16}$ | $\left.\begin{array}{l} 4^2.0^6.0^{16} \\ -2^2.2^6.0^{16} \end{array}\right\}$ | 28 | 6 |
| | $\left.\begin{array}{l} 1^8.-3.1^{15} \\ 1^8.3.-1.-1^{15} \end{array}\right\}$ | 256 | 3 |
| | $\left.\begin{array}{l} 2^4.0^4.2^4.0^{12} \\ 0^4.2^4.-2^4.0^{12} \end{array}\right\}$ | 1 120 | 9 |
| | $\left.\begin{array}{l} 3.-1.1^6.-1^6.1^{10} \\ -1.3.1^6.1^6.-1^{10} \end{array}\right\}$ | 896 | 5 |
| | Total | 2 300 | 1 |

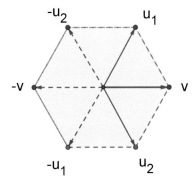

Figure 11.1 One of the 2300 hexagonal sublattices containing a given 2-vector $v$

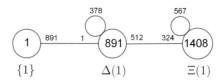

Figure 11.2 The action of $Co_2$ on the 2300 hexagonal lattices

them modulo 11 as shown. We conclude that $Co_2$ acts transitively on the 2300 sublattices.

We assert that the action of $Co_2$ on these 2300 sublattices is rank 3 with suborbits of lengths 1, 891 and 1408, and incidence diagram as shown in Figure 11.2. In order to see this, first note that any two of these sublattices will span a 3-dimensional sublattice, since the fixed 2-vector lies in both of them. If we take the fixed sublattice to be that spanned by $v = (4, 4, 0, 0^{21})$, $u_1 = (4, 0, 4, 0^{21})$ and $u_2 = (0, 4, -4, 0^{21})$, then we see that the vector $u_3 = (0, 4, 4, 0^{21})$ has inner products 16, 16, 0 with $v, u_1, u_2$ respectively. In the terminology of Conway (1971) this gives a lattice of type 222–2224, which is to say a 2-vector $u_3$ has been adjoined to our fixed 222-lattice such that $v - u_3$ and $u_1 - u_3$ are 2-vectors, and $u_2 - u_3$ is a 4-vector. We can now classify all 2-vectors $u_3$ that have this property:

| vector shape | calculation | number |
|---|---|---|
| $(0, 4, 4, 0^{21})$ | 1 | 1 |
| $(4, 0, 0, \pm 4, 0^{20})$ | $21 \times 2$ | 42 |
| $(2, 2, 2, \pm 2^5, 0^{16})$ | $21 \times 2^4$ | 336 |
| $(3, 1, 1, \pm 1^{21})$ | $2^9$ | 512 |
| Total | | 891 |

We further note that a vector $u_3$ of shape $(2, 2, 2, 2^5, 0^{16})$ has inner products 16, 8, 8 with our three fixed vectors $v, u_1$ and $u_2$, and that the same is true of $v - u_3$. So such vectors come in pairs generating the same 3-dimensional sublattice. We may now perform a similar calculation for such vectors:

| vector shape | calculation | number of pairs |
|---|---|---|
| $(2, 2, 0, \pm 2^6, 0^{15})$ | $56 \times 2^5/2$ | 896 |
| $(3, 1, -1, \pm 1^{21})$ $(1, 3, 1, \pm 1^{21})$ | $2^9$ | 512 |
| Total | | 1408 |

## 11.5 The Conway Group $Co_2$

We wish to use the Conway element to verify that these orbits of vectors under the action of those elements of $N$ that visibly fix $v, u_1$ and $u_3$ fuse into single orbits. We make them explicit in the MOG, thus

$$v := \begin{array}{|c|} \hline 4 \cdot \cdot \cdot \cdot \cdot \\ 4 \cdot \cdot \cdot \cdot \cdot \\ \cdot \cdot \cdot \cdot \cdot \cdot \\ \cdot \cdot \cdot \cdot \cdot \cdot \\ \hline \end{array} ; u_1 := \begin{array}{|c|} \hline 4 \cdot \cdot \cdot \cdot \cdot \\ \cdot \cdot \cdot \cdot \cdot \cdot \\ 4 \cdot \cdot \cdot \cdot \cdot \\ \cdot \cdot \cdot \cdot \cdot \cdot \\ \hline \end{array} ; u_2 := \begin{array}{|c|} \hline \cdot \cdot \cdot \cdot \cdot \cdot \\ 4 \cdot \cdot \cdot \cdot \cdot \\ -4 \cdot \cdot \cdot \cdot \cdot \\ \cdot \cdot \cdot \cdot \cdot \cdot \\ \hline \end{array}.$$

Now let $\pi$ denote the permutation of $M_{24}$ that fixes the last two columns of the MOG pointwise, and interchanges the 1st and 4th, 2nd and 3rd rows of the first four columns, as displayed in Figure 11.3. Then the element $\xi_T \pi$, where $T$ is the first column of the MOG, fixes $v, u_1$ and $u_2$ and can thus be used to fuse the suborbits described earlier. We have $\xi_T \pi$:

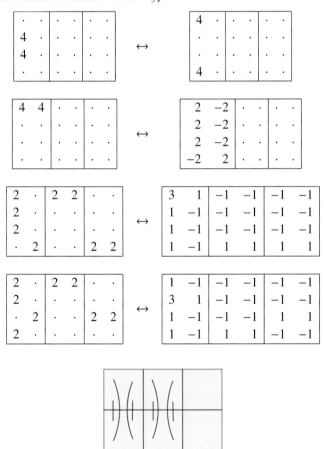

Figure 11.3 The permutation $\pi$

Thus verifying that the suborbits do indeed fuse in the way claimed. Moreover, the fact that one of the suborbits in the 891 orbit has just one member enables us to verify that the incidence graph of the permutation action of $Co_2$ on these 2300 points is as shown in Figure 11.2 with very little additional work. A brief explanation of how graph diagrams like this one can be used to decompose the permutation representation into its irreducible components is given in Chapter 12.

### 11.5.1 $Co_2$ Using MAGMA

First note that $-\xi_T$, the negative of the Conway element, fixes the vector $-4v_\infty + v_\Omega$, as does the permutation $\alpha$ of order 23 used in the construction of $M_{24}$. Over $\mathbb{Z}_3$ these elements become $-xim$ and $am$ in Section 11.4 and they, in fact, generate $Co_2$:

```
> co2:=sub<gl24|-xim,am>;
> #co2;
42305421312000
> u1:=Vector(24,[z3!1:i in [1..23]] cat [z3!0]);
> #(u1^co2);
1
> u2:=Vector(24,[z3!0] cat [z3!1:i in [1..23]]);
> #(u2^co2);
4600
```

In the above $u_1$ represents $-4v_\infty + v_\Omega$ (modulo 3) and $u_2$ represents $-4v_1 + v_\Omega$, a 2-vector having inner product 16 with $u_1$. Then $u_1 - u_2 = -4v_\infty + 4v_1$, the other edge of the 222-triangle defined by $u_1$ and $u_2$, which also has inner product 16 with $u_1$. Thus the 4600 2-vectors that are images of $u_2$ under the action of $Co_2$ fall into 2300 pairs. We first obtain the **CosetAction** of $Co_2$ on the stabilizer of $u_2$; thus $f4600$ below is the homomorphism mapping $24 \times 24$ matrices of $Co_2$ onto permutations of degree 4600, and $gp4600$ is the resulting permutation group. We now wish to obtain the action on these 2300 pairs such as $\{u_2, u_1 - u_2\}$, so we ask which other point is fixed by the stabilizer of $u_2$. It turns out to be labelled 1201; thus this point corresponds to $-4v_\infty + 4v_1$. We obtain the stabilizer of $\{u_2, u_1 - u_2\}$ and the **CosetAction** of $Co_2$ over it as $gp2300$.

```
> stu2:=Stabilizer(co2,u2);
> f4600,gp4600,k:=CosetAction(co2,stu2);
> Degree(gp4600);
```

```
4600
> Fix(f4600(stu2));
{ 1, 1201 }
> stpr:=Stabilizer(gp4600,{1,1201});
> Index(gp4600,stpr);
2300
> f2300,gp2300,kk:=CosetAction(gp4600,stpr);
```

Now that we have the group $Co_2$ as permutations of degree 2300, we can investigate its action. We find that it is of rank three and, in order to confirm the intersection numbers of the graph of valence 891 as presented in Figure 11.2, we ask for the number of points joined to 1 and to a point $j$ in the 891-orbit. We find that there are 378 of them, and that the remaining 512 joins of $j$ are in the 1408-orbit, thus confirming the parameters in Figure 11.2.

```
> st1:=Stabilizer(gp2300,1);
> oost1:=Orbits(st1);
> [#oost1[i]:i in [1..3]];
[ 1, 891, 1408 ]
> 1^gp2300.1 in oost1[2];
true
> #(oost1[2] meet oost1[2]^gp2300.1);
378
> #(oost1[3] meet oost1[2]^gp2300.1);
512
```

## 11.6 The Conway Group $Co_3$

The 3-vector $u = v_\Omega + 4v_\infty$ is visibly fixed by the copy of $M_{23}$ stabilizing $\infty$. We consider the ways in which $u$ can be expressed as the sum of two 2-vectors. Note that if $u_1, u_2 \in \Lambda_2$ such that $u_1 + u_2 \in \Lambda_3$, then

$$(u_1 + u_2) \cdot (u_1 + u_2) = 2.16 + 2.16 + 2u_1 \cdot u_2 = 3.16 \Rightarrow u_1 \cdot u_2 = -8,$$

and so

$$(u_1 + u_2) \cdot u_1 = 2.16 - 8 = 24.$$

Then we see that the ways in which $u$ can be so expressed are given by:

| $5.1^{23}$ | Number |
|---|---|
| $\begin{cases} 4.4.0^{22} \\ 1.-3.1^{22} \end{cases}$ | 23 |
| $\begin{cases} 3.(-1)^7.1^{16} \\ 2.2^7.0^{16} \end{cases}$ | 253 |
| Total | 276 |

So we see that $Co_3$, the stabilizer of a 3-vector in $\cdot O$, can be thought of as a permutation group on 276 letters. Under the action of $M_{23}$ these 276 fall into two orbits, one of which corresponds one to one with the 23 points of $\Omega \setminus \{\infty\}$ and the other corresponds to the *heptads* of the Steiner system $S(4, 7, 23)$, which is to say the 253 octads that contain $\infty$.

### 11.6.1 $Co_3$ from the Matrix Representation of $\cdot O$

We could demonstrate transitivity on this set of 276 letters by working out the lengths of the orbits of the stabilizer of a 3-vector of a different shape, such as $2^{12}.0^{12}$ that is fixed by $M_{12}$ and demonstrating numerically that they must fuse. However, in this instance we prefer to produce an element of $\cdot O$ that fixes $u$ and lies outside the monomial subgroup $N \cong 2^{12} : M_{24}$. We may readily check that the element $\eta$ in Figure 11.4 has this property.

| 5 | 1 | 1 | 1 | 1 | 1 |
|---|---|---|---|---|---|
| 1 | 1 | 1 | 1 | 1 | 1 |
| 1 | 1 | 1 | 1 | 1 | 1 |
| 1 | 1 | 1 | 1 | 1 | 1 |

$\xi_T \Rightarrow$

| -1 | -1 | -1 | -1 | -1 | -1 |
|---|---|---|---|---|---|
| 3 | -1 | -1 | -1 | -1 | -1 |
| 3 | -1 | -1 | -1 | -1 | -1 |
| 3 | -1 | -1 | -1 | -1 | -1 |

$\xi_{T'} \Rightarrow$

| 1 | 5 | 1 | 1 | 1 | 1 |
|---|---|---|---|---|---|
| 1 | 1 | 1 | 1 | 1 | 1 |
| 1 | 1 | 1 | 1 | 1 | 1 |
| 1 | 1 | 1 | 1 | 1 | 1 |

$\beta \Rightarrow$

| 5 | 1 | 1 | 1 | 1 | 1 |
|---|---|---|---|---|---|
| 1 | 1 | 1 | 1 | 1 | 1 |
| 1 | 1 | 1 | 1 | 1 | 1 |
| 1 | 1 | 1 | 1 | 1 | 1 |

We already have $N$ and the Conway element $\xi_T$ as $24 \times 24$ matrices over $\mathbb{Z}_3$, and so in order to obtain a matrix representing $\eta$ we simply require a permutation of $M_{24}$ that takes $T$, the first column of the MOG, to $T'$. Such an element is given by $d$ below.

```
> d:=m24!(24,23)(6,9)(11,20)(16,10)(13,18)(14,2)(21,22)(8,7);
> dm:=PermutationMatrix(z3,[i^d:i in [1..24]]);
> eta:=gl24!xim*dm^-1*xim*dm*bm;
> Order(eta);
```

## 11.6 The Conway Group Co$_3$

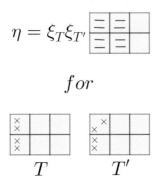

Figure 11.4 An element outside M$_{23}$ fixing our 3-element

```
2
> #sub<gl24|eta,am>;
495766656000
```

We readily see that $\eta$ fuses the two orbits above and may now ask MAGMA for the order of the subgroup of ·O it generates together with the 23-cycle of M$_{24}$ that we have labelled $\alpha$. This indeed verifies that

$$|\text{Co}_3| = \frac{|\cdot\text{O}|}{|\Lambda_3|}.$$

We now wish to use this matrix representation of Co$_3$ to obtain a permutation representation on 276 letters. The vector $u_3 = 4v_\infty + 4v_0$ is one of the 552 2-vectors having inner product 16 with $u$, and modulo 3 it has shape $0^{22}.1^2$.

```
> u3:=Vector(24,[z3!0:i in [1..22]] cat [z3!1:i in [1..2]]);
> co3:=sub<gl24|eta,am>;
> #(u3^co3);
552
```

This enables us to obtain a permutation action on 552 letters by using the **CosetAction** command giving the action of Co$_3$ on the cosets of the stabilizer of $u_3$. Here the function mapping a 24 × 24 matrix of Co$_3$ to a permutation of degree 552 is given by $f552$; the image group is denoted by $gp552$ and the kernel of the action is given by $kk$. As we know, these 552 vectors fall into 276 pairs and we want the action on those pairs. So we first identify the pair containing 1 that turns out to be $\{1, 219\}$. We may now simply ask for the stabilizer of this pair and take the action on the cosets of that stabilizer.

```
> stu3:=Stabilizer(co3,u3);
> f552,gp552,kk:=CosetAction(co3,stu3);
```

146                    The Conway Group ·O

```
> Fix(Stabilizer(gp552,1));
{ 1, 219 }
> stprb:=Stabilizer(gp552,{1,219});
> f276,gp276,kkk:=CosetAction(gp552,stprb);
> Degree(gp276);
276
> #gp276;
495766656000
```

Having verified that the resulting permutation group $gp276$ does indeed have the correct order, we may now confirm other properties of $Co_3$.

```
> st1a:=Stabilizer(gp276,1);
> #(2^st1a);
275
> oost12:=Orbits(Stabilizer(gp276,[1,2]));
> #oost12;
4
> [#oost12[i]:i in [1..4]];
[ 1, 1, 112, 162 ]
>
```

We first confirm that the stabilizer of a point in the 276 point action acts transitively on the remaining 275 points, and so $Co_3$ acts doubly transitively on 276 points. We then ask for the orbit lengths of the stabilizer of two points and find that the stabilizer of a point in $Co_3$ is a rank 3 permutation group on 275 points with suborbits of lengths 275 = 1 + 112 + 162. It is, of course, McL : 2, the automorphism group of the McLaughlin sporadic simple group, see the ATLAS, Conway et al. (1985, pages 100 and 134).

### 11.6.2 $Co_3$ Directly as Permutations of Degree 276

As we have seen, $Co_3$ is best regarded as a doubly transitive permutation group of degree 276 acting on the 23 points of $\Omega \setminus \{\infty\}$ together with the 253 heptads of the Steiner system S(4, 7, 23) defined on that set. It is of interest to construct the group directly, using our element $\eta$ of Figure 11.4 but without first obtaining the matrix representation. Accordingly we construct $xx276$, a sequence of length 276, labelled $xx276$, consisting of the points followed by the heptads.

```
> s23:=Sym(23);
> m23:=sub<s23|
> (1,2,3,4,5,6,7,8,9,10,11,12,13,14,15,16,17,18,19,20,21,22,23),
> (11,1)(22,2)(4,20)(18,10)(16,14)(8,17)(13,21)(12,7)>;
```

## 11.6 The Conway Group $Co_3$

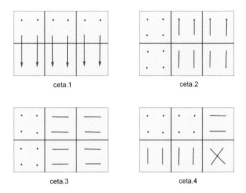

Figure 11.5 Generators for the centralizer in $M_{23}$ of $\eta$

```
> #m23;
10200960
> xx276:=[{i}:i in [1..23]] cat Setseq(({23,3,6,9,19,15,5}^m23));
```

We now require a **procedure**, labelled **perm276**, which will convert a permutation of $M_{23}$ into a permutation of degree 276 acting on $xx276$. This is applied to our usual generator $\alpha$ and to $\gamma$, which interchanges the 3rd and 4th, 5th and 6th columns of the MOG whilst fixing the rows, to obtain $M_{23}$ acting on the 276 points and heptads. We now have to construct our element $\eta$ and to facilitate this we notice that the permutations shown in Figure 11.5 commute with $\eta$; they generate a subgroup *ceta* of order 48 that can more conveniently be generated by the two elements $dp$ and $ep$, as defined below.

```
> s276:=Sym(276);
> procedure perm276(uu,xx276,~uup);
procedure> uus:=[1: i in [1..276]];
procedure> for i in [1..276] do
procedure|for> for j in [1..276] do
procedure|for|for> if xx276[i]^uu eq xx276[j] then
procedure|for|for|if> uus[i] := j;
procedure|for|for|if> end if;
procedure|for|for> end for;end for;
procedure> uup:=s276!uus;
procedure> end procedure;
> perm276(m23.1,xx276,~ap);
> perm276(m23.2,xx276,~bp);
> #sub<s276|ap,bp>;
10200960

> ceta:=sub<m23|(3,6,9)(19,15,5)(4,16,13)(20,14,21)(18,8,12)(10,17,7),
> (11,4)(1,20)(22,18)(2,10)(16,13)(14,21)(8,12)(17,7),
```

148                    The Conway Group ·O

```
> (11,1)(22,2)(4,20)(18,10)(16,14)(8,17)(13,21)(12,7),
> (6,9)(15,5)(16,13)(14,21)(22,2)(18,10)(8,7)(12,17)>;
> #ceta;
48

> perm276(ceta.1*ceta.3,xx276,~dp);
> perm276(ceta.2*ceta.4,xx276,~ep);
> ceta276:=sub<s276|dp,ep>;
> #ceta276;
48
```

As we saw on page 144, the element $\eta$ has order 2 and so we shall produce a set $pr$ containing all the pairs that are interchanged by $\eta$. We shall then treat each of these pairs as a transposition in the symmetric group $S_{276}$ and multiply them together to obtain $\eta$. To do this we need a representative of each of the orbits of $ceta$ on these pairs; we then take the set of all images of those representatives. The MAGMA instruction **Index(xx276,heptad)** identifies for a given heptad the position in $xx276$ it occupies. For instance, the heptad consisting of the first brick of the MOG (less $\infty$) is in position 24 and the point 0 is in position 23; since $\eta$ interchanges them, $\{23, 24\}$ is one of the pairs:

thus

```
> pr:={{23,Index(xx276,{23,3,6,9,19,15,5})}};
> pr:=pr join ({3,19}^ceta276);
> #pr;
4
> pr:=pr join ({11,Index(xx276,{23,11,20,14,21,22,2})}^ceta276);
> #pr;
12
> pr:=pr join ({22,Index(xx276,{23,11,1,18,8,12,2})}^ceta276);
> #pr;
20
> acteta:=[[{23,6,9,4,1,12,17},{23,15,5,11,20,12,17}],
> [{23,3,19,11,1,4,20},{23,3,19,8,12,17,7}],
> [{23,3,15,11,2,13,7},{23,6,19,4,16,8,18}],
> [{3,6,9,11,4,16,13},{3,6,9,1,20,14,21}],
> [{3,6,19,16,1,22,17},{6,9,5,16,1,8,2}],
> [{3,6,5,11,1,12,7},{3,6,5,13,21,22,2}],
> [{3,11,4,22,2,12,17},{19,20,4,16,13,12,17}],
> [{3,11,1,13,14,18,22},{19,16,21,18,10,17,7}],
> [{3,19,15,11,16,18,12},{9,5,15,20,21,22,8}],
```

## 11.6 The Conway Group $Co_3$

```
> [{3,5,15,22,18,7,17},{3,5,15,8,12,2,10}]];
> for i in [1..10] do
for> pr:=pr join ({Index(xx276,acteta[i][1]),
>Index(xx276,acteta[i][2])}^ceta276);
for> end for;
> #pr;
132
> test:=Id(s276);
> for tt in pr do
for> test:=test*s276!(Setseq(tt)[1],Setseq(tt)[2]);
for> end for;
> etap:=test;
> #sub<s276|ap,bp,etap>;
495766656000
```

From the above we see that there are 20 pairs involving points, three of which map points to points. There remain a further 10 orbits of *ceta* on pairs in which heptads are mapped to heptads. Representatives for each of these are given below in *acteta*; in each case the action of $\eta$ must be worked out as in:

| 2 | 2 | . | 2 | . | . |
|---|---|---|---|---|---|
| . | . | 2 | . | . | . |
| 2 | . | . | . | . | 2 |
| 2 | . | . | . | 2 | . |

$\xi_T \Rightarrow$

| 1 | 1 | −1 | 1 | −1 | −1 |
|---|---|----|---|----|----|
| 3 | −1 | 1 | −1 | −1 | −1 |
| 1 | −1 | −1 | −1 | −1 | 1 |
| 1 | −1 | −1 | −1 | 1 | −1 |

$\xi_{T'} \Rightarrow$

| 2 | 2 | . | 2 | . | . |
|---|---|---|---|---|---|
| . | . | 2 | . | . | . |
| 2 | . | . | . | . | 2 |
| 2 | . | . | . | 2 | . |

$\beta \Rightarrow$

| 2 | 2 | 2 | . | . | . |
|---|---|---|---|---|---|
| . | . | . | 2 | . | . |
| . | 2 | . | . | . | 2 |
| . | 2 | . | . | 2 | . |

where $\beta$ denotes the involutory permutation in the definition of $\eta$, see Figure 11.4. Thus

$\eta$ :

| ∞ | × | . | × | . | . |
|---|---|---|---|---|---|
| . | . | × | . | . | . |
| × | . | . | . | . | × |
| × | . | . | . | × | . |

$\leftrightarrow$

| ∞ | × | × | . | . | . |
|---|---|---|---|---|---|
| . | . | . | × | . | . |
| . | × | . | . | . | × |
| . | × | . | . | × | . |

The remaining nine cases listed in *acteta* below are left as an exercise in applying Conway elements to vectors in the Leech lattice.

We see that *pr* contains 132 pairs, which means that $\eta$ has $276 - 2 \times 132 = 12$ fixed points and so belongs in class 2B of $Co_3$, as we see from the character table in the ATLAS Conway et al. (1985, page 135). Indeed, it is an interesting exercise to identify the 12 heptads that are fixed by $\eta$.

## 11.7 The Finite Simple Groups

Although this book is mainly concerned with the Mathieu group $M_{24}$ and the Conway group ·O that grows out of it, it seems appropriate to include a brief description of the set of all finite simple groups, and in particular to give a short history of those that are termed *sporadic*. The Introduction to the ATLAS (Conway et al., 1985) contains a more detailed synopsis, written by John Conway, of the various infinite families of finite simple groups, and the body of the ATLAS includes a description of a representative of each of these families together with details of each of the 26 sporadics. The book *The Finite Simple Groups* (Wilson, 2009a) gives an in-depth modern account of all these groups together with a comprehensive bibliography.

### 11.7.1 A Brief Description of the Infinite Families of Simple Groups

A group is said to be *simple* if it contains no normal subgroups other than itself and the trivial group, which is to say that the only homomorphic images it possesses are isomorphic copies of itself and the trivial group. Every finite group possesses a *composition series*, a nested sequence of subgroups each of which is normal in the one above, such that the quotient of successive terms is simple. In this sense the simple groups are the building blocks of finite groups; they can be thought of as playing the same role as that played by prime numbers in number theory. Indeed, in an abelian group every subgroup is normal, and so every non-trivial element in a simple abelian group must generate. Thus the only simple abelian groups are cyclic groups of prime order.

The alternating groups $A_n$ are simple for $n \geq 5$; indeed it was the simplicity of $A_5$ that enabled Galois to prove that – unlike quadratics, cubics and quartics – there can be no systematic way of solving the general quintic equation by taking successive $n$th roots. So group theory has its origins in Galois theory. Every alternating group possesses an outer automorphism of order 2 that is realized in the corresponding symmetric group; thus

$$A_n : 2 \cong S_n \text{ for all } n \geq 5.$$

This is the full outer automorphism group of $A_n$ except for $n = 6$ when $S_6$ acts on 6 letters in two non-permutation identical ways and possesses an outer automorphism interchanging the two, see Section 9.12.

The classical groups consist of the linear, unitary, symplectic and orthogonal groups. We shall only give detailed information about the classical groups that

### 11.7.2 The Classical Groups

#### The Linear Groups

The set of all non-singular $n \times n$ matrices over $\mathbb{F}_q$, the field of order $q = p^r$ for $p$ prime, is the *general linear group* in $n$ dimensions over $\mathbb{F}_q$; it is denoted by $\mathrm{GL}_n(q)$. The columns of such a matrix form a basis for the underlying $n$-dimensional vector space over $\mathbb{F}_q$ and so the order of this group is given by

$$|\mathrm{GL}_n(q)| = (q^n - 1)(q^n - q) \cdots (q^n - q^{n-1}).$$

This group is not, in general, simple but if we 'top and tail' it by (a) restricting to matrices of determinant 1 to obtain the *special linear group* $\mathrm{SL}_n(q)$, of index $(q - 1)$ in $\mathrm{GL}_n(q)$, and then by (b) factoring out its centre, which consists of the remaining $(q - 1, n)$ scalar matrices, we obtain the *projective special linear group* $\mathrm{PSL}_n(q)$ that is simple except for small values of $n$ and $q$. So we have

$$|\mathrm{PSL}_n(q)| = (q^n - 1)(q^n - q) \cdots (q^n - q^{n-1})/(q - 1)(q - 1, n).$$

In fact $\mathrm{PSL}_2(2) \cong S_3$ and $\mathrm{PSL}_2(3) \cong A_4$; otherwise $\mathrm{PSL}_n(q)$ is simple for all $n \geq 2$. Following Artin's convention we denote the simple group $\mathrm{PSL}_n(q)$ by $L_n(q)$. The linear groups that occur most frequently in this book have $n = 2$ and we see that

$$|L_2(q)| = (q^2 - 1)(q^2 - q)/(q - 1)(q - 1, n) = (q + 1)q(q - 1)/(q - 1, 2).$$

$L_2(q)$ acts naturally on the $(q^2 - 1)/(q - 1) = q + 1$ 1-dimensional subspaces of the underlying 2-dimensional space, generators for which may be taken as

$$\begin{pmatrix} 1 \\ 0 \end{pmatrix} \sim \infty \text{ and } \begin{pmatrix} \lambda \\ 1 \end{pmatrix} \sim \lambda \text{ for } \lambda \in \mathbb{F}_q.$$

Thus $\mathrm{GL}_2(q)$ may be thought of as permuting the points of the *projective line*

$$P_1(q) = \{\infty\} \cup \mathbb{F}_q.$$

Scalar matrices fix all 1-dimensional spaces and so lie in the kernel of the homomorphism from $\mathrm{GL}_2(q)$ into $S_{q+1}$. Furthermore we have for $ad - bc \neq 0$,

$$\begin{pmatrix} a & b \\ c & d \end{pmatrix} \begin{pmatrix} x \\ 1 \end{pmatrix} = \begin{pmatrix} ax + b \\ cx + d \end{pmatrix} \sim \frac{ax + b}{cx + d} \in \mathbb{F}_q \text{ if } cx + d \neq 0,$$

and

$$\begin{pmatrix} a & b \\ c & d \end{pmatrix} \begin{pmatrix} 1 \\ 0 \end{pmatrix} = \begin{pmatrix} a \\ c \end{pmatrix} \sim \frac{a}{c} \in \mathbb{F}_q \text{ if } c \neq 0,$$

with the image being $\infty$ if $cx+d = 0$ in the first case, and if $c = 0$ in the second. With the interpretations implied by the above, the action of $\mathrm{PGL}_2(q)$ on $\mathrm{P}_1(q)$ is given by the *linear fractional maps*

$$x \mapsto \frac{ax+b}{cx+d} \text{ for } ad - bc \neq 0.$$

Note that if $p$ is odd then the order of the *projective general group* is given by

$$|\mathrm{PGL}_2(q)| = (q^2 - 1)(q^2 - q)/(q - 1) = 2|\mathrm{L}_2(q)|.$$

In general, the multiplicative subgroup of $\mathbb{F}_q$ is cyclic of order $q - 1$. Suppose that $\lambda \in \mathbb{F}_q$ is a *primitive element*, which is to say it has multiplicative order $q - 1$. Then the linear fractional map

$$\beta : x \mapsto \lambda x$$

has order $q - 1$ as a permutation. So, if $p$ is an odd prime then this is an odd permutation and if $p = 2$, it is an even permutation. The subgroup of index $q + 1$ fixing $\infty \in \mathrm{P}_1(q)$ consists of all *affine* transformations

$$\{x \mapsto ax + b : 0 \neq a, b \in \mathbb{F}_q\},$$

an elementary abelian $p$-group of order $q = p^r$, extended by the cyclic group $\langle \beta^2 \rangle$ in the odd case and $\langle \beta \rangle$ when $p = 2$. Galois himself knew that $q + 1$ was the lowest degree for a faithful permutation representation of $\mathrm{L}_2(q)$ except for the cases $q = 5, 7$ or $11$ when $\mathrm{L}_2(q)$ has an action on $q$ letters. In these cases we have

(i) $\mathrm{L}_2(5)$ possesses a subgroup isomorphic to $A_4$ and $\mathrm{L}_2(5) \cong A_5 \cong \mathrm{L}_2(4)$;
(ii) $\mathrm{L}_2(7)$ possesses a subgroup isomorphic to $S_4$ and $\mathrm{L}_2(7) \cong \mathrm{L}_3(2)$;
(iii) $\mathrm{L}_2(11)$ possesses a subgroup isomorphic to $A_5$.

In the case $q = 7$ the group $\mathrm{L}_3(2)$ acts on the 7 points and 7 lines of the Fano plane, see Section 2.2. In Chapter 14 we see how the exceptional isomorphisms in the cases $q = 5$ and $q = 7$ give rise to the Mathieu groups $M_{12}$ and $M_{24}$, respectively.

In the case $q = 11$ the group $\mathrm{L}_2(11)$ preserves the beautiful $(11, 5, 2)$ *biplane* that consists of 11 lines with five points on each line such that every pair of points lie together on precisely two lines. These can be taken as the set of quadratic residues $Q = \{1, 3, 9, 5, 4\}$ and its translates under $n \mapsto n+1$ modulo 11. This action on 11 points, in this case 11 involutions in some larger group, was used in Curtis (1993) and Curtis (2007b, page 138) to define and construct by hand the smallest Janko simple group $J_1$, see Janko (1965, 1966).

## 11.7 The Finite Simple Groups

A final isomorphism that should be mentioned here is that

$$L_4(2) \cong A_8;$$

thus $A_8$ has a permutation action on the $15 = 2^4 - 1$ non-zero vectors in a 4-dimensional vector space over $\mathbb{Z}_2$. The point stabilizer of this action has shape $2^3 : L_2(7)$: an elementary abelian group of order $2^3$ generated by the seven images of the permutation $(\infty\ 0)(1\ 3)(2\ 6)(4\ 5)$ under conjugation by $L_2(7)$, extended by $L_2(7) = \langle (0\ 1\ 2\ 3\ 4\ 5\ 6), (\infty\ 0)(1\ 6)(2\ 3)(4\ 5)\rangle$. This subgroup conveniently exhibits the isomorphism $L_3(2) \cong L_2(7)$.

The classic text on linear groups is Leonard E. Dickson's book that was first published in 1901, a modern edition of which is Dickson (2007). This book includes a description of the maximal subgroups of $L_2(q)$; more modern expositions of the maximal subgroups of classical groups are given by Bray et al. (2013), King (2005) and Kleidman and Liebeck (1990). Wilson's book *The Finite Simple Groups* (Wilson, 2009a), as mentioned earlier, is an excellent source for information on all of these groups.

We obtain the other classical groups – the unitary, symplectic and orthogonal groups, see Weyl (1953) – by considering the subgroups of certain general linear groups that preserve some additional algebraic structure. Standard algebra texts such as Jacobson (1980, Chapter 6) contain elementary approaches to these groups, as does the more recent Burness and Giudici (2016); perhaps the definitive text on the subject is *La Géometrie des Groupes Classiques* (Dieudonné, 1963).

### The Unitary Groups

We shall describe the unitary groups in some detail here as it is the group $U_3(3)$ that is used in Chapter 15 to initiate the Thompson chain of groups that culminates in the largest Conway simple group $Co_1$. The explicit matrices in $U_3(3)$ we need are given in Section 15.3.1. In order to construct finite unitary groups, we consider the finite field $\mathbb{F}_{q^2}$, of order $q^2$. The multiplicative subgroup of $\mathbb{F}_{q^2}$ is cyclic of order $q^2 - 1$ and so if $\lambda \in \mathbb{F}_{q^2}$ then $\lambda^{q^2} = \lambda$. Thus the automorphism

$$\lambda \mapsto \lambda^q = \bar{\lambda} \text{ say, satisfies } \bar{\bar{\lambda}} = (\lambda^q)^q = \lambda^{q^2} = \lambda$$

and has order 2. We can then define the *general unitary group* by

$$GU_n(q) = \{A \in GL_n(q^2) \mid A\bar{A}^t = I_n\}, \qquad (11.1)$$

where, if $A = (a_{ij})$, then the *conjugate* matrix $\bar{A}$ is defined by $\bar{A} = (\overline{a_{ij}})$, and $A^t$ denotes the *transpose* of $A$. Thus $GU_n(q)$ is the set of all $n \times n$ matrices with entries in $\mathbb{F}_{q^2}$ the transpose of whose conjugate is their inverse. Note that the entries in $GU_n(q)$ are in $\mathbb{F}_{q^2}$ rather than its subfield $\mathbb{F}_q$. Thus the matrices in $U_3(3)$ have entries in $\mathbb{F}_9$. We see that for $A, B \in GU_n(q)$,

$$\overline{AB}^t = (\bar{A}\bar{B})^t = \bar{B}^t\bar{A}^t = B^{-1}A^{-1} = (AB)^{-1},$$

and so $AB \in GU_n(q)$ and $GU_n(q)$ is indeed a subgroup. Note that if $A \in GU_n(q)$ then

$$A\bar{A}^t = I_n \implies \sum_k a_{ik}\overline{a_{jk}} = \delta_{ij},$$

where $\delta_{ii} = 1, \delta_{ij} = 0$ if $i \neq j$, and so the rows (or columns) of $A$ form an orthonormal basis with respect to the inner product

$$\underline{x} \cdot \underline{y} = \sum x_i \overline{y_i}. \qquad (11.2)$$

In general, if $\underline{x}$ and $\underline{y}$ are any $n$-dimensional row vectors written over $\mathbb{F}_{q^2}$, that is to say $1 \times n$ matrices $x$ and $y$, then

$$\underline{x} \cdot \underline{y} = x\overline{y}^t.$$

If the matrix $A$ preserves the inner product of vectors in the sense that the images of two vectors after multiplication by $A$ have the same inner product as the original vectors, then we have

$$\underline{x}A \cdot \underline{y}A = xA \, \overline{(yA)}^t = xA\overline{A}^t \overline{y}^t = x\overline{y}^t = \underline{x} \cdot \underline{y} \text{ for all } \underline{x}, \underline{y} \Leftrightarrow A\overline{A}^t = I_n.$$

So the set of unitary matrices as defined earlier consists of those matrices preserving the inner product defined by (11.2). As with the linear groups, we restrict to matrices of determinant 1 to obtain the *special unitary group* $SU_n(q)$, and then factor out the centre to obtain the $PSU_n(q)$, the *projective special unitary group*. For $n \geq 2$ $PSU_n(q)$ is simple except for

$$PSU_2(2) \cong S_3; \ PSU_2(3) \cong A_4; \ PSU_3(2) \cong 3^2 : Q_8,$$

where $Q_8$ denotes the *quaternion group* of order 8. Again following Artin we adopt the shorter version $U_n(q)$ for the simple group.

## The Order of $U_n(q)$

We have seen that the order of $GU_n(q)$ is precisely the number of ordered orthonormal bases of an $n$-dimensional vector space over $\mathbb{F}_{q^2}$ with respect to the inner product (11.2). In order to calculate this number, we recall some elementary facts about a finite field $\mathbb{F}$. Firstly the additive order of the identity element must be a prime $p$, otherwise $\mathbb{F}$ would contain divisors of zero, and so $k_0$, the subfield generated by the identity, is a copy of $\mathbb{Z}_p$, the integers modulo

## 11.7 The Finite Simple Groups

$p$. Regarding $\mathbb{F}$ as a vector space over this *prime subfield* $k_0$, we see that every element of $\mathbb{F}$ can be written uniquely as

$$\sum_{i=1}^{m} \xi_i u_i$$

for some $m$ where $\{u_1, u_2, \ldots, u_m\}$ is a basis for $\mathbb{F}$ over $k_0$ and the $\xi_i \in k_0$. Thus $|\mathbb{F}| = p^m$.

In fact, it can be shown that there is a finite field of order $p^m$ for any prime $p$ and any natural number $m$, see for instance Jacobson (1985, page 287), and that any two finite fields of the same order are isomorphic. Finite fields are nested in the sense that if $m = rs$ is a composite number, then the field $\mathbb{F}_{p^m}$ contains subfields of orders $p^r$ and $p^s$. Moreover the multiplicative subgroup of a finite field is cyclic and so, in our case, if we let

$$\mathbb{F}_q \cong k \leq K \cong \mathbb{F}_{q^2},$$

then for any $\lambda \in K$,

$$\lambda \bar{\lambda} = \lambda \lambda^q = \lambda^{q+1}$$

has order dividing $q - 1$ and so lies in $k$. Every non-zero element of $k$ is the $(q+1)$th power of the same number of elements of $K$ since

$$\lambda^{q+1} = \mu^{q+1} \Rightarrow (\lambda \mu^{-1})^{q+1} = 1 \Rightarrow \lambda = v^i \mu \quad \text{for some} \quad i,$$

where $v \in K$ is a generator for the unique cyclic multiplicative subgroup of $K$ of order $q + 1$. The *norm* of $\underline{x}$, an $n$-dimensional vector over $\mathbb{F}_{q^2}$, is defined to be

$$\underline{x} \cdot \underline{x} = x\bar{x}^t = \sum x_i \bar{x}_i = \sum x_i^{q+1} \in k,$$

and, since $\lambda \underline{x} \cdot \lambda \underline{x} = \lambda^{q+1} \underline{x} \cdot \underline{x}$ we see that for any non-zero element of $k$ there is the same number of vectors in $\mathbb{F}_{q^2}^n$ of that norm. Vectors that have norm 0 are known as *isotropic*. We seek an ordered basis of vectors of norm 1. We count such vectors, see for example Wilson (2009a, page 65), using induction. If $z_n$ denotes the number of isotropic vectors in $\mathbb{F}_{q^2}^n$ and $y_n$ denotes the number of vectors of norm 1, then the total number of vectors is given by

$$q^{2n} = 1 + z_n + (q - 1)y_n.$$

Suppose $V_{n+1}$ is an $(n + 1)$-dimensional space over $\mathbb{F}_{q^2}$ and $V_n$ is its $n$-dimensional subspace with $(n + 1)$th coordinate 0, then $V_{n+1}$ has $z_n$ isotropic vectors with 0 in the final position. If a vector of $V_n$ has non-zero norm $\lambda \in k$, then we may complete it to an isotropic vector of $V_{n+1}$ by placing a $(q+1)$th root of $-\lambda$ in the final position. Thus we see that

$$z_{n+1} = z_n + (q+1)(q-1)y_n.$$

Eliminating $y_n$ we obtain a recurrence relation for $z_n$:

$$z_{n+1} = -qz_n + (q+1)(q^{2n} - 1),$$

which may be solved to obtain

$$z_n = (q^n - (-1)^n)(q^{n-1} + (-1)^n).$$

Note that we may readily check that this function does satisfy the recurrence relation and that $z_0 = z_1 = 0$ as required, so it is indeed the unique solution. Recall that we are seeking an ordered orthonormal basis so, simplifying the expression for $y_n$, we see that the first column may be chosen in

$$y_n = q^{n-1}(q^n - (-1)^n) \text{ ways.}$$

The orthogonal complement of the first column is an $(n-1)$-dimensional space, which thus contains $y_{n-1}$ vectors of norm 1. So the number of ordered orthonormal bases is

$$|\text{GU}_n(q)| = \prod_{i=1}^{n} q^{i-1}(q^i - (-1)^i) = q^{\binom{n}{2}} \prod_{i=1}^{n}(q^i - (-1)^i).$$

If $A \in \text{GU}_n(q)$ then $A\bar{A}^t = I_n$ and so, if $|A| = \delta$ then $\delta\bar{\delta} = \delta^{q+1} = 1$. Every $(q+1)$th root of unity is possible and so the order of the special unitary group is given by

$$|\text{SU}_n(q)| = |\text{GU}_n(q)|/(q+1).$$

The centre of this group consists of scalar matrices $\lambda I_n$ such that $\lambda^n = 1$ and $\lambda\bar{\lambda} = \lambda^{q+1} = 1$, which is satisfied by $(n, q+1)$, the highest common factor of $n$ and $q+1$, values of $\lambda$. Thus the order of the projective special unitary group $U_n(q)$ that is simple apart from the cases mentioned earlier is given by

$$|U_n(q)| = q^{\binom{n}{2}} \prod_{i=2}^{n}(q^i - (-1)^i)/(n, q+1).$$

In particular, putting $q = 3$, we have $U_3(3) = \text{SU}_3(3)$ and

$$|U_3(3)| = 3^3(3^3 + 1)(3^2 - 1)/(4, 3) = 27 \cdot 28 \cdot 8 = 6048.$$

### Subgroups of $\mathrm{GL}_n(q)$ Preserving a Form

Putting this into a more general context, the function $f$ from $V \times V$ into $\mathbb{F}_{q^2}$ defined by

$$f(x, y) = x \, \bar{y}^t,$$

satisfies, for $\lambda_1, \lambda_2 \in \mathbb{F}_{q^2}$,

$$f(\lambda_1 x_1 + \lambda_2 x_2, y) = \lambda_1 f(x_1, y) + \lambda_2 f(x_2, y)$$

but

$$f(x, \lambda_1 y_1 + \lambda_2 y_2) = \overline{\lambda_1} f(x, y_1) + \overline{\lambda_2} f(x, y_2).$$

It is *linear* in $x$ but *semilinear* in $y$. In these circumstances we say it is a *sesquilinear* (one-and-a-half linear) form. Moreover, it is *conjugate-symmetric* in that

$$f(y, x) = \overline{f(x, y)}.$$

Associated with $f$ is the *Hermitian* form $F : V \mapsto F$ defined by

$$F(x) = f(x, x).$$

The Hermitian form $F$ is said to be *non-singular* if there exist no non-zero vector $x_0 \in V$ such that $f(x_0, y) = 0$ for all $y \in V$, where $f$ is the associated conjugate-symmetric sesquilinear form. In these circumstances a basis of $V$ may be chosen so that $F$ takes the form implied by (11.2), namely

$$F(\bar{x}) = x_1 \overline{x_1} + x_2 \overline{x_2} + \cdots + x_n \overline{x_n}.$$

So, the *general unitary group* $\mathrm{GU}_n(q)$, which is defined to be the subgroup of the general linear group $\mathrm{GL}_n(q^2)$ preserving a given non-singular Hermitian form, may be taken to be the set of matrices given earlier in (11.1).

The symplectic and orthogonal groups are defined in an analogous manner. Thus a form $f : V \times V \mapsto F$ is *alternating* if

$$f(x, y) = -f(y, x) \text{ for all } x, y \in V.$$

Note that if the characteristic of $F$ is not 2, then this implies that $f(x, x) = 0$ for all $x \in V$. If $\mathrm{char} F = 0$ then we must add this as an additional condition. If an alternating form is linear in $x$ so that

$$f(\lambda_1 x_1 + \lambda_2 x_2, y) = \lambda_1 f(x_1, y) + \lambda_2 f(x_2, y),$$

then it is also linear in $y$ and so it is *bilinear*. An alternating bilinear form is also known as a *symplectic form*. If

$$\{x \in V \mid f(x, y) = 0 \text{ for all } y \in V\} = \{0\},$$

then the form is said to be *non-singular*. The rank of a symplectic form is necessarily even, say $n = 2m$, and a basis may be chosen so that

$$f(x, y) = x_1 y_{m+1} + x_2 y_{m+2} + \cdots + x_m y_{2m} - x_{m+1} y_1 - x_{m+2} y_2 - \cdots - x_{2m} y_m.$$

The subgroup of $\mathrm{GL}_{2m}(q)$ preserving a non-singular symplectic form is the *symplectic group* $\mathrm{Sp}_{2m}(q)$. Its elements will necessarily have determinant 1, and so the special and general symplectic groups coincide. Factoring out the centre in the usual manner, we obtain the *projective symplectic group* that is simple except for small cases and so we denote it by $S_{2m}(q)$, the exceptions being

$$\mathrm{PSp}_2(2) \cong S_3, \quad \mathrm{PSp}_2(3) \cong A_4 \quad \text{and} \quad \mathrm{PSp}_4(2) \cong S_6.$$

The order of the projective symplectic group is given by

$$|S_{2m}(q)| = q^{m^2}(q^{2m} - 1)(q^{2m-2} - 1) \cdots (q^2 - 1)/(q - 1, 2).$$

A form

$$f : V \times V \mapsto \mathbb{F}_q,$$

which is linear in both arguments $x$ and $y$ and which, in addition, satisfies

$$f(x, y) = f(y, x),$$

is a *symmetric bilinear form*. A *quadratic form* $F$ is a function from $V$ into $\mathbb{F}_q$ that satisfies

$$F(\lambda x + \mu y) = \lambda^2 F(x) + \lambda \mu f(x, y) + \mu^2 F(y),$$

where $f(x, y)$ is the *associated symmetric bilinear form*. As in the previous cases, the *kernel* of $f$ is defined to be the set of all $x$ such that $f(x, y) = 0$ for all $y \in V$, but now we must also define the *kernel of $F$* to be those vectors in the kernel of $f$ for which $F(x) = 0$. For $F$ a quadratic form the dimension of whose kernel is zero, we define the *general orthogonal group* $\mathrm{GO}_n(q, F)$ to be the set of all elements of $\mathrm{GL}_n(q)$ that preserve $F$. The situation is more complicated than in the previous cases, however, as 'topping and tailing' no longer necessarily results in a simple group, although $\mathrm{PSO}_n(q, F)$ does contain a subgroup that is usually simple. It is this subgroup, which has index at most 2 in $\mathrm{PSO}_n(q, F)$, that we label $O_n(q, F)$. Moreover, the parity of $n$ plays a significant role as there is just one equivalence class of non-singular quadratic forms when $n$ is odd and two when $n$ is even. When $n = 2m$ is even these are distinguished by the labels $O_{2m}^+(q)$ and $O_{2m}^-(q)$. We have

$$n = 2m+1 | O_n(q) |$$
$$= q^{m^2}(q^{2m}-1)(q^{2m-2}-1)\cdots(q^2-1)/(2,q-1),$$
$$n = 2m | O_n^+(q) |$$
$$= q^{m(m-1)}(q^m-1)(q^{2m-2}-1)(q^{2m-4}-1)\cdots(q^2-1)/(4,q^m-1),$$
$$n = 2m | O_n^-(q) |$$
$$= q^{m(m-1)}(q^m+1)(q^{2m-2}-1)(q^{2m-4}-1)\cdots(q^2-1)/(4,q^m+1).$$

A useful synopsis of these groups is given in the ATLAS (Conway et al., 1985, page xi), and a thorough treatment is provided in *The Finite Simple Groups* (Wilson, 2009a, Chapter 3).

### 11.7.3 The Chevalley Groups

A remarkable breakthrough in the theory of finite groups came in 1955 when Chevalley found a way of defining bases of the complex simple Lie algebras such that the structure constants are all rational integers, see Chevalley (1955). This means that these algebras and the associated Lie groups, the *adjoint Chevalley groups*, can be defined over any finite field. A wonderfully accessible account of this process can be found in *Simple Groups of Lie Type* (Carter, 1972). It turns out that all the classical groups mentioned earlier may be constructed in this manner, along with many others. The fundamental root systems of the simple Lie algebras are displayed in the *Dynkin diagrams* that are shown in Figure 11.6. In these diagrams, each node stands for a vector, if two nodes are joined by a single edge then the corresponding vectors have the same length and are at an angle of $2\pi/3$ with one another; if they are not connected then they are at right angles to one another. When nodes are joined by an edge with an arrowhead on it, the roots have lengths in the ratio $\sqrt{2}:1$ or $\sqrt{3}:1$ for a double and triple edge respectively, with the arrow pointing from the longer to the shorter root; the pairs of corresponding roots are at angles $3\pi/4$ and $5\pi/6$ respectively.

It turns out that the linear groups $L_{n+1}(q)$ correspond to the adjoint Chevalley groups $A_n(q)$; the symplectic groups $S_{2m}(q)$ correspond to $C_m(q)$; orthogonal groups of odd rank $O_{2m+1}(q)$ correspond to $B_m(q)$; and in the even case $O_{2m}^+(q)$ corresponds to $D_m(q)$. Fitting the remaining two families of classical groups into this elegant structure had to await a further development.

In 1959 Steinberg developed a procedure that not only brought the two remaining families of classical groups into the Chevalley fold, but also produced

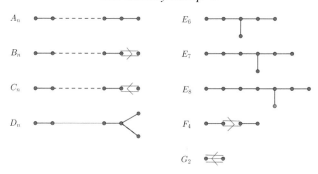

Figure 11.6 The Dynkin diagrams

two new families of finite simple groups. Several of the Dynkin diagrams in Figure 11.6 whose roots all have the same length exhibit obvious symmetries, namely: $A_n$ can be reflected through its centre; the two 'antennae' of a $D_n$ can be interchanged; and $E_6$ can be reflected through its trivalent vertex. All these *graph automorphisms* have order two, but the diagram $D_4$ possesses three edges meeting at a trivalent point and so possesses an automorphism of order 3. The fields of order $\mathbb{F}_{q^2}$ and $\mathbb{F}_{q^3}$ possess automorphisms of order 2 and 3 respectively, namely $x \mapsto x^q$. Steinberg found that extending the field $\mathbb{F}_q$ to $\mathbb{F}_{q^2}$ when the graph automorphism has order 2 and to $\mathbb{F}_{q^3}$ when it has order 3 we obtain new groups; namely the elements of the group $X_n(q^r)$, for $X = A, D$ or $E$, that are fixed by the quotient of the graph automorphism by the field automorphism. This process was referred to as *twisting* and the resulting groups are the *twisted Chevalley groups*. It turned out that

$$^2A_n(q) \cong U_{n+1}(q) \ (n \geq 2) \quad \text{and} \quad ^2D_n(q) \cong O^-_{2n}(q),$$

and that $^3D_4(q)$ and $^2E_6(q)$ were new simple groups. The preceding superscript denotes the order of the graph automorphism and we talk of *doubly twisted* or *triply twisted* $X_n(q)$.

But, if the lengths of the vectors were ignored, it was clear that some of the other Dynkin diagrams possessed automorphisms, namely $B_2 = C_2$, $F_4$ and $G_2$. Suzuki (1964) and Ree (1961, 1960) extended twisting to these diagrams to obtain new groups that are named after them. However, in order to cope with roots of different lengths they needed to restrict the characteristic $p$ of the underlying field; explicitly, for the double-edge diagrams $B_2$ and $F_4$ the characteristic had to be 2, and for the triple-edge diagram $G_2$ the characteristic had to be 3. Furthermore, the order of the field had to be an odd power of $p$.

We thus obtained new twisted groups

$$^2B_2(2^{2m+1}) \quad \text{Suzuki group} \quad Sz(q),$$
$$^2F_4(2^{2m+1}) \quad \text{Ree group} \quad R_1(q),$$
$$^2G_2(3^{2m+1}) \quad \text{Ree group} \quad R_2(q).$$

In practice the subscript on the Ree groups is usually omitted, as the field order $q$ identifies the family. All these groups are simple except for the smallest representative of each family. We have

$$^2B_2(2) \cong 5:4 \quad = \langle x, y \mid x^5 = y^4 = 1, x^y = x^2 \rangle,$$
a *Frobenius group* of order 20, and
$$^2G_2(3) \cong L_2(8):3, \quad \text{the linear group extended by its field automorphism.}$$

The group $^2F_4(2)$ is not simple but its derived group $^2F_4(2)'$, which has index 2, is a simple group, something that occurs in no other family; it is referred to as the *Tits group*.

### 11.7.4 The Sporadic Simple Groups

Until the 1960s the only known finite simple groups that did not fit into one of the infinite families were the five Mathieu groups $M_{11}, M_{12}, M_{22}, M_{23}$ and $M_{24}$ and it was tacitly assumed that there were no more. Then, in 1964, Janko discovered a new simple group in which the centralizer of an involution had shape $2 \times A_5$. This came as a great shock followed by great excitement as more and more *sporadic* finite simple groups were discovered in various different ways. For instance, by seeking a simple group having a certain internal structure: specifying the shape of the centralizer of an involution, as Janko had done with the group now referred to as $J_1$, or whose Sylow 2-subgroup mimicked that of a known group. Alternatively a combinatorial object – a graph, a code, a block design or a lattice, say – was constructed using a known group, and then the new object was shown to have far more symmetries than we had originally built in. This, of course, is precisely what happened with the Conway group ·O where $M_{24}$ was used to construct the Leech lattice whose group of symmetries was a much larger group. In a sense the discovery of ·O provided evidence that the number of new groups was not inexhaustible: although the Leech lattice provided three new groups itself, most of the newly discovered groups were involved in ·O in some way.

A fresh approach to discovering new groups was introduced by Bernd Fischer who mimicked the symmetric groups by seeking groups generated by a class of involutions any two of which had product of order 2 or 3. These he termed

3-*transposition* groups. This resulted in three new groups $Fi_{22}$, $Fi_{23}$ and $Fi_{24}$. The last of these is not simple but its derived group of index 2 is simple; the 3-transpositions lie outside the simple group. This approach had proved so successful that it was extended to groups generated by a class of involutions any two of which had product of low but less restricted order. This led Fischer to conjecture the existence of a very large new group, and eventually Robert Griess constructed what is now known as the *Fischer–Griess Monster* M. With around $8 \times 10^{53}$ elements, this group does indeed have massive order but, of course, each of the infinite families has members larger than M and the smallest of the $E_8$ family has order around $10^{70}$. However, what is truly monstrous about M is the fact that its lowest degree complex matrix representation is in 196 883 dimensions, and the smallest number of letters on which it can act faithfully as a permutation group is around $2 \times 10^{20}$.

With the discovery of M, the stream of new sporadic simple groups almost dried up but Janko was not yet finished; in 1972, see Janko (1976), he announced a further new group, thus becoming the discoverer of the first and the last sporadic simple groups to be discovered after Mathieu. This fascinating group contains $M_{24}$ and can be constructed using $M_{24}$ as control subgroup as in Chapter 14.

The table in the ATLAS, see Conway et al. (1985, page 238), shows which sporadic groups are involved in one another. It can be seen that all but six of the 26 are involved in M. For this reason these six groups (three of the four groups found by Janko, $J_1$, $J_3$ and $J_4$ and other groups named after their discoverers, see Rudvalis, 1973, O'Nan, 1976 and Lyons, 1972) are known as the *pariahs*.

Conway produced a simplified construction of M, see Conway (1985), and determined to understand where the group came from. An important development in this direction came with the investigation of the so-called *Y-diagrams*, see Conway et al. (1985). The projective plane of order 3, see Section 2.2, consists of 13 points and 13 lines, each consisting of four points, such that every pair of points lie together on a unique line, and every two lines intersect in a unique point. You can draw a bipartite graph with 26 nodes, the points and lines, in which each line is joined to the four points that lie on it. It turns out that if you interpret the 26 vertices of this graph as involutions that commute if they are disjoint and have product of order 3 if they are joined, then the resulting group has the wreath product $M \wr 2$, the so-called *Bimonster*, as a homomorphic image. Many subgroups of this configuration are described in the ATLAS, Conway et al. (1985, page 232), together with defining relations.

## 11.8 The Classification of Finite Simple Groups

In what must rank as one of the greatest intellectual achievements of the 20th century, the Classification of Finite Simple Groups states that any finite simple group appears in the list of groups described briefly earlier. The possibility that such an accomplishment was even feasible seemed remote until relatively recently. I well remember in the late 1960s sitting at a table with John Thompson and John Conway while they were discussing whether there was an infinite number of sporadic simple groups (which would, of course, render classification impossible) or a large but finite number. Their views differed but I noticed that some months later their positions had reversed. In fact, it was impossible to say at that stage.

The driving force behind the classification was Daniel Gorenstein supported by many mathematicians but most significantly by Richard Lyons and Ron Solomon. Sadly Gorenstein died suddenly in 1992 but the other two, assisted by many others including Inna Capdeboscq, picked up the mantle and the massive proof is now available in a dozen volumes. The argument is extremely subtle in places and a particularly troublesome part, known as the *quasi-thin case*, was dealt with separately by Michael Aschbacher and Steve Smith, the former of whom had for many years led the way in the classification project.

## 11.9 A Word about Moonshine

The next development again involved John McKay who had originally pointed Conway at the Leech lattice. He had noticed that the power series expansion of an important function in number theory had a coefficient close to the minimal irreducible degree of M. Specifically, the *modular function j* has $q$-expansion given by

$$j(\tau) = \frac{1}{q} + 744 + 196\,884q + 21\,493\,760q^2 + 864\,299\,970q^3 + \cdots,$$

where $q = e^{2\pi i \tau}$, and McKay pointed out to Conway, Thompson and others that the coefficient of $q$ was just one more than the degree of the lowest dimensional representation of M. Initially this observation was dismissed as a numerological fluke, but when it became clear that the other coefficients were also simple numerical combinations of the irreducible degrees of M, people realized that something deep was going on. Conway coined the expression *Monstrous Moonshine* to describe the unexplained connection with number theory, and he and Norton conjectured a link between M and elliptic modular functions, see Conway and Norton (1979).

At this point Richard Borcherds joined the fray, see Borcherds (1998), and in an extraordinary piece of work in which he linked the Monster, elliptic curves and string theory, through *vertex algebras*, he verified the Moonshine conjectures. For this profound breakthrough Borcherds was awarded the Fields Medal. He writes 'I consider Conway's calculation of the automorphism group of the 26-dimensional Lorentzian lattice $\Pi_{25,1}$, see Conway (1983) and Borcherds et al. (1984), to be one of his deepest and most fundamental results: it underlies several areas such as monstrous moonshine and infinite dimensional generalized Kac–Moody algebras. Most of my published papers depend in some way on this result of Conway's which shows that the automorphism group of $\Pi_{25,1}$ is the product of $-1$ and an extension $R \cdot \Lambda \cdot \text{Aut}(\Lambda)$, where $R$ is the reflection group, $\Lambda$ is the Leech lattice, and $\text{Aut}(\Lambda)$ is the Conway group'. This space is described briefly in Section 10.2.3 and for further details the reader is referred to Conway and Sloane (1988, page 522). Borcherds goes on to state that Conway's result explains how below dimension 26 lattices are somehow manageable, but that beyond that point things 'go wild'. 'For example, the even, positive definite, unimodular lattices consist of 1 in dimension 8, 2 in dimension 16, 24 in dimension 24, but more than a billion in dimension 32'.

# 12

# Permutation Actions of $M_{24}$

## 12.1 Background

For an introduction to the representations and characters of finite groups, see James and Liebeck (1993) or Collins (1990). For more advanced reading, see Isaacs (1976) and Curtis and Reiner (1962).

In this chapter we shall give a brief account of how representations and characters occur in connection with permutations. A group $G$ is said to *act* on the finite set $\Omega$ if there is a homomorphism $\phi$ from $G$ into $S_\Omega$, the symmetric group on $\Omega$. If this homomorphism has trivial kernel, then we say that the action is *faithful*, which will usually be the case in the examples we shall be considering. We can associate a matrix representation of degree $n = |\Omega|$ with such a permutation action by constructing a vector space over, say, the complex numbers $\mathbb{C}$ with basis $\{\Omega\}$. Thus

$$V = \left\{ \sum_{\omega \in \Omega} \lambda_\omega \omega \mid \lambda_\omega \in \mathbb{C} \right\};$$

here, $V$ is known as the (complex) *permutation module* associated with the permutation action of $G$. The elements of $G$ will permute the basis elements and so the matrices defining their action will have a single entry 1 in each row, a single entry 1 in each column and zeros elsewhere; they will be *monomial*. The trace of such a matrix, its *character value*, will be the number of 1s on the main diagonal; in other words, it will be $|\text{Fix}_\Omega(g)|$, the number of $\omega \in \Omega$ that are mapped to themselves by $g\phi$. The function $\pi$ defined by

$$\pi : g \mapsto |\text{Fix}_\Omega(g)|$$

is the *permutation character* of $G$ on $\Omega$. Corresponding to the trivial representation of $G$, which maps every element of $G$ to the identity, is the trivial character $1_G$, which thus has character value 1 on every element. We readily obtain a formula, usually referred to as Burnside's Lemma, for $\mathrm{orb}(G,\Omega)$, the number of orbits of $G$ in its action on $\Omega$. Thus

**Theorem 12.1** *Let the group $G$ act on the finite set $\Omega$ with permutation character $\pi$ and trivial character $1_G$, then*

$$\langle \pi, 1_G \rangle = \frac{1}{|G|} \sum_{g \in G} |\mathrm{Fix}_\Omega(g)| = \mathrm{orb}(G, \Omega).$$

*Proof* By the definition of the inner product of characters, we have

$$\langle \pi, 1_G \rangle = \frac{1}{|G|} \sum_{g \in G} \pi(g) \cdot 1 = \frac{1}{|G|} \sum_{g \in G} |\mathrm{Fix}_\Omega(g)|.$$

We now count the set

$$X = \{(\omega, g) \mid \omega \in \Omega, g \in G, \text{ such that } \omega g = \omega\}$$

in two different ways. Suppose that $\mathrm{orb}(G, \Omega) = r$, and that the $r$ orbits of $G$ on $\Omega$ are $\Omega_i$ for $i = 1, \ldots, r$; so that

$$\Omega = \Omega_1 \cup \Omega_2 \cup \cdots \cup \Omega_r.$$

Let $\omega_i$ be a representative of the orbit $\Omega_i$ for $i = 1, \ldots, r$. Then we have

$$|\Omega_i| = |\omega_i^G| = |G : G_{\omega_i}|,$$

where $G_{\omega_i}$ denotes the stabilizer in $G$ of $\omega_i$. Since elements in the same orbit have stabilizers of the same order, we have:

$$\begin{aligned}
|X| &= \sum_{g \in G} |\mathrm{Fix}_\Omega(g)| \\
&= \sum_{\omega \in \Omega} |G_\omega| = \sum_{i=1}^{r} |G_{\omega_i}| \cdot |\omega_i^G| \\
&= \sum_{i=1}^{r} |G_{\omega_i}| \cdot |G : G_{\omega_i}| \\
&= \sum_{i=1}^{r} |G| = r|G| = \mathrm{orb}(G, \Omega) \cdot |G|,
\end{aligned}$$

as required. □

Note that this result is intuitively what we would expect as the 'sum' of all elements in an orbit is fixed by every element in $G$ and so we see $r$ copies of the trivial representation in the permutation representation. The theorem is saying that there are these and no more.

Now if $G$ acts on two sets $\Omega$ and $\Delta$ then we can think of it acting on the Cartesian product $\Omega \times \Delta$ in the obvious manner:

$$(\omega, \delta) g = (\omega g, \delta g). \text{ for } \omega \in \Omega, \delta \in \Delta.$$

## 12.1 Background

Then $(\omega, \delta)$ is fixed by $g$ if, and only if, both $\omega$ and $\delta$ are, and so

$$|\text{Fix}_{\Omega \times \Delta}(g)| = |\text{Fix}_{\Omega}(g)| \times |\text{Fix}_{\Delta}(g)|.$$

If $\pi$ and $\sigma$ are the two corresponding permutation characters, then we have

$$\langle \pi, \sigma \rangle = \frac{1}{|G|} \sum_{g \in G} |\text{Fix}_{\Omega}(g)| \cdot |\text{Fix}_{\Delta}(g)| = \frac{1}{|G|} \sum_{g \in G} |\text{Fix}_{\Omega \times \Delta}(g)|,$$

and so we have proved

**Theorem 12.2** *If $G$ acts on $\Omega$ and $\Delta$ with corresponding permutation characters $\pi$ and $\sigma$, then we have*

$$\langle \pi, \sigma \rangle = \text{orb}(G, \Omega \times \Delta),$$

*the number of orbits on the Cartesian product $\Omega \times \Delta$.*

If the actions on $\Omega$ and $\Delta$ are transitive, then this number is simply the number of orbits of the stabilizer of a point in one set on the points of the other set. In particular, we have proved

**Theorem 12.3** *If $G$ acts transitively on $\Omega$ and $\Delta$ and if $\omega \in \Omega$ and $\delta \in \Delta$, then*

$$\text{orb}(G_\omega, \Delta) = \text{orb}(G_\delta, \Omega).$$

**Note** In this case, if $G_\omega$ has orbits of lengths $n_1 \leq n_2 \leq \cdots \leq n_r$ on $\Delta$ with $n_1 + n_2 + \cdots + n_r = |\Delta|$, and $G_\delta$ similarly has orbits of lengths $m_1 \leq m_2 \leq \cdots \leq m_r$ on $\Omega$ with $m_1 + m_2 + \cdots + m_r = |\Omega|$, then

$$n_1|\Omega| = m_1|\Delta|, \quad n_2|\Omega| = m_2|\Delta|, \ldots, n_r|\Omega| = m_r|\Delta|,$$

these being the orbit lengths on $\Omega \times \Delta$ in increasing order. Thus the $m_i$ and the $n_i$ are in the same ratio to one another; that is,

$$m_1 : m_2 : \cdots : m_r = n_1 : n_2 : \cdots : n_r.$$

For instance, in Figures 12.4 and 12.5 we see that the orbits of the trio stabilizer on sextets are of lengths

$$7 : 84 : 336 : 1344;$$

multiplying these by $3795/1771 = 15/7$, we obtain

$$15 : 180 : 720 : 2880,$$

the lengths of the orbits of the sextet stabilizer acting on trios.

| Class | $|C_G(x)|$ | cycle shape | $\pi_{\text{octads}}$ | $\pi_{\text{dua}}$ | $\pi_{\text{sextets}}$ | $\pi_{\text{trios}}$ |
|---|---|---|---|---|---|---|
| 1A | $|G|$ | $1^{24}$ | 759 | 1288 | 1771 | 3795 |
| 2A | 21504 | $1^8.2^8$ | 71 | 56 | 91 | 99 |
| 2B | 7680 | $2^{12}$ | 15 | 48 | 51 | 75 |
| 3A | 1080 | $1^6.3^6$ | 21 | 10 | 16 | 15 |
| 3B | 504 | $3^8$ | 0 | 7 | 7 | 15 |
| 4A | 384 | $2^4.4^4$ | 7 | 8 | 11 | 19 |
| 4B | 128 | $1^4.2^2.4^4$ | 11 | 4 | 7 | 7 |
| 4C | 96 | $4^6$ | 3 | 4 | 7 | 7 |
| 5A | 60 | $1^4.5^4$ | 4 | 3 | 1 | 0 |
| 6A | 24 | $1^2.2^2.3^2.6^2$ | 5 | 2 | 4 | 3 |
| 6B | 24 | $6^4$ | 0 | 3 | 3 | 3 |
| 7A | 42 | $1^3.7^3$ | 3 | 0 | 0 | 1 |
| B** | 42 | $1^3.7^3$ | 3 | 0 | 0 | 1 |
| 8A | 16 | $1^2.2.4.8^2$ | 1 | 2 | 1 | 1 |
| 10A | 20 | $2^2.10^2$ | 0 | 3 | 1 | 0 |
| 11A | 11 | $1^2.11^2$ | 0 | 1 | 0 | 0 |
| 12A | 12 | $2.4.6.12$ | 1 | 2 | 2 | 1 |
| 12B | 12 | $12^2$ | 0 | 1 | 1 | 1 |
| 14A | 14 | $1.2.7.14$ | 1 | 0 | 0 | 1 |
| B** | 14 | $1.2.7.14$ | 1 | 0 | 0 | 1 |
| 15A | 15 | $1.3.5.15$ | 1 | 0 | 1 | 0 |
| B** | 15 | $1.3.5.15$ | 1 | 0 | 1 | 0 |
| 21A | 21 | $3.21$ | 0 | 0 | 0 | 1 |
| B** | 21 | $3.21$ | 0 | 0 | 0 | 1 |
| 23A | 23 | $1.23$ | 0 | 0 | 0 | 0 |
| B** | 23 | $1.23$ | 0 | 0 | 0 | 0 |

Figure 12.1 The class list for $M_{24}$ showing some permutation characters

In the case when $G$ acts transitively on $\Omega = \Delta$, this result tells us that $\langle \pi, \pi \rangle$ gives the number of orbits of $G_\omega$ on $\Omega$; this number is known as the *rank* of the permutation action of $G$ on $\Omega$. Clearly, one orbit on $\Omega \times \Omega$ consists of the diagonals

$$D = \{(\omega, \omega) \mid \omega \in \Omega\},$$

and so $G$ is doubly transitive on $\Omega$ if, and only if, this number is 2. In this case we must have that $\pi$ is the sum of the trivial character and an irreducible.

We do not include character tables here but refer the reader to the ATLAS, see Conway et al. (1985). However, in Figure 12.1 we give the class list of $M_{24}$ together with the cycle shapes of elements of each class acting on 24 letters and the permutation characters on the 759 octads, the 1288 dua (pairs of complementary dodecads), the 1771 sextets and the 3795 trios.

## 12.2 Orbitals and Graphs

Suppose now that $G$ acts transitively but not doubly transitively on the set $\Omega$ and that

$$A = \{(\omega, \omega') \mid \omega, \omega' \in \Omega\}$$

is an *orbital* of $G$, that is to say an orbit of $G$ in its action on $\Omega \times \Omega$. Then

$$\bar{A} = \{(\omega', \omega) \mid (\omega, \omega') \in A\}$$

is also an orbital of $G$ and $|A| = |\bar{A}|$. The orbital $A'$ is known as the orbital *paired* with $A$; and if $A = A'$ then we say that $A$ is *self-paired*. We may define a directed graph on $\Omega$ by

$$\omega \to \omega' \iff (\omega, \omega') \in A.$$

If $A = A'$, which will always be the case if there is no other orbital of length $|A|$ (which will hold for the permutation actions of $M_{24}$ that we shall consider here), then this graph becomes undirected. However, before considering in detail these graphs and the wealth of information they contain, we shall investigate the orbitals in greater generality, see for instance Bannai and Ito (1984, page 45) and Cameron (1994, page 63).

We denote the orbitals of our transitive group $G$ by $R_i$ for $i \in [0, 1, \ldots, r-1]$, where $R_0$ denotes the diagonal orbital $\{(\omega, \omega) : \omega \in \Omega\}$. Thus $\langle \pi, \pi \rangle = r$. For convenience we denote the orbital paired with $R_i$ by $R_{i'}$. To each orbital $R_i$ we associate an $n \times n$ matrix $A_i$, whose rows and columns correspond to the points of $\Omega$ such that

$$(A_i)_{\alpha\beta} = \begin{cases} 1 \text{ if } (\alpha, \beta) \in R_i, \\ 0 \text{ otherwise.} \end{cases}$$

These *adjacency matrices* have $|R_i|/n$ 1s in each row and each column; for convenience we denote this number by $n_i$ it being the number of points in the $i$th orbit of the stabilizer of a point in $\Omega$. Moreover, if $R_i$ is self-paired, then the corresponding $A_i$ will be symmetric. We see that $A_0 = I_n$, the $n \times n$ identity matrix, and that $A_{i'} = A_i^t$, the transpose of $A_i$. Since every $(\alpha, \beta)$ lies in one and only one orbital, we see that

$$A_0 + A_1 + \cdots + A_{r-1} = J,$$

the all 1s matrix.

Now suppose that $(\alpha, \beta) \in R_k$ and consider the set

$$Y = \{\gamma \in \Omega \mid (\alpha, \gamma) \in R_i, (\gamma, \beta) \in R_j\}.$$

The cardinality of this set is independent of the choice of $\alpha$ and $\beta$ since if $(\alpha', \beta') \in R_k$, then there exists $g \in G$ such that $\alpha' = \alpha^g, \beta' = \beta^g$; and so $(\alpha^g, \gamma^g) \in R_i$ and $(\gamma^g, \beta^g) \in R_j$ for all $\gamma \in Y$. Denote the number $|Y|$ by $p_{ij}^k$. Then we have

$$A_i A_j = \sum_k p_{ij}^k A_k. \qquad (12.1)$$

Since the positions in which the $A_i$ have their non-zero entries are disjoint, these $r$ matrices are linearly independent when regarded as vectors in an $n^2$-dimensional space, and the above product formula shows that they span an $r$-dimensional subalgebra:

$$\mathcal{U} = \langle A_0, A_1, \ldots, A_{r-1} \rangle.$$

Now let $\rho$ denote the permutation representation that maps each element $g \in G$ to an $n \times n$ permutation matrix whose $\alpha\beta$th entry is $\delta_{\alpha^g, \beta}$ (where $\delta_{ij}$ denotes the Kronecker delta function that takes the value 1 if $i = j$ and 0 otherwise). Let $X = (X_{ij})$ be any complex $n \times n$ matrix. Then

$$(\rho(g)) X (\rho(g)^{-1})_{\alpha\beta} = \sum_{r,s} \delta_{\alpha^g s} X_{st} \delta_{t^{g-1} \beta} = X_{\alpha^g \beta^g}.$$

So $X$ commutes with $\rho(g)$ for all $g \in G$ if, and only if, its entries are constant on orbitals. In other words, if $X$ is a linear combination of the basis elements of $\mathcal{U}$. Thus $\mathcal{U}$ consists of all $n \times n$ matrices that commute with $\rho(g)$ for all $g \in G$; it is the *centralizer algebra* of $\rho$.

Complete reducibility of representations of $G$ ensures that there exists a change of basis, achieved by conjugation by a non-singular matrix $B$, such that

$$B^{-1} \rho(G) B = \begin{bmatrix} I_{e_1} \otimes M_1 & 0 & \cdots & 0 \\ 0 & I_{e_2} \otimes M_2 & \cdots & 0 \\ \vdots & \vdots & \ddots & \vdots \\ 0 & 0 & \cdots & I_{e_k} \otimes M_k \end{bmatrix},$$

where the $M_i$ are the irreducible representations involved in $\rho$, and $M_i$ has dimension $f_i$ and occurs $e_i$ times in the decomposition. The notation $A \otimes B$ where $A$ is an $m \times n$ matrix and $B$ is an $r \times s$ matrix denotes the *Kronecker* or

## 12.2 Orbitals and Graphs

*tensor product* that is defined to be the $mr \times ns$ matrix given by

$$A \otimes B = \begin{bmatrix} a_{11}B & \cdots & a_{1n}B \\ \vdots & \ddots & \vdots \\ a_{m1}B & \cdots & a_{mn}B \end{bmatrix}.$$

So, in the case when $A$ is an $e \times e$ unit matrix, this is simply $e$ copies of $B$ down the main diagonal and 0 matrices elsewhere.

The tensor product has the property that whenever the matrices involved are such that the usual products $AC$ and $BD$ can be formed, then

$$(A \otimes B)(C \otimes D) = AC \otimes BD.$$

In particular

$$(I_e \otimes M)(H \otimes I_f) = H \otimes M = (H \otimes I_f)(I_e \otimes M),$$

when $H$ is any $e \times e$ matrix and $M$ is an $f \times f$ matrix. So the centralizer algebra then consists of all matrices of the form

$$\begin{bmatrix} H_1 \otimes I_{f_1} & 0 & \cdots & 0 \\ 0 & H_2 \otimes I_{f_2} & \cdots & 0 \\ \vdots & \vdots & \ddots & \vdots \\ 0 & 0 & \cdots & H_k \otimes I_{f_k} \end{bmatrix}, \quad (12.2)$$

where $H_i$ is an *arbitrary* $e_i \times e_i$ matrix. Thus $\mathcal{U}$ is isomorphic to a direct sum of complete matrix algebras and we have

$$\dim \mathcal{U} = \sum_1^k e_i^2 = r.$$

Since $G$ is transitive, the trivial representation occurs just once and so $e_1 = f_1 = 1$. But a complete matrix algebra can only be commutative if it has dimension 1; in this case $H_i = (h_i)$ will be a $1 \times 1$ matrix and the term $H_i \otimes I_{f_i} = h_i I_{f_i}$ will be a scalar $f_i \times f_i$ matrix with $h_i$ down the diagonal. Thus, since diagonal matrices commute with one another, we see that

**Theorem 12.4** *The centralizer algebra of a transitive permutation representation $\rho$ of a group $G$ is commutative if, and only if, no irreducible representation occurs more than once in the decomposition of $\rho$ into a sum of irreducibles.*

In this case the permutation representation is said to be *multiplicity free*. We have seen that when the orbitals are all self-paired, we have $A_i = A_i^t$ for all $i = 1, \ldots, r$, and so every matrix in $\mathcal{U}$ is symmetric. Thus

$$A_i A_j = A_i^t A_j^t = (A_j A_i)^t = A_j A_i,$$

and so the centralizer algebra $\mathcal{U}$ is commutative. This will be the case for all the permutation representations of $M_{24}$ that we are considering here, so they will all be multiplicity free. In particular, we shall have $e_i = 1$ for all $i \in [0, \ldots, r-1]$ and the centralizer algebra as in Equation (12.2) will consist of diagonal matrices. Thus the space $V$ decomposes into the sum of $k$ eigenspaces of dimensions $f_1 = 1, f_2, \ldots, f_k$. But we now have

$$\dim \mathcal{U} = \sum_1^k e_i^2 = \sum_1^k 1 = k = r,$$

so the number of constituents in the decomposition is indeed the rank of the permutation representation:

$$\mathcal{U} = W_0 \oplus W_1 \oplus \cdots \oplus W_{r-1}.$$

(Note that, since $W_0$ is the unique trivial representation, we label the $r$ irreducible representations from $1, \ldots, r-1$.) Indeed, since the $A_i$ are commuting, real, symmetric matrices they may be simultaneously diagonalized by an orthogonal matrix; so the spaces $W_i$ are mutually orthogonal to one another under the standard inner product.

## 12.3 Graph Diagrams

In presenting the diagram of one of the associated graphs, as in Figures 12.2, 12.3, 12.4, and 12.5, we shall always choose $R_1$ to be the shortest orbit on $\Omega \times \Omega$ other than $D$. Thus, for $\omega \in \Omega$ we let $\Delta(\omega)$ be the shortest non-trivial orbit of $G_\omega$, the stabilizer in $G$ of $\omega$, on $\Omega$ and join $\omega$ to the points of $\Delta(\omega)$. For example, in the action of $M_{24}$ on the 759 octads, we join each octad to the 30 octads disjoint from it. There are $\binom{8}{4} \times 4 = 280$ octads intersecting our fixed octad in four points, and a further $\binom{8}{2} \times 16 = 448$ intersecting it in two points, the relevant numbers being read off from the Todd triangle Figure 3.1. A diagram of the resulting graph can be seen at the top of Figure 12.2 where we see the orbits of the octad stabilizer (O) and note that, for instance, an octad $W$ intersecting our fixed octad $U$ in four points is disjoint from three octads disjoint from $U$, three octads intersecting $U$ in four points and 24 octads intersecting $U$ in two points. The remaining three diagrams in Figure 12.2 show how the duum stabilizer (D), the sextet stabilizer (S) and the trio stabilizer (T) act on the octads. More detailed descriptions of the various orbits, describing how these four stabilizing subgroups act on one another, are given in Table 12.1. Note that O, D, S and T preserve partitions of the 24 points of $\Omega$ into $8 + 16$, $12^2$, $4^6$, $8^3$ points respectively. In Table 12.1 we see how the members of a particular orbit cut across the partition preserved by the relevant stabilizing subgroup. Thus,

## 12.3 Graph Diagrams

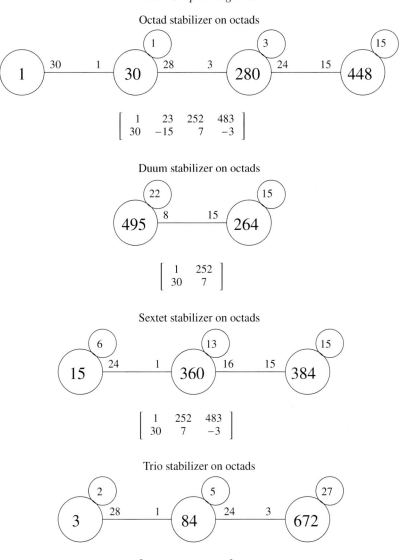

Figure 12.2 The action of some maximal subgroups on octads

for instance, in the 720-orbit of S acting on trios, each of the three octads of the trio has two points in four of the tetrads of the sextet and 0 in the other two.

The intersections in which members of a particular orbit cut the partition

Table 12.1 Orbits of some maximal subgroups of $M_{24}$ acting on the cosets of one another

|   | Octads | Dua | Sextets | Trios |
|---|---|---|---|---|
| O | $8.0$ (1)<br>$0.8$ (30)<br>$4.4$ (280)<br>$2.6$ (448) | $6.6/2.10$ (448)<br>$4.8/8.4$ (840) | $(4.0)^2 (0.4)^4$ (35)<br>$(3.1) (1.3)^5$ (840)<br>$(2.2)^4 (0.4)^2$ (896) | $(8.0) (0.8)^2$ (15)<br>$(4.4)^2 (0.8)$ (420)<br>$(4.4) (2.6)^2$ (3360) |
| D | $4.4$ (495)<br>$2.6$ (264) | $12.0/0.12$ (1)<br>$4.8/8.4$ (495)<br>$6.6/6.6$ (792) | $(2.2)^6$ (396)<br>$(4.0)^2 (2.2)^4$ (495)<br>$(3.1)^3 (1.3)^3$ (880) | $(4.4)^3 r$ (495)<br>$(4.4)^3 nr$ (1320)<br>$(6.2) (2.6) (4.4)$ (1980) |
| S | $4^2.0^4$ (15)<br>$0^2.2^4$ (360)<br>$3.1^5$ (384) | $2^6/2^6$ (288)<br>$4.0.2^4/0.4.2^4$ (360)<br>$3^3.1^3/1^3.3^3$ (640) | $(4.0^5)^6$ (1)<br>$(2^2.0^4)^6$ (90)<br>$(3.1.0^4)^2 (1^4.0^2)^4$ (240)<br>$(2.1^2.0^3)^4 (1^4.0^2)^2$ (1440) | $(4^2.0^4)^3$ (15)<br>$(4^2.0^4) (0^2.2^4)^2$ (180)<br>$(2^4.0^2)^3$ (720)<br>$(3.1^5)^2 (0^2.2^4)$ (2880) |
| T | $8.0^2$ (3)<br>$4^2.0$ (84)<br>$4.2^2$ (672) | $4^3/4^3 r$ (168)<br>$4^3/4^3 nr$ (448)<br>$6.4.2/2.4.6$ (672) | $(4.0^2)^6$ (7)<br>$(4.0^2)^2 (0.2^2)^4$ (84)<br>$(2^2.0)^6$ (336)<br>$(3.1.0) (1.3.0) (1^2,2)^4$ (1344) | $(8.0^2)^3$ (1)<br>$(8.0^2) (0.4^2)^2$ (42)<br>$(4^2.0)^3$ (56)<br>$(4.2^2)^2 (0.4^2)$ (1008)<br>$(4.2^2)^3$ (2688) |

## 12.3 Graph Diagrams

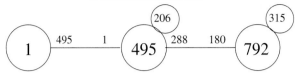

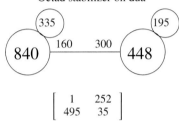

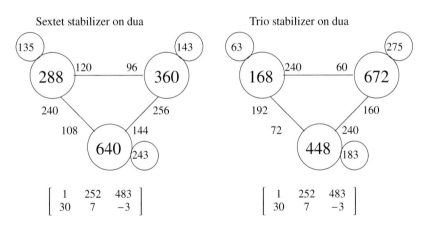

Figure 12.3 The action of some maximal subgroups on dua

define the orbit uniquely except in one case: D acting on trios, and the corresponding T acting on dua. [Note that these correspond to a single orbit in the action on the Cartesian product Dua × Trios.] Our symbols $r$ and $nr$ refer to the fact that in one case the tetrads in which the duum intersects the octads of the trio are refinements of the trio, in the other they are not. Examples of the

## Sextet stabilizer on sextets

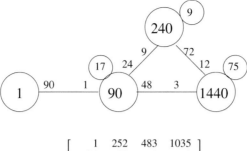

## Octad stabilizer on sextets

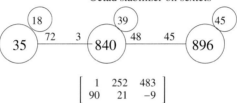

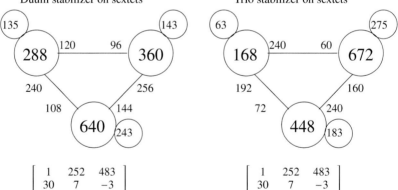

Figure 12.4 The action of some maximal subgroups on sextets

two cases in the MOG trio are respectively

$(r)$ and $(nr)$.

## 12.4 Eigenvalues and Invariant Subspaces

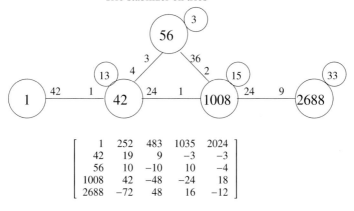

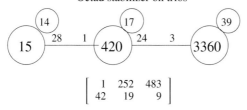

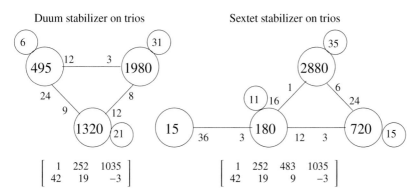

Figure 12.5 The action of some maximal subgroups on trios

Graphs of this nature related to other sporadic simple groups are given in Praeger and Soicher (1997).

## 12.4 Eigenvalues and Invariant Subspaces

Let $G$ act transitively on the set $\Omega$; then we have seen that the number of orbits that $G_\omega$, the stabilizer of a point $\omega \in \Omega$, has on $\Omega$ is given by the inner product of the permutation character with itself, which is to say the *norm* of the permutation character. If this number is $r$ and the permutation representation is multiplicity free, then the permutation module over the complex numbers $\mathbb{C}$ must decompose into the direct sum of $r$ $G$-invariant irreducible submodules. We now explain how the degrees of these irreducible constituents may be deduced from the graphs exhibited in Figures 12.2–12.5. Firstly we define

$$R_i(\omega) = \{\omega' \mid (\omega, \omega') \in R_i\}$$

and fixing $\omega_0 \in \Omega$ we let

$$u_i = \sum_{\omega \in R_i(\omega_0)} \omega;$$

so $u_0 = \omega_0$. We view the matrix $A_i$ as representing a linear transformation from $V$ to $V$ defined by

$$\alpha_i : \omega \mapsto \sum_{\omega' \in R_i(\omega)} \omega';$$

thus $\omega \in \Omega$ is mapped to the sum of those points that are joined to $\omega$ in the $i$th graph. Then, as with the matrices,

$$\alpha_i \alpha_j = \sum_{k=0}^{r-1} p_{ij}^k \alpha_k.$$

We consider the action of the $\alpha_i$ on $W = \langle u_0, u_1, \ldots, u_{r-1} \rangle$, the $r$-dimensional subspace of $V$ spanned by the $u_i$, thus

$$\alpha_1(u_0) = u_1; \quad \alpha_1(u_1) = p_{11}^0 u_0 + p_{11}^1 u_1 + \sum_{k>1} p_{11}^k u_k.$$

And more generally, since $p_{01}^i = p_{i1}^0 = 0$ for $i \neq 1$, $p_{01}^1 = n_1$ and $p_{11}^0 = 1$, we have

$$\alpha_1(u_i) = \sum_k p_{i1}^k u_k.$$

Thus the $r \times r$ matrix $P_1$ whose $ij$th entry is $p_{i1}^j$ gives the action of $\alpha_1$ on $W$, and the $r$ eigenvalues of $P_1$ will be eigenvalues for the action of $\alpha_1$ on $V$.

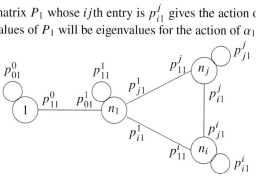

## 12.4 Eigenvalues and Invariant Subspaces

We have concentrated on $\alpha_1$ simply because the smallest non-diagonal orbital $R_1$ gives the graph of smallest valency. In the general case, using the orbital $R_k$, the parameters related to two suborbits are as indicated in

Since the number of edges from the $i$th suborbit to the $j$th equals the number of edges from the $j$th suborbit to the $i$th, we have

$$n_i p^i_{jk} = n_j p^j_{ik} \text{ for all } i, j, k,$$

and

$$\sum_i p^j_{ik} = n_k \text{ the valency of the } k\text{th graph.}$$

The action of $\alpha_k$ is given by

$$\alpha_k(u_i) = \sum_j p^j_{ik} u_j,$$

and the matrix $P_k$ whose $ij$th entry is $p^j_{ik}$ give the eigenvalues of $\alpha_k$. Since the $\alpha_i$ are commuting linear transformations they have the same eigenspaces, although their eigenvalues will in general be different. We demonstrate these ideas by consideration of $M_{24}$ acting on the 759 octads and choose the minimal graph in which octads are joined if they are disjoint. A diagram of this graph is shown at the top of Figure 12.2. We have

$$P_1 = \begin{pmatrix} p^0_{01} & p^1_{01} & p^2_{01} & p^2_{01} \\ p^0_{11} & p^1_{11} & p^2_{11} & p^2_{11} \\ p^0_{21} & p^1_{21} & p^2_{21} & p^2_{21} \\ p^0_{31} & p^1_{31} & p^2_{31} & p^2_{31} \end{pmatrix} = \begin{pmatrix} 0 & 1 & 0 & 0 \\ 30 & 1 & 3 & 0 \\ 0 & 28 & 3 & 15 \\ 0 & 0 & 24 & 15 \end{pmatrix},$$

from which we can deduce that the eigenvalues of $A_1$ and $\alpha_1$ are 30, −3, −15 and 7. We do not yet know the dimensions of the associated eigenspaces, which are of course the degrees of the irreducible representations in the permutation representation. However, we do know that the trivial character occurs once in the representation and that

$$\alpha_1 : \sum_{\omega \in \Omega} \omega \mapsto n_1 \left( \sum_{\omega \in \Omega} \omega \right).$$

In this case $n_1 = 30$ and so the eigenspace corresponding to 30 has dimension 1. Similarly $A_2$ and $A_3$ have 1-dimensional eigenspaces corresponding to eigenvalues 280 and 448, respectively. If $f_1 = 1, f_2, f_3, f_4$ are the dimensions of the eigenspaces, then we certainly have

$$f_1 + f_2 + f_3 + f_4 = 759.$$

But the matrix $A_1$ has zeros on the main diagonal and so when diagonalized it still has trace 0. Thus

$$30 f_1 + (-3) f_2 + (-15) f_3 + (7) f_4 = 30 - 3 f_2 - 15 f_3 + 7 f_4 = 0.$$

The same is true for $A_2$ and $A_3$ and so, if we obtain their eigenvalues we shall have four linear equations in $f_2, f_3$ and $f_4$. We can facilitate this process by making use of the matrix product formula in Equation (12.1). Thus we have

$$A_i A_1 = \sum p_{i1}^k A_k$$

and from the graph diagram we can read off

$$\begin{aligned} A_1^2 &= 30I + A_1 + 3A_2, \\ A_2 A_1 &= 28 A_1 + 3 A_2 + 15 A_3, \\ A_3 A_1 &= 24 A_2 + 15 A_3. \end{aligned}$$

From which we deduce that

$$\begin{aligned} A_1^3 &= 30 A_1 + A_1^2 + 3(28 A_1 + 3 A_2 + 15 A_3) \\ &= A_1^2 + 114 A_1 + 3(A_1^2 - 30 I - A_1) + 45 A_3 \\ &= 4 A_1^2 + 111 A_1 - 90 I + 45 A_3. \end{aligned}$$

These give expressions for $A_2$ and $A_3$ as polynomials in $A_1$ and so, if $\lambda$ is an eigenvalue for $A_1$, then

$$\frac{1}{3}(\lambda^2 - \lambda - 30) \text{ and } \frac{1}{45}(\lambda^3 - 4\lambda^2 - 111\lambda + 90)$$

are the corresponding eigenvalues for $A_2$ and $A_3$, respectively. We thus obtain all eigenvalues:

$$\begin{array}{c|cccc} A_0 & 1 & 1 & 1 & 1 \\ A_1 & 30 & -3 & 7 & -15 \\ A_2 & 280 & -6 & 4 & 70 \\ A_3 & 448 & 8 & -12 & -56 \end{array}. \qquad (12.3)$$

Indeed, acting $A_1$ once more on the expression for $A_3$ above we may obtain a quartic in $A_1$, namely

$$A_1^4 - 19 A_1^3 - 171 A_1^2 + 2115 A_1 + 30.315 = (A_1 - 30)(A_1 + 3)(A_1 - 7)(A_1 + 15)$$

## 12.4 Eigenvalues and Invariant Subspaces

whose roots are the four eigenvalues of $A_1$. Now if $\mathbf{f} = (f_0, f_1, f_2, f_3)^t$ is the column vector whose entries are the degrees of the irreducible constituents of the permutation representation on octads, then $P$, the $4 \times 4$ matrix in Equation (12.3), satisfies

$$P\mathbf{f} = (759, 0, 0, 0)^t = \mathbf{a}, \text{ say,}$$

and so

$$\mathbf{f} = P^{-1}\mathbf{a} = \frac{\operatorname{adj}P}{|P|}\mathbf{a} = \frac{1}{|P|}\begin{pmatrix} |Q_{00}| & \cdot & \cdot & \cdot \\ -|Q_{10}| & \cdot & \cdot & \cdot \\ |Q_{20}| & \cdot & \cdot & \cdot \\ -|Q_{30}| & \cdot & \cdot & \cdot \end{pmatrix}\begin{pmatrix} 759 \\ 0 \\ 0 \\ 0 \end{pmatrix},$$

where $Q_{ij}$ is the $3 \times 3$ matrix obtained by removing the $i$th row and the $j$th column from $P$. Hence

$$f_i = (-1)^i 759 |Q_{i0}|/|P|.$$

We find $|P| = -759.880$, $|Q_{00}| = -880$, $|Q_{10}| = 880.483$, $|Q_{20}| = -880.252$ and $|Q_{30}| = 880.23$ and so $\mathbf{f} = (1\,48\,325\,223)^t$.

A delightful alternative formula for these degrees, which involves left and right eigen*vectors* for the same eigenvalue, is given in Cameron (1999, page 74). Thus if $p_i(j)$ denotes the eigenvalue of $A_i$ on the common eigenspace $W_j$ then $f_j$ the dimension of $W_j$, is given by

$$f_j = \frac{n}{\sum_i |p_i(j)|^2/n_i},$$

where $n_i$ is the valency of the graph afforded by $A_i$. Thus, in the above example using the eigenvalues given in Equation (12.3), we have

$$f_2 = \frac{759}{1 + \frac{(-3)^2}{30} + \frac{(-6)^2}{280} + \frac{8^2}{448}} = \frac{759.70}{70 + 21 + 9 + 10} = 483,$$

and similarly for $f_3$ and $f_4$. A proof of this formula is given in Cameron and van Lint (1991, page 202).

As we have seen, when the group $G$ acts transitively on $\Omega$ and on $\Delta$ then the number of orbits of $G_\omega$, the stabilizer of a point $\omega \in \Omega$ on $\Delta$ equals the number of orbits of $G_\delta$, the stabilizer of a point $\delta \in \Delta$ on $\Omega$, and this number equals the inner product of the associated permutation characters. In the multiplicity-free case with which we are dealing, this number is the number of irreducible characters the two permutation characters have in common. As an example of how this works, consider the action of the sextet stabilizer on octads, the diagram of whose graph is given in Figure 12.2. If we let $w_1$ be the sum of the terms in the 15-orbit, $w_2$ the sum of terms in the 360-orbit and $w_3$ be the sum of terms in the 384-orbit, then we have

$$\alpha_1(w_1) = 6w_1 + w_2,$$
$$\alpha_1(w_2) = 24w_1 + 13w_2 + 15w_3,$$
$$\alpha_1(w_3) = 16w_2 + 15w_3,$$

giving the matrix

$$\begin{pmatrix} 6 & 1 & 0 \\ 24 & 13 & 15 \\ 0 & 16 & 15 \end{pmatrix}$$

whose eigenvectors are eigenvectors of $\alpha_1$. One eigenvalue is, of course, $n_1 = 30$, the valency of the graph, and we may calculate the other two to be 7 and $-3$. Thus the two permutation representations have the irreducible 252-dimensional representation and the irreducible 483-dimensional representation in common. The following mnemonic often facilitates calculations involving the parameters of our graph diagrams.

### 12.4.1 Calculating the Eigenvalues of $P_i$

The following approach was helpful in working out the eigenvalues listed in Figures 12.2–12.5. It uses the way the well-known determinant formula

$$|A| = |(a_{ij})| = \sum_{\sigma \in S_n} \text{sgn}(\sigma) a_{1\sigma(1)} a_{2\sigma(2)} \cdots a_{n\sigma(n)},$$

where $S_n$ denotes the symmetric group n n letters and $\text{sgn}(\sigma) = +1$ if $\sigma$ is even and $-1$ if $\sigma$ is odd, relates to the $r \times r$ matrices $P_k$ whose entries are parameters in the graph diagrams. Thus we consider the eigenvalue equation

$$|P_k - xI_r| = \begin{vmatrix} p^0_{0k} - x & p^1_{0k} & \cdots & p^i_{0k} & \cdots & p^j_{0k} & \cdots & p^{r-1}_{0k} \\ p^0_{1k} & p^1_{1k} - x & \cdots & p^i_{1k} & \cdots & p^j_{1k} & \cdots & p^{r-1}_{1k} \\ \vdots & \vdots & \ddots & \vdots & \ddots & \vdots & \ddots & \vdots \\ p^0_{ik} & p^1_{ik} & \cdots & p^i_{ik} - x & \cdots & p^j_{ik} & \cdots & p^{r-1}_{ik} \\ \vdots & \vdots & \ddots & \vdots & \ddots & \vdots & \ddots & \vdots \\ p^0_{jk} & p^1_{jk} & \cdots & p^i_{jk} & \cdots & p^j_{jk} - x & \cdots & p^{r-1}_{jk} \\ \vdots & \vdots & \ddots & \vdots & \ddots & \vdots & \ddots & \vdots \\ p^0_{(r-1)k} & p^1_{(r-1)k} & \cdots & p^i_{(r-1)k} & \cdots & p^j_{(r-1)k} & \cdots & p^{r-1}_{(r-1)k} - x \end{vmatrix}$$

$= 0.$

## 12.4 Eigenvalues and Invariant Subspaces

In the resulting $r$th degree equation in $x$, the sum of the roots is the trace of the matrix, which is to say the sum of the *loops* in the $R_k$-graph. We, of course, have given the $R_1$-graph in our diagrams. We think of this as the number of ways in which you can go from a point in a suborbit back to the same suborbit, a cycle of length 1. Thus in the rank 4 action of the octad stabilizer on octads, if the eigenvalues are 30, $\lambda$, $\mu$, $\nu$ then

$$30 + \lambda + \mu + \nu = 1 + 3 + 15 = 19.$$

The sum of the products of roots in pairs is then given by

$$\sum \lambda\mu = \sum_{i<j}(p^j_{ik}p^j_{jk} - p^i_{ik}p^i_{jk}),$$

the first terms of which correspond to elements of shape $1^2$ in the symmetric group $S_2$ (go from a point in suborbit $i$ back to suborbit $i$ and from suborbit $j$ back to suborbit $j$); the second terms correspond to a 2-cycle in $S_2$ (the number of ways you can go from a point of suborbit $i$ to suborbit $j$ and back again). The second type are, of course, odd permutations and so they carry a minus sign. In the same graph we thus have

$$30(\lambda+\mu+\nu)+\mu\nu+\nu\lambda+\lambda\mu = (3.15+15.1+1.3)-(30.1+28.3+24.15) = -411.$$

The sum of products of roots in threes is then given by:

$$\sum_{\lambda,\mu,\nu} \lambda\mu\nu = \sum_{\substack{r,s,t \\ \text{distinct}}} \begin{vmatrix} p^r_{rk} & p^s_{rk} & p^t_{rk} \\ p^r_{sk} & p^s_{sk} & p^t_{sk} \\ p^r_{tk} & p^s_{tk} & p^t_{tk} \end{vmatrix}$$

$$= \sum_{\substack{r,s,t \\ \text{distinct}}} p^r_{rk}p^s_{sk}p^t_{tk} - \sum_{\substack{r,s,t \\ \text{distinct}}} p^r_{rk}p^t_{sk}p^s_{tk} + \sum_{\substack{r,s,t \\ \text{distinct}}} p^s_{rk}p^t_{sk}p^r_{tk}.$$

The first terms correspond to the element $1^3$ in $S_3$; the second terms correspond to transpositions $1.2$ in $S_3$, which are again odd permutations and so carry a negative sign; and the third terms correspond to 3-cycles $r \mapsto s \mapsto t \mapsto r$. Note that we must also count $r \mapsto t \mapsto s \mapsto r$. So with the octad graph we have

$$30(\mu\nu + \nu\lambda + \lambda\mu) + \lambda\mu\nu = 1.3.15 - (1.24.15 + 3.30.1 + 15.30.1 + 15.28.3)$$
$$= -2115,$$

there being no triangles in the diagram.

The other symmetric functions follow the same pattern. Thus the product of all four eigenvalues in our example is given by

$$30\lambda\mu\nu = 30.1.24.15 - 30.1.3.15 = 30.315,$$

which correspond to permutations in $S_4$ of shapes $2^2$ and $1^2.2$, all other shapes contributing 0. Thus the eigenvalues $\lambda, \mu, \nu$ are the roots of

$$x^3 + 11x^2 - 81x - 315 = (x+15)(x+3)(x-7) = 0.$$

Again, consider the action of the sextet stabilizer on octads whose diagram also appears in Figure 12.2. If the three eigenvalues here are 30, $\lambda$ and $\mu$, then $\lambda+\mu = (6+13+15)-30$ and $30\lambda\nu = (6.13.15-6.16.15-15.24.1) = -30.21$, and so $\lambda$ and $\mu$ are roots of the quadratic

$$x^2 - 4x - 21 = (x-7)(x+3).$$

Thus the two non-trivial irreducibles that the permutation character of the sextet stabilizer has in common with that of the octad stabilizer are those corresponding to eigenvalues 7 and $-3$, namely those of degrees 252 and 483.

### The Rank 3 Case

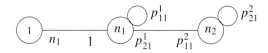

In the rank 3 case let the three eigenvalues be $n_1, \lambda$ and $\mu$. Then we immediately have that

$$\lambda + \mu = p_{11}^1 + p_{21}^2 - n_1$$

and

$$\lambda\mu = -p_{21}^2,$$

so we can essentially read them off from the diagram. But if $r$ and $s$ are the degrees of the eigenspaces, we have

$$1 + r + s = n = 1 + n_1 + n_2 \text{ and } n_1.1 + \lambda r + \mu s = 0,$$

the second equation being the trace of $A_1$. As an example of this, consider the action on dua whose diagram appears in Figure 12.3. We have that $\lambda$ and $\mu$ are roots of

$$x^2 - (206 + 315 - 495)x - 315 = x^2 - 26x - 315 = (x-35)(x+9) = 0,$$

and

$$1 + r + s = 1288, \quad 495.1 + 35r - 9s = 0,$$

## 12.4 Eigenvalues and Invariant Subspaces

giving $r = 252$ and $s = 1035$. This process is particularly useful for working out the sum of the roots and the product of the roots, where very few terms of the symmetric group $S_r$ contribute. We illustrate this with the rank 5 graph of the action of $M_{24}$ on trios:

**Example 12.5** Obtain the eigenvalues and dimensions of the eigenspaces for the action of $M_{24}$ on trios.

*Solution* Let the eigenvalues be 42, $\lambda$, $\mu$, $\nu$, $\rho$. In the following calculations the summations are over the four unknown eigenvalues, thus

$$42 + \lambda + \mu + \nu + \rho = 64 \Rightarrow \sum \lambda = 22,$$

and

$$42\left(\sum \lambda\right) + \sum \lambda\mu = 936 \Rightarrow \sum \lambda\mu = 936 - 42.22 = 12.$$

Further,

$$42\left(\sum \lambda\mu\right) + \sum \lambda\mu\nu = -270 \Rightarrow \sum \lambda\mu\nu = -270 - 42.12 = -774.$$

The arithmetic has been omitted in these calculations but in the calculation of the product of the roots we give all details. We note that of the cycle shapes $1^5$, $1^3.2$, $1^2.3$, $1.4$, $1.2^2$, $5$, $2.3$ only $1^3.2$ and $1.2^2$ contribute and we have

$$42\lambda\mu\nu\rho = -42.1.3.15.33 + 3.42.1.24.9 + 33.42.1.36.2 = 42.9.9.19.$$

Thus $\lambda, \mu, \nu, \rho$ are the roots of

$$x^4 - 22x^3 + 12x^2 + 774x + 1539 = (x - 19)(x - 9)(x + 3)^2,$$

giving eigenvalues for the $R_1$ graph of 42, 19, 9, and $-3$ repeated. In the notation used above, these are the eigenvalues $p_i(1)$; that is to say the eigenvalues of $A_1$ acting on the eigenspace $W_i$ for $i = 0, 1, \ldots, 4$.

Since the $A_j$ have common eigenspaces, we may read off the relationship between the various eigenvalues for the same eigenspace from the graph. To ease notation for now only we let $\lambda_i$ denote $p_i(k)$, the eigenvalue of $A_i$ acting on the fixed eigenspace $W_k$. Then $A_1^2 = 42I + 13A_1 + 3A_2 + A_3$ tells us that $\lambda_1^2 = 42 + 13\lambda_1 + 3\lambda_2 + \lambda_3$. Similarly we have

$$\begin{aligned}
\lambda_1\lambda_2 &= 4\lambda_1 + 3\lambda_2 + 2\lambda_3, \\
\lambda_1\lambda_3 &= 24\lambda_1 + 36\lambda_2 + 15\lambda_3 + 9\lambda_4, \\
\lambda_1\lambda_4 &= 24\lambda_3 + 33\lambda_4.
\end{aligned}$$

Thus

$$\lambda_2 = (2\lambda_1^2 - 22\lambda_1 - 84)/(3 + \lambda_1),$$
$$\lambda_3 = \lambda_1^2 - 13\lambda_1 - 3\lambda_2 - 42,$$
$$\lambda_4 = 24\lambda_3/(\lambda_1 - 33).$$

These equations enable us to complete the full eigenvalue table $(p_i(j))$ below apart from the $\lambda_3$ and $\lambda_4$ entries when $\lambda_1 = -3$. However, the map $A_0 + A_1 + A_2 + A_3 + A_4 = J$, the all 1s matrix whose nullity is $n - 1$. It maps $W_0$ to itself and since $W_0^\perp = W_1 + W_2 + W_3 + W_4$ is preserved by $J$, we have

$$\sum_k p_i(k) = 0$$

for $i > 0$. We thus obtain

| $p_i(j)$ | $A_0$ | $A_1$ | $A_2$ | $A_3$ | $A_4$ | $f_i$ |
|---|---|---|---|---|---|---|
| $W_0$ | 1 | 42 | 56 | 1008 | 2688 | 1 |
| $W_1$ | 1 | 19 | 10 | 42 | $-72$ | 252 |
| $W_2$ | 1 | 9 | $-10$ | $-48$ | 48 | 483 |
| $W_3$ | 1 | $-3$ | 10 | $-24$ | 16 | 1035 |
| $W_4$ | 1 | $-3$ | $-4$ | 18 | $-12$ | 2024 |

where the degrees $f_i$ have been worked out either by solving the linear equations or by using the Cameron–van Lint formula given above. □

# 13

# Natural Generators of the Mathieu Groups

The approach of this book has been to construct the Steiner system $S(5, 8, 24)$ on a 24-element set $\Omega$, prove that it is unique up to re-labelling of its points and define the Mathieu group $M_{24}$ to be the set of all permutations of $\Omega$ that preserve the system. Our construction enabled us to deduce the order of $M_{24}$ and that it acts quintuply transitively on $\Omega$.

Whilst we firmly believe that the best way to understand and study this remarkable group is through its action on the Steiner system, an alternative approach is to obtain the group directly by writing down permutations generating it, and then deduce the Steiner system as the fixed point sets of a class of involutions. What follows is a delightfully natural way of writing down such a set of generators.

It turns out that the Octern subgroup On, the maximal subgroup of $M_{24}$ isomorphic to the linear group $L_2(7)$, affords such an approach.

Now $L_2(7) \cong L_3(2)$ acts on the seven points and the seven lines of the Fano plane, and both the stabilizer of a point and the stabilizer of a line are subgroups isomorphic to $S_4$. However, these copies of $S_4$ are *not* conjugate within $L_2(7)$. [They are, of course, conjugate within Aut $L_3(2) \cong PGL_2(7)$ when points and lines are interchanged.] Indeed, the behaviour within $M_{24}$ of representatives of these two classes is completely different: one class has trivial centralizer whilst the other centralizes an involution of $M_{24}$. This is demonstrated by the following self-explanatory MAGMA code that takes the generators of the Octern group as given in Figure 7.3. We have identified two non-conjugate copies of the Klein fourgroup $V_4$ within the centralizer of an involution in On, and taken their normalizers.

```
> s24:=Sym(24);
> m24:=sub<s24|
> (1,2,3,4,5,6,7,8,9,10,11,12,13,14,15,16,17,18,19,20,21,22,23),
> (24,23)(3,19)(6,15)(9,5)(11,1)(4,20)(16,14)(13,21)>;
```

```
> #m24;
244823040
> e2:=m24!
> (24,20)(23,11)(8,22)(3,15)(13,16)(10,18)(2,9)(1,5)
> (4,17)(19,12)(6,14)(7,21);
> e7:=m24!(23,3,6,19,9,5,15)(1,4,16,20,13,21,14)(2,18,8,10,12,7,17);
> oct:=sub<m24|e7,e2>;
> #oct;
168
> ce2:=Centralizer(oct,e2);
> s4subs := {};
> for tt in oct do
for> if Order(tt) eq 3 and #sub<oct|ce2,tt> eq 24 then
for|if> s4subs := s4subs join {sub<oct|ce2,tt>};
for|if> end if;
for> end for;
> #s4subs;
2
> s4subss:=Setseq(s4subs);
> [#Centralizer(m24,s4subss[i]):i in [1..2]];
[ 2, 1 ]
> s4:=s4subss[1];
> t0:=Centralizer(m24,s4).1;
> #(t0^oct);
7
> #sub<m24|(t0)^oct>;
244823040
```

The resulting involution $t_0$ must have seven images, $\mathcal{T} = \{t_0, t_1, \ldots, t_6\}$, under conjugation by On and if we let $H = \langle \mathcal{T} \rangle$ then we see that

$$N_{M_{24}}(H) \geq \langle \text{On}, t_0 \rangle.$$

But On is maximal in $M_{24}$ and $t_0 \notin$ On, so $1 \neq H$ must be normal in the simple group $M_{24}$. Thus $H = M_{24}$ and $\mathcal{T}$ is a set of generators.

We now give a number of ways in which the generators $t_i$ may be written down as permutations on 24 letters, using just a knowledge of $L_2(7)$.

## 13.1 The Combinatorial Approach

The linear group $L \cong L_3(2)$ of order 168 permutes the seven non-zero vectors of a 3-dimensional vector space over $\mathbb{Z}_2$ that form the points of the *Fano plane*, see Figures 2.1 and 2.2. It possesses two classes of 7-cycles, each containing 24 elements; we take the conjugacy class containing the 7-cycle (0 1 2 3 4 5 6) as our set $\Omega$, and so

$$\Omega = \{(a_0\ a_1\ a_2\ a_3\ a_4\ a_5\ a_6) \mid \{a_i, a_{i+1}, a_{i+3}\} \in \mathcal{L} \text{ for } i = 0, 1, \ldots, 6\},$$

## 13.1 The Combinatorial Approach

where $\mathcal{L}$ denotes the set of lines and the subscripts are to be read modulo 7. These 7-cycles are displayed in Table 13.1. Clearly $L$ acts transitively on this set; it will turn out to be a copy of the Octern group in the copy of $M_{24}$ that we are about to construct. Indeed, we can already see that an element of order three with which $L$ commutes is the squaring map:

$$\xi : w \mapsto w^2 \text{ for } w \in \Omega.$$

The element of $L_3(2)$ fixing the columns and the top row of Figure 13.1 whilst cycling the other rows downwards corresponds to $(0\ 1\ 2\ 3\ 4\ 5\ 6)^3 = (0\ 3\ 6\ 2\ 5\ 1\ 4)$ and the element from Figure 7.3 that acts as $(\infty\ 0)(1\ 6)(2\ 3)(4\ 5)$ on the eight terns corresponds to $(1\ 4)(3\ 5)$ here. We now write all the elements of $\Omega$ so that they start with 0 as in Table 13.1 and define

$$t_0 : (0\ u\ v\ w\ x\ y\ z) \mapsto (0\ u\ v\ w\ x\ y\ z)^{(v\ z)(x\ y)} = (0\ u\ z\ w\ y\ x\ v).$$

Note that the permutation $(v\ z)(x\ y)$ is in $L$ and so this does indeed define a one-to-one map from $\Omega$ onto $\Omega$, a permutation of cycle shape $2^{12}$. Do note though that the element we are conjugating by depends on the 7-cycle we are conjugating, so it is certainly not an element of $L_3(2)$. We may now repeat the process to obtain a permutation $t_1$ by starting the 24 7-cycles of $\Omega$ with 1 rather than 0 and conjugating by the equivalent permutation $(v\ z)(x\ y)$. In this manner we obtain seven elements $t_0, \ldots, t_6$ of cycle shape $2^{12}$ that are displayed in Table 13.2.

*These seven involutions generate the Mathieu group $M_{24}$*

The curious reader will wonder why we have chosen to conjugate these 7-cycles of $\Omega$ by the permutation $(v\ z)(x\ y)$. The short answer is so as to get the familiar copy of $M_{24}$ with which we are working throughout this book. However, we could equally well have chosen to conjugate $(0\ u\ v\ w\ x\ y\ z)$ by $(u\ w)(x\ y)$ or $(u\ w)(v\ z)$, these being the other two involutions that fix every line through 0 in the Fano plane Figure 2.2. In this way we should obtain two further copies of $M_{24}$. Indeed, there are three further ways of obtaining copies of $M_{24}$ containing the particular Octern group $L_3(2)$ with which we have started. These correspond not to the points of the Fano plane, but to its lines. Explicitly we choose a line, $\{1, 2, 4\}$ say, and write every 7-cycle of $\Omega$ so that $\{1, 2, 4\}$ appear in the 1st, 2nd and 4th positions, in some order. Thus every 7-cycle will appear as

$$(\alpha\ \beta\ w\ \gamma\ x\ y\ z) \text{ where } \{\alpha, \beta, \gamma\} = \{1, 2, 4\}.$$

We then conjugate this element by one of $(w\ x)(y\ z)$, $(w\ y)(x\ z)$ or $(w\ z)(x\ y)$ but, of course, we must always conjugate by the same one. In this way we obtain three more fixed point free involutions that we can label $[124]_1$, $[124]_2$ and

Table 13.1 Labelling of the 24 7-cycles in an orbit of $L_3(2)$ with the points of $\Omega$

| | | | | | |
|---|---|---|---|---|---|
| ∞  | (0 1 2 3 4 5 6) | 11 | (0 2 4 6 1 3 5) | 22 | (0 4 1 5 2 6 3) |
| 0  | (0 2 1 6 4 5 3) | 20 | (0 4 2 5 1 3 6) | 8  | (0 1 4 3 2 6 5) |
| 15 | (0 1 6 3 5 4 2) | 16 | (0 5 1 4 6 2 3) | 18 | (0 6 5 2 1 3 4) |
| 5  | (0 5 3 4 2 6 1) | 4  | (0 2 5 6 3 1 4) | 2  | (0 3 2 1 5 4 6) |
| 9  | (0 5 2 4 3 1 6) | 1  | (0 3 5 1 2 6 4) | 17 | (0 2 3 6 5 4 1) |
| 19 | (0 6 4 2 3 1 5) | 14 | (0 3 6 1 4 5 2) | 7  | (0 4 3 5 6 2 1) |
| 6  | (0 5 6 4 1 3 2) | 21 | (0 1 5 3 6 2 4) | 12 | (0 6 1 2 5 4 3) |
| 3  | (0 4 6 5 3 1 2) | 13 | (0 3 4 1 6 2 5) | 10 | (0 6 3 2 4 5 1) |

$[124]_3$. For a fixed subscript, the seven elements obtained, corresponding to the seven lines, generate a further copy of $M_{24}$.

In this manner we have obtained six copies of $M_{24}$ containing the original copy of $L_3(2)$ acting transitively on 24 points. At the end of this chapter we shall show that these are the only copies of $M_{24}$ that contain this copy of $L_3(2)$ as a maximal subgroup.

Table 13.2 The seven involutory generators for $M_{24}$ acting on 7-cycles in $L_3(2)$

| | |
|---|---|
| $t_1$ | (1 12)(2 22)(3 24)(4 8)(5 10)(6 18)(7 17)(9 15)(11 20)(13 23)(14 16)(19 21) |
| $t_2$ | (1 5)(2 18)(3 15)(4 13)(6 14)(7 23)(8 22)(9 10)(11 21)(12 20)(16 17)(19 24) |
| $t_3$ | (1 11)(2 6)(3 19)(4 9)(5 24)(7 15)(8 10)(12 22)(13 18)(14 20)(16 23)(17 21) |
| $t_4$ | (1 18)(2 3)(4 21)(5 19)(6 13)(7 12)(8 9)(10 14)(11 16)(15 20)(17 22)(23 24) |
| $t_5$ | (1 20)(2 17)(3 21)(4 7)(5 23)(6 24)(8 19)(9 14)(10 16)(11 13)(12 15)(18 22) |
| $t_6$ | (1 19)(2 20)(3 17)(4 15)(5 12)(6 23)(7 13)(8 18)(9 24)(10 22)(11 14)(16 21) |
| $t_0$ | (1 13)(2 14)(3 20)(4 11)(5 16)(6 9)(7 22)(8 21)(10 12)(15 24)(17 23)(18 19) |

**Carrying this out in MAGMA**

It is a possibly therapeutic and not too time-consuming process to work out the action of these generators on $\Omega$ by hand. However, it is interesting and informative to see how this can be carried out in MAGMA. Note that as usual in the MAGMA-code below 0 has been replaced by 23 and ∞ by 24 in the action on $\Omega$, and 0 has been replaced by 7 as a point of the Fano plane. The sequence $ts$ lists the 7-cycles of $\Omega$ in numerical order, rather than the order in Table 13.1 where the 7-cycles are arranged so as to facilitate reading off generators of the Octern group. The function $f$ conjugates by $(v\ z)(x\ y)$ and the seven generators

## 13.1 The Combinatorial Approach

are listed in $gens = [t_1, t_2, \ldots, t_0]$. Note that we use the fact that

$$(v\ z)(x\ y) : a = (i\ u\ v\ w\ x\ y\ z) \mapsto (i\ u\ z\ w\ y\ x\ v) = (i\ i^a\ i^{a^6}\ i^{a^3}\ i^{a^5}\ i^{a^4}\ i^{a^2})$$

in defining the function $f$ below. In this code, and in subsequent calculations for $M_{12}$, permutations are first obtained as sequences of length $n$ with the image of $i$ in the $i$th position; the sequence is then coerced into membership of $S_n$ as a product of disjoint cycles.

```
> s7:=Sym(7);
> ts:=[s7|
> (1, 2, 6, 4, 7, 3, 5), (1, 5, 4, 6, 7, 3, 2),
> (1, 2, 7, 4, 6, 5, 3), (1, 4, 7, 2, 5, 6, 3),
> (1, 7, 5, 3, 4, 2, 6), (1, 3, 2, 7, 5, 6, 4),
> (1, 7, 4, 3, 5, 6, 2), (1, 4, 3, 2, 6, 5, 7),
> (1, 6, 7, 5, 2, 4, 3), (1, 7, 6, 3, 2, 4, 5),
> (1, 3, 5, 7, 2, 4, 6), (1, 2, 5, 4, 3, 7, 6),
> (1, 6, 2, 5, 7, 3, 4), (1, 4, 5, 2, 7, 3, 6),
> (1, 6, 3, 5, 4, 2, 7), (1, 4, 6, 2, 3, 7, 5),
> (1, 7, 2, 3, 6, 5, 4), (1, 3, 4, 7, 6, 5, 2),
> (1, 5, 7, 6, 4, 2, 3), (1, 3, 6, 7, 4, 2, 5),
> (1, 5, 3, 6, 2, 4, 7), (1, 5, 2, 6, 3, 7, 4),
> (1, 6, 4, 5, 3, 7, 2), (1, 2, 3, 4, 5, 6, 7)];
> f:=func<i,tt|s7!
> (i,i^tt,i^(tt^6),i^(tt^3),i^(tt^5),i^(tt^4),i^(tt^2))>;
> s24:=Sym(24);
> gens:=[Id(s24):i in [1..7]];
> for i in [1..7] do
for> test:=[1:i in [1..24]];
for> for j in [1..24] do
for|for> for k in [1..24] do
for|for|for> if f(i,ts[j]) eq ts[k] then
for|for|for|if> test[j] := k;
for|for|for|if> end if;
for|for|for> end for;
for|for> end for;
for> gens[i]:=s24!test;
for> end for;
> gens;
[
(1, 12)(2, 22)(3, 24)(4, 8)(5, 10)(6, 18)
(7, 17)(9, 15)(11, 20)(13, 23)(14, 16)(19, 21),

(1, 5)(2, 18)(3, 15)(4, 13)(6, 14)(7, 23)
(8, 22)(9, 10)(11, 21)(12, 20)(16, 17)(19, 24),

(1, 11)(2, 6)(3, 19)(4, 9)(5, 24)(7, 15)
(8, 10)(12, 22)(13, 18)(14, 20)(16, 23)(17, 21),

(1, 18)(2, 3)(4, 21)(5, 19)(6, 13)(7, 12)
(8, 9)(10, 14)(11, 16)(15, 20)(17, 22)(23, 24),
```

```
 (1, 20)(2, 17)(3, 21)(4, 7)(5, 23)(6, 24)
 (8, 19)(9, 14)(10, 16)(11, 13)(12, 15)(18, 22),

 (1, 19)(2, 20)(3, 17)(4, 15)(5, 12)(6, 23)
 (7, 13)(8, 18)(9, 24)(10, 22)(11, 14)(16, 21),

 (1, 13)(2, 14)(3, 20)(4, 11)(5, 16)(6, 9)
 (7, 22)(8, 21)(10, 12)(15, 24)(17, 23)(18, 19)
]
> #sub<s24|gens>;
244823040
```

Finally, we ask for the order of the subgroup of $S_{24}$ generated by the seven elements in **gens**, confirming that they generate a copy of $M_{24}$.

### 13.1.1 The Geometric Approach

The Klein[1] map $\kappa$ is an extraordinary geometric object analogous to the regular dodecahedron that we shall investigate in Section 13.1.2; but instead of having 12 pentagonal faces meeting three at each vertex, $\kappa$ has 24 heptagonal faces meeting likewise three at each vertex. Thus it has $(24 \times 7)/3 = 56$ vertices and $(24 \times 7)/2 = 84$ edges. Euler's formula

$$V - E + F = 2(1 - g),$$

where $V$ denotes the number of vertices, $E$ the number of edges, $F$ the number of faces and $g$ the genus of the lowest genus surface on which it can be drawn, then gives us

$$24 - 84 + 56 = 2(1 - g),$$

and so $g = 3$.

In order to construct this object we take the 24 7-cycles in a conjugacy class of $L_3(2)$, as in $\Omega$ above, and let the numbers $0, 1, \ldots, 6$ correspond to colours. Thus

$$
\begin{aligned}
0 &\sim \text{grey} \\
1 &\sim \text{purple} \\
2 &\sim \text{green} \\
3 &\sim \text{red} \\
4 &\sim \text{black} \\
5 &\sim \text{blue} \\
6 &\sim \text{yellow.}
\end{aligned}
$$

---

[1] I am indebted to Norman Biggs for pointing out to me that the graph I had spent several weeks constructing was actually the Klein map.

## 13.1 The Combinatorial Approach

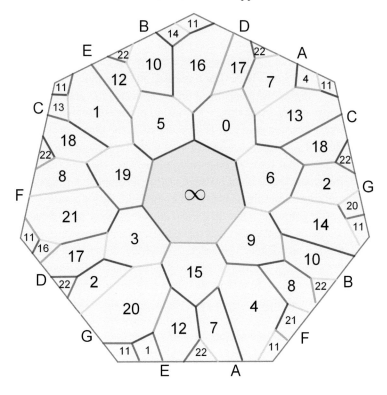

Figure 13.1 Generators for $M_{24}$ on the faces of the Klein map

We then take 24 heptagons and colour the *edges* of each of them with these seven colours going clockwise in the order dictated by the 7-cycle. We must now arrange these 24 heptagons together, three at each vertex, so that the colours match up: a red edge alongside a red edge and so on. In Curtis (2007b) we found all possible ways that we can do this so that the resulting configuration has all the symmetries of the $L_3(2)$ we started with. There are in fact just two ways in which this can be accomplished and the one which interests us here is the Klein map shown in Figure 13.1. This diagram requires a little explanation.

If we take a torus, that is to say a doughnut-shaped surface, we can cut it and bend it straight so as to form a cylinder. We can then cut the cylinder along its length and fold it out so as to form a rectangle. Reversing this process we can take a rectangle and identify a pair of opposite edges, that is to say we understand that any line that strikes one of these edges continues onto our rectangle on the opposite edge. We then also identify the other pair of edges and adopt the same convention. In this way our rectangle affords a diagram of the torus in the plane in which every point is visible. To say that a surface has genus $n$ means that it is equivalent to a sphere with $n$ *handles*, which is to say a sphere with $n$ holes in it. The torus has one hole and is thus a surface of genus 1. The higher the genus the more complicated the graphs that can be drawn on it without crossing edges. For instance, the celebrated Four Colour Theorem states that any map that can be drawn in the plane, or equivalently on the surface of a sphere, can be coloured with four colours in such a way that no two faces that have an edge in common have the same colour. This means, in particular, that it is not possible to draw a complete 5-graph $K_5$ in the plane without edges crossing. However, in Figure 13.2 we see a map with seven faces drawn (diagrammatically) on a torus such that every pair of faces have an edge in common. Replacing faces by vertices, joined if they have an edge in common, this is equivalent to a complete 7-graph $K_7$, drawn with no crossings.

Now Euler's formula has told us that we require a surface of genus 3 in order to exhibit the Klein map. To do this we take a 14-gon and identify its 14 edges in pairs, just as we identified the four edges of our rectangle in two pairs to give a surface of genus 1; in Figure 13.1 every edge of the heptagon in the diagram represents two edges of the surface. These are labelled $A, B, \ldots, G$. The 24 faces are labelled with the points of $\Omega$, which are both elements of $P_1(23)$ and 7-cycles that now correspond to heptagons with coloured edges. Thus the face labelled $\infty$ corresponds to

(0 1 2 3 4 5 6) ~ (grey purple green red black blue yellow).

Suppose, for instance, we are on face 7 and we leave the surface on the top right boundary edge labelled $A$ with a blue edge on our right and a grey edge on our left, then we re-enter the surface through the boundary edge labelled $A$ on the base line. Again, if we leave through boundary edge $B$ top left with a yellow edge on our left and a black one on our right, then we re-enter through boundary edge $B$ bottom right.

The connection between the colouring and this classic map is that when the group $L_3(2)$ of rotational symmetries of $\kappa$ acts on the 84 edges, it has seven blocks of imprimitivity with 12 edges in each of them. Each block is coloured with one of the seven colours. These blocks are readily described geometrically. For instance, using the red block, proceed along the red edge with face 5 on your left and $\infty$ on your right, turn left at the end, then at the next junction the right branch is also labelled red; proceed along this edge with 16 on your left and 0 on your right, turn left onto the green edge leaving through the boundary edge $D$ re-entering bottom left, bearing right onto the red edge between 21 on the left and 17 on the right ... and so on. Note that we could have coloured

## 13.1 The Combinatorial Approach

the map by going right then left, rather than left then right as we have done in Figure 13.1.

The permutations that generate $M_{24}$ may now be read off Figure 13.1:

*Choose a colour and interchange all pairs of faces that are separated by an edge of that colour.*

Of course, every face has an edge of each colour, and so this defines a permutation of the 24 faces of cycle shape $2^{12}$. One now sees that the grey edges give rise to our permutation $t_0$ above, the purple edges to $t_1$ and so on.

Referring again to Figure 13.1, for a particular colour, grey say, instead of interchanging two faces that are separated by a grey edge we could interchange two face that are *joined* by a grey edge. Thus $\infty$ is interchanged with 14; 0 is interchanged with 10 and so on, giving rise to the permutation

$$s_0 = (\infty\ 14)(0\ 10)(5\ 22)(19\ 4)(3\ 9)(15\ 2)(6\ 20)(1\ 8)(21\ 13)(18\ 11)(7\ 16)(12\ 17),$$

which corresponds to the combinatorial map

$$s_0 : (0\ u\ v\ w\ x\ y\ z) \mapsto (0\ u\ v\ w\ x\ y\ z)^{(u\ w)(v\ z)} = (0\ w\ z\ u\ x\ y\ v).$$

As mentioned above, the seven permutations $\{s_0, s_1, \ldots, s_6\}$ generate a different copy of $M_{24}$. Now $t_0$ and $s_0$ commute with one another and so their product is a further involution. It is the third point-like generating set. Thus

$$t_0 s_0 : (0\ u\ v\ w\ x\ y\ z) \mapsto (0\ u\ v\ w\ x\ y\ z)^{(u\ w)(x\ y)}.$$

The remaining three copies are obtained from the other colouring that is achieved by applying a reflective symmetry to $\kappa$. For instance, if we apply the reflection in the vertical line through the centre of $\kappa$, which is to say

$$(\infty)(15)(16)(11)(18)(22)(0\ 5)(6\ 19)(3\ 9)(7\ 12)(4\ 20)(10\ 17)(8\ 2)(1\ 13)(14\ 21),$$

then $t_0$ becomes the permutation defined combinatorially by

$$[013]_3 : (\alpha\ \beta\ w\ \gamma\ x\ y\ z) \mapsto (\alpha\ \beta\ w\ \gamma\ x\ y\ z)^{(w\ z)(x\ y)}$$
$$= (\alpha\ \beta\ z\ \gamma\ y\ x\ w),$$

where $\{\alpha, \beta, \gamma\} = \{0, 1, 3\}$.

### More about the Klein Map

As we have seen, the faces of the Klein map correspond to 7-cycles in a conjugacy class of the linear group $L_3(2)$. This class is closed under squaring, that is to say that a 7-cycle is conjugate in the group to its square. In the map a face, its square and its 4th power form a triangle of faces at maximum distance

Figure 13.2 A map requiring seven colours on a surface of genus 1

from one another. To move from a face to its square we leave it by any one of the seven edges radiating out from it, so from ∞ we might move along the blue edge to face 16, then the face corresponding to its square is the one opposite the point of entry, namely 11. To move to the 4th power we reverse the process, so from ∞ we move to an adjoining face, say 19, and proceed along the opposite edge, which is yellow in this case, to face 22. Thus

$$[\infty, 11, 22] \sim [(0\ 1\ 2\ 3\ 4\ 5\ 6), (0\ 2\ 4\ 6\ 1\ 3\ 5), (0\ 4\ 1\ 5\ 2\ 6\ 3)]$$

as in the top line of Figure 13.1. Note that in Figure 7.3 the element of order three that centralizes the Octern group in $S_{24}$ cycles the first column in a different sense from the other seven columns, as is borne out in Figure 13.1 where the square of face 0 is face 8 and so on.

### Connection with the Klein Quartic

The Klein map exhibited in Figure 13.1 was discovered by Felix Klein in 1878 in connection with the celebrated quartic curve named after him, see Klein (1878), an equation for which may be taken as

$$F(x, y, z) = xy^3 + yz^3 + zx^3 = 0. \tag{13.1}$$

Solutions to this equation are triples of complex numbers $(x, y, z)$ that satisfy the equation, and so correspond to 1-dimensional subspaces of $\mathbb{C}^3$. Clearly cycling the three coordinates preserves the curve, and so the transformation given by

$$\beta = \begin{pmatrix} 0 & 1 & 0 \\ 0 & 0 & 1 \\ 1 & 0 & 0 \end{pmatrix}$$

## 13.1 The Combinatorial Approach

maps points on the curve to points on the curve. If we let $\eta = e^{2\pi i/7}$, a complex 7th root of unity, then the matrix

$$\alpha = \begin{pmatrix} \eta & 0 & 0 \\ 0 & \eta^2 & 0 \\ 0 & 0 & \eta^4 \end{pmatrix}$$

also preserves the curve. We see that $\alpha^\beta = \alpha^4$ and so $\langle \alpha, \beta \rangle \cong 7:3$, a Frobenius group of order 21. In fact the curve has far more symmetries than these. In order to see this, we follow Baker (1935), and let[2] $a = \eta + \eta^{-1}$, $b = \eta^2 + \eta^{-2}$ and $c = \eta^4 + \eta^{-4}$. We may then readily check that these complex numbers, which are the roots of the cubic $x^3 + x^2 - 2x - 1 = 0$ satisfy

$$a^2 = b+2, \qquad b^2 = c+2, \qquad c^2 = a+2,$$
$$bc + c + 1 = 0, \quad ca + a + 1 = 0, \quad ab + b + 1 = 0,$$
$$a^{-1} = a + b, \qquad b^{-1} = b + c, \qquad c^{-1} = c + a,$$

and using these relations we can check that the matrix

$$\gamma = d^{-1} \begin{pmatrix} 1 & a^{-1} & b \\ a^{-1} & b & 1 \\ b & 1 & a^{-1} \end{pmatrix}$$

also preserves the curve, where $d = -(1 + a^{-1} + b)$.

The famous Hurwitz bound states that a group preserving a surface of genus $g$ cannot have order exceeding $84(g-1)$. The surface in our case has $g = 3$ and so the maximal order of a group of linear transformations preserving the curve is 168. In fact we can show that the matrices $\alpha$ and $\gamma$ above satisfy

$$\alpha^7 = \gamma^2 = (\alpha\gamma)^3 = [\alpha, \gamma]^4 = 1,$$

which is a familiar presentation for the group $L_3(2)$. Since the group is simple and has order 168, and since the matrices are non-trivial, we must have

$$\langle \alpha, \gamma \rangle = \langle \alpha, \beta, \gamma \rangle \cong L_3(2).$$

So Klein followed by Baker, in seeking the group preserving the curve $F = 0$, had constructed a 3-dimensional complex representation of the group $L_3(2)$. Turning this process around we see that they had shown that $F = 0$ is a degree 4 *invariant* of $L_3(2)$.

Now associated with a curve $F = 0$ is its *Hessian*, which is the determinant

---

[2] In ATLAS notation, see Conway et al. (1985), we should have $a = b7 = (-1 + i\sqrt{7})/2$, $b = a*2$, $c = a*4$.

of second-degree partial derivatives:

$$\mathcal{H}(x, y, z) = \begin{vmatrix} \frac{\partial^2 F}{\partial x^2} & \frac{\partial^2 F}{\partial x \partial y} & \frac{\partial^2 F}{\partial x \partial z} \\ \frac{\partial^2 F}{\partial y \partial x} & \frac{\partial^2 F}{\partial y^2} & \frac{\partial^2 F}{\partial y \partial z} \\ \frac{\partial^2 F}{\partial z \partial x} & \frac{\partial^2 F}{\partial z \partial y} & \frac{\partial^2 F}{\partial z^2} \end{vmatrix}.$$

It is shown in, for example, Hilton (1920, p. 100) that a curve meets its Hessian in either *points of inflexion* or *multiple points*. Since our curve possesses no multiple points, all intersections of $F = 0$ with $\mathcal{H} = 0$ are points of inflexion. Now $\mathcal{H}$ is homogeneous of degree 4 and so all of its second derivatives are quadratic, and the determinant above is homogeneous of degree 6; thus $F = 0$ and $\mathcal{H} = 0$ intersect in $4 \times 6 = 24$ points all of which are points of inflexion. These correspond to the faces of $\kappa$ in Figure 13.1.

Now $L_3(2)$ possesses two further invariants of interest in this context, one of degree 14 and the other of degree 21, giving rise to $14 \times 4 = 56$ and $21 \times 4 = 84$ special points on the Klein quartic, respectively. The former, which correspond to the 56 vertices of $\kappa$, are the *points of contact of bitangents*. A quartic curve can have 28 tangents that touch it twice, a configuration that gives rise to the *group of the 28 bitangents* see for example Conway et al. (1985, page 46), so there are 56 of them. The latter, which correspond to the 84 edges of $\kappa$ are the so-called *sextactic points*. Any five points in the plane have a unique conic passing through them and so at any point of a higher-order differentiable curve there is a conic making 5-point contact with the curve. In the same way that some tangents to a curve make 3-point contact and are thus points of inflexion, some of these conics make 6-point contact: the 84 sextactic points. The invariant of degree 21 has been worked out by Miles Reid (2023), using MAGMA and he finds that it factorizes into the product of 21 linear terms; thus these 84 sextactic points fall into 21 collinear sets of four points. These can be seen clearly from the colouring of the edges in Figure 13.1 where $\{17/21, 3/19, \infty/5, 0/16\}$ is one such.

We saw that on $\kappa$ there are triples of faces at maximum distance from one another that correspond to a 7-cycle, its square and its 4th power. It is of interest to ask what these triples correspond to geometrically. A straight line should cut a quartic curve four times and so the tangent at a point of inflexion, which has made 3-point contact at that point, must cut the curve once more. Now our element $\alpha$ fixes just three points, namely (1 0 0), (0 1 0) and (0 0 1), and since it must preserve the set of 24 points of inflexion and since $24 \equiv 3$ modulo 7, these must themselves be points of inflexion. But the stabilizer of the point of inflexion (1 0 0) which is $\langle \alpha \rangle$, must fix the tangent at that point and thus fix the point where that tangent cuts the curve again. So that point can only be one of the remaining two points fixed by $\langle \alpha \rangle$. Repeating the process, the tangent at this second point of inflexion must cut the curve at the remaining point of inflexion, and the tangent at this third point of inflexion must pass through the original point. In this way we see that the 24 points of inflexion fall into eight triangles, each with a sense of rotation; these are the *triangles of inflexion* and they correspond, of course, to the sets $\{\lambda, \lambda^2, \lambda^4\}$ for $\lambda$ a 7-cycle in our chosen conjugacy class of $L_3(2)$.

### 13.1.2 The Algebraic Explanation

We can, in fact, construct these seven generators directly from the abstract group $L_3(2)$ as follows. For consider $L \cong L_3(2)$ acting transitively on 24 points; its point stabilizer being a cyclic group of order 7. If $H \leq L$ with $H \cong S_4$ then $H$ must act regularly on the 24 points (since no non-trivial element of $H$ fixes a point), and so is acting as, say, the *right* regular representation of $H$. We now ask what is the centralizer of $H$ in the symmetric group $\Sigma \cong S_{24}$. Certainly the left regular representation will commute with the right regular representation since, if $L_x$ denotes multiplication by $x$ on the left and $R_y$ multiplication by $y$ on the right, then

$$(L_x R_y)(g) = L_x(R_y(g)) = L_x(gy) = x(gy) = (xg)y = R_y(xg) = R_y(L_x(g))$$
$$= (R_y L_x)(g) \quad \text{for all } g \in H.$$

Moreover, the only permutation that fixes a point and commutes with a transitive group is the identity, and so

$$C_\Sigma(H) = K \cong S_4,$$

the *left* regular representation. It turns out that the involutions in $K$ that provide generators for a copy of $M_{24}$ are in the Klein fourgroup, see Curtis (2007b, p. 26), and so there are three possibilities. There were also two choices for which class of $S_4$ we started with, and so if $d$ denotes the number of ways in which a transitive copy of $L_3(2)$ acting on 24 points can be extended to a copy of $M_{24}$ then $d = 2 \times 3 = 6$. To confirm this we count the set

$$\{(L, M) \mid L \cong L_3(2), M \cong M_{24}, L \leq M \leq \Sigma, L \text{ transitive}\}$$

in two ways making use of the fact that $N_\Sigma(L) \cong 3 \times \text{PGL}_2(7)$. Firstly note that a transitive copy of $L_3(2)$ in $M_{24}$ must be generated by an element of order 7, cycle shape $1^3 \cdot 7^3$, and an involution of cycle shape $2^{12}$. The only maximal subgroup that such a subgroup could be contained in other than the Octern group is the trio stabilizer. But a transitive subgroup of the trio group would have to act transitively on the three octads of the trio, and the simple group $L_3(2)$ can have no such action. So any transitive copy of $L_3(2)$ in $M_{24}$ must be a copy of the Octern group. We then have

$$|\Sigma : N_\Sigma(L)| \times d = |\Sigma : N_\Sigma(M)| \times |M : N_M(L)|,$$

and so

$$d = \frac{24!}{|M|} \times \frac{|M|}{|L|} \times \frac{6|L|}{24!} = 6.$$

## 13.2 The Mathieu Group $M_{12}$

What follows is an analagous treatment of the smaller Mathieu group $M_{12}$. It may seem more natural to deal with the smaller group first, and this is indeed the order in which we dealt with them in Curtis (2007b), however we have chosen this arrangement here for two reasons. Firstly, this was the order in which we discovered these sets of generators. Having thoroughly investigated the seven involutory generators of $M_{24}$, we decided to look for an analogy in $M_{12}$ and the emergence of the five generators of order three and their connection with the dodecahedron emerged effortlessly. Secondly, this is after all a book about $M_{24}$ so it seems right to lead with its construction.

Our construction of generators for $M_{24}$ made use of the fact that the linear group $L \cong L_2(7) \cong L_3(2)$ *factorizes* as $L = HK$, where $H \cong S_4$ and $K \cong C_7$. In the action of $L$ on 24 points, which may be taken as a conjugacy class of 7-cycles in $L_3(2)$, $H$ acts regularly.

Analogously, with $L \cong L_2(5) \cong A_5$ we have $L = HK$ where $H \cong A_4$ and $K \cong C_5$. In the action on 12 points, which may be taken as a conjugacy class of 5-cycles in $A_5$, $H$ acts regularly.

### 13.2.1 The Combinatorial Approach

The alternating group $A_5$ possesses two conjugacy classes of 5-cycles, a given 5-cycle being conjugate to its inverse but not to its square or its cube. Let $A \cong A_5$ acting naturally on the set $Y = \{1, 2, 3, 4, 5\}$, and let $a = (1\ 2\ 3\ 4\ 5) \in A$, then we may take the two classes to be

$$\Lambda = \{a^g \mid g \in A\} \text{ and } \bar\Lambda = \{(a^2)^g \mid g \in A\}.$$

## 13.2 The Mathieu Group $M_{12}$

Table 13.3 Labelling the 24 5-cycles with elements of $P_1(11)$

| | $\Lambda$ | | | | $\bar{\Lambda}$ | | |
|---|---|---|---|---|---|---|---|
| $\infty$ | (1 2 3 4 5) | 0 | (1 5 4 3 2) | $\bar{\infty}$ | (1 3 5 2 4) | $\bar{0}$ | (1 4 2 5 3) |
| 1 | (1 3 2 5 4) | 2 | (1 4 5 2 3) | $\bar{1}$ | (1 2 4 3 5) | $\bar{2}$ | (1 5 3 4 2) |
| 9 | (1 5 2 4 3) | 7 | (1 3 4 2 5) | $\bar{9}$ | (1 2 3 5 4) | $\bar{7}$ | (1 4 5 3 2) |
| 4 | (1 3 5 4 2) | 8 | (1 2 4 5 3) | $\bar{4}$ | (1 5 2 3 4) | $\bar{8}$ | (1 4 4 3 2 5) |
| 3 | (1 5 3 2 4) | 6 | (1 4 2 3 5) | $\bar{3}$ | (1 3 4 5 2) | $\bar{6}$ | (1 2 5 4 3) |
| 5 | (1 4 3 5 2) | X | (1 2 5 3 4) | $\bar{5}$ | (1 3 2 4 5) | $\bar{X}$ | (1 5 4 2 3) |

We shall define permutations on the set $\Lambda$ and eventually extend this to permutations on $\Lambda \cup \bar{\Lambda}$. In Table 13.3 we write out all of the 5-cycles so that they begin with 1, and label $\Lambda$ using the projective line

$$P_1(11) = \{\infty, 0, 1, \ldots, X\} = \mathbb{Z}_{11} \cup \{\infty\},$$

where as usual X stands for 10. The elements in $\bar{\Lambda}$ are then labelled with the set $\{\bar{\infty}, \bar{0}, \ldots, \bar{X}\}$ with the convention that if the 5-cycle $\lambda \in \Lambda$ is labelled $n$, then $\lambda^2$ is labelled $\bar{n}$. Elements of $A$ permute $\Lambda$ (and $\bar{\Lambda}$) by conjugation and so we obtain an embedding of $A$ in the symmetric group $S_{12}$ and indeed, since $A_5$ is simple, in the alternating group $A_{12}$. We shall denote the permutation of $\Lambda$ that is induced by a permutation $\pi \in A$ by $\hat{\pi}$. Thus $\hat{a} = (1\ 9\ 4\ 3\ 5)(2\ 7\ 8\ 6\ X)$.

We now define a new permutation of the 12 points of $\Lambda$ that we shall denote by $s_1$. For $(1\ w\ x\ y\ z) \in \Lambda$ we define

$$s_1 : (1\ w\ x\ y\ z) \mapsto (1\ w\ x\ y\ z)^{(x\ y\ z)} = (1\ w\ y\ z\ x).$$

This is certainly a well-defined function from $\Lambda$ to $\Lambda$ as we have conjugated by an even permutation. Moreover, since $s_1^3$ acts as the identity permutation $s_1$ possesses an inverse, namely $s_1^2$, and so is a permutation. It does not fix any 5-cycle and so it has cycle shape $3^4$ on $\Lambda$. Indeed we soon see that

$$s_1 \sim (\infty\ 8\ X)(0\ 3\ 9)(1\ 4\ 7)(2\ 6\ 5).$$

The permutations $s_i$ for $i = 2, \ldots, 5$ are defined in an analogous manner by writing all the 5-cycles beginning with $i$ and then conjugating $(i\ w\ x\ y\ z)$ by $(x\ y\ z)$ as before. We then have the delightful fact that

$$\langle \hat{a}, s_1 \rangle = \langle s_1, s_2, \ldots, s_5 \rangle \cong M_{12}.$$

If we are not concerned with the order in which the 12 5-cycles are labelled, we may simply take $\Lambda$ to be the set of all conjugates of $a = (1\ 2\ 3\ 4\ 5)$ in the

Table 13.4 Action of the five symmetric generators on $\Lambda \cup \bar{\Lambda}$

$$
\begin{align}
s_1 &= (\infty\ 8\ X)(0\ 3\ 9)(1\ 4\ 7)(2\ 6\ 5)(\bar{\infty}\ \bar{3}\ \bar{5})(\bar{0}\ \bar{8}\ \bar{7})(\bar{1}\ \bar{6}\ \bar{9})(\bar{2}\ \bar{4}\ \bar{X}) \\
s_2 &= (\infty\ 6\ 2)(0\ 5\ 4)(9\ 3\ 8)(7\ X\ 1)(\bar{\infty}\ \bar{5}\ \bar{1})(\bar{0}\ \bar{6}\ \bar{8})(\bar{9}\ \bar{X}\ \bar{4})(\bar{7}\ \bar{3}\ \bar{2}) \\
s_3 &= (\infty\ X\ 7)(0\ 1\ 3)(4\ 5\ 6)(8\ 2\ 9)(\bar{\infty}\ \bar{1}\ \bar{9})(\bar{0}\ \bar{X}\ \bar{6})(\bar{4}\ \bar{2}\ \bar{3})(\bar{8}\ \bar{5}\ \bar{7}) \\
s_4 &= (\infty\ 2\ 8)(0\ 9\ 5)(3\ 1\ X)(6\ 7\ 4)(\bar{\infty}\ \bar{9}\ \bar{4})(\bar{0}\ \bar{2}\ \bar{X})(\bar{3}\ \bar{7}\ \bar{5})(\bar{6}\ \bar{1}\ \bar{8}) \\
s_5 &= (\infty\ 7\ 6)(0\ 4\ 1)(5\ 9\ 2)(X\ 8\ 3)(\bar{\infty}\ \bar{4}\ \bar{3})(\bar{0}\ \bar{7}\ \bar{2})(\bar{5}\ \bar{8}\ \bar{1})(\bar{X}\ \bar{9}\ \bar{6})
\end{align}
$$

alternating group. Note that, as in the $M_{24}$ case:

$$(x\ y\ z) : a := (i\ v\ x\ y\ z) \mapsto (i\ v\ y\ z\ x) = (i\ i^a\ i^{a^3}\ i^{a^4}\ i^{a^2}).$$

The construction of $M_{12}$ is then achieved in MAGMA by

```
> A:=Alt(5);
> a:=A!(1,2,3,4,5);
> ts:=Setseq(a^A);
> f:=func<i,tt|A!(i,i^tt,i^(tt^3),i^(tt^4),i^(tt^2))>;
> s12:=Sym(12);
> gens:=[Id(s12):i in [1..5]];
> for i in [1..5] do
for> test:=[1:m in [1..12]];
for> for j in [1..12] do
for|for> for k in [1..12] do
for|for|for> if f(i,ts[j]) eq ts[k] then
for|for|for|if> test[j]:=k;
for|for|for|if> end if;
for|for|for> end for;
for|for> end for;
for> gens[i]:=s12!test;
for> end for;
> #sub<s12|gens>;
95040
>
```

As we have seen above, $M_{12}$ possesses an outer automorphism of order 2 that interchanges the two dodecads of the fixed duum, and we can now realize this outer automorphism interchanging $\Lambda$ and $\bar{\Lambda}$, which are playing the roles of the

## 13.2 The Mathieu Group $M_{12}$

two dodecads. We extend the action of $s_i$ to $\Lambda \cup \bar{\Lambda}$ by defining

$$s_i : (i\ w\ x\ y\ z) \mapsto (i\ w\ x\ y\ z)^{(x\ y\ z)} = (i\ w\ y\ z\ x) \text{ if } (i\ w\ x\ y\ z) \in \Lambda,$$
$$\mapsto (i\ w\ x\ y\ z)^{(z\ y\ x)} = (i\ w\ z\ x\ y) \text{ if } (i\ w\ x\ y\ z) \in \bar{\Lambda},$$

which results in the permutations displayed in Table 13.4. Note that we shall need two functions **f1** and **f2** in the MAGMA code, one for $\Lambda$ and one for $\bar{\Lambda}$.

Now let $S \cong S_5$ be the symmetric group acting on the set $Y$, when odd elements of $S$ interchange the two classes $\Lambda$ and $\bar{\Lambda}$ by conjugation. Even elements of $S$ permute the $s_i$ in the natural way, but odd elements permute and *invert*. Thus

$$s_i^{\hat{\pi}} = s_{i^\pi} \text{ if } \pi \in A; \quad s_i^{\hat{\pi}} = s_{i^\pi}^{-1} \text{ if } \pi \in S \setminus A.$$

These statements could be proved directly by applying generators for $A$ and $S$ to the permutations in Table 13.4, but they are proved formally in Curtis (2007b, p. 5). We obtain $M_{12}$ acting non-permutation identically on two sets of size 12 with the following MAGMA code:

```
> S:=Sym(5);
> A:=DerivedGroup(S);
> a:=S!(1,2,3,4,5);
> ts:=Setseq(a^A) cat Setseq((a^2)^A);
> s24:=Sym(24);
> f1:=func<i,tt|A!(i,i^tt,i^(tt^3),i^(tt^4),i^(tt^2))>;
> f2:=func<i,tt|A!(i,i^tt,i^(tt^4),i^(tt^2),i^(tt^3))>;
> gens:=[Id(s24): r in [1..5]];
> for i in [1..5] do
for> test:=[1:s in [1..24]];
for> for j in [1..12] do
for|for> for k in [1..12] do
for|for|for> if f1(i,ts[j]) eq ts[k] then
for|for|for|if> test[j]:=k;
for|for|for|if> end if;
for|for|for> end for;
for|for> end for;
for> for j in [13..24] do
for|for> for k in [13..24] do
for|for|for> if f2(i,ts[j]) eq ts[k] then
for|for|for|if> test[j] :=k;
for|for|for|if> end if;
for|for|for> end for;
for|for> end for;
```

```
for> gens[i]:=s24!test;
for> end for;
> #sub<s24|gens>;
95040
```

In order to obtain the outer automorphism of $M_{12}$, which interchanges the two sets of size 12, $\Lambda$ and $\bar{\Lambda}$, we must adjoin an odd permutation of $S_5$ acting on the 24 5-cycles. We choose the transposition (4 5) and achieve its action with the following MAGMA code:

```
> dum45:=[1:i in [1..24]];
> for i in [1..24] do
for> for j in [1..24] do
for|for> if ts[i]^S!(4,5) eq ts[j] then
for|for|if> dum45[i]:=j;
for|for|if> end if;
for|for> end for;
for> end for;
> t45:=s24!dum45;
> #sub<s24|gens,t45>;
190080
>
```

So far we have chosen to conjugate our 5-cycles of form $(i\ w\ x\ y\ z)$ by the 3-cycle $(x\ y\ z)$. In fact we could equally well have chosen $(y\ z\ w)$, $(z\ w\ x)$ or $(w\ x\ y)$ and conjugated every 5-cycle by it. In this way we should have obtained four distinct copies of $M_{12}$ which, as we shall see later, are the only ways in which $A_4$ acting transitively on 12 points can be extended to a copy of $M_{12}$. The behaviour of these four copies is investigated in more detail in Curtis (2007b) but we mention here that they are cycled by the 'squaring' transformation

$$\tau : \lambda \mapsto \lambda^2 \text{ for } \lambda \in \Lambda \cup \bar{\Lambda}.$$

As we see from Table 13.3, this element acts as

$$\tau \sim \quad (\infty\ \bar{\infty}\ 0\ \bar{0})(1\ \bar{1}\ 2\ \bar{2})(9\ \bar{9}\ 7\ \bar{7})$$
$$(4\ \bar{4}\ 8\ \bar{8})(3\ \bar{3}\ 6\ \bar{6})(5\ \bar{5}\ X\ \bar{X}).$$

It commutes with the $\hat{A}$ and cycles the four copies of $M_{12}$ that contain it.

### 13.2.2 The Geometric Approach

We readily see that the group of rotational symmetries of the regular dodecahedron has order 60: We may rotate until any one of the 12 faces is on the base

## 13.2 The Mathieu Group $M_{12}$

giving a factor of 12, there then remains a cyclic group of order 5 fixing this base. The 20 vertices of the dodecahedron then fall into five blocks of size 4 (in two different ways) such that each block forms the vertices of a regular tetrahedron. This can be seen in Figure 13.3 where in either diagram if we choose a red vertex and move along any edge from it, bear right at the first junction and left at the second we come to another red vertex. Thus the four red vertices are all equidistant from one another, and similarly for the other colours. The other partition into five tetrahedra is obtained by reflecting through the centre of the dodecahedron and is displayed in Figure 13.4; here one moves from one red vertex to another by turning first left and then right. Indeed, if we take the four vertices of one of these tetrahedra together with their reflections in the centre of the dodecahedron, we obtain the eight vertices of a cube. The five cubes obtained in this way are permuted as the alternating group $A_5$ by the group of rotational symmetries of the dodecahedron, as are the two sets of five tetrahedra. In Figures 13.3 and 13.4 the vertices have been coloured with five colours, one for each of the five tetrahedra, and so each face has a vertex of each colour. If we let these colours correspond to the numbers $1, \ldots, 5$ as

$$
\begin{array}{rcl}
\text{black} & \sim & 1 \\
\text{yellow} & \sim & 2 \\
\text{red} & \sim & 3 \\
\text{blue} & \sim & 4 \\
\text{green} & \sim & 5,
\end{array}
$$

then we see that, reading around the faces in a clockwise sense, the faces correspond to the 5-cycles of $\Lambda$ as labelled in Table 13.3 in the first diagram of Figure 13.3 and to those of $\bar{\Lambda}$ in the second. We saw in the combinatorial approach to $M_{12}$ that there are four ways of extending a copy of $A_5$ acting transitively on 12 letters to a copy of $M_{12}$, and this will be explained further in the algebraic approach that follows. So how do these copies of $M_{12}$ manifest themselves on the faces of the dodecahedron? Of course the version that most interests us is the one that generates the usual copy. Since $M_{12}$ contains $L_2(11)$ acting on the 12-point projective line $P_1(11)$, we label the faces $\{\infty, 0, \ldots, \times\}$. We obtained $s_i$ by conjugating each 5-cycle $(i\ w\ x\ y\ z)$ by $(x\ y\ z)$, so how do we see this on the faces of the dodecahedron? When $i = 1$ we must refer to the black vertices and observe in Figure 13.3 that the three edges from each of them connects to three faces; thus, for instance, the top right black vertex in the first diagram of Figure 13.3 is joined to faces $\infty$, 8 and X. Rotate these triples of faces *clockwise* to obtain

$$s_1 \sim (\infty\ 8\ X)(0\ 3\ 9)(1\ 4\ 7)(2\ 6\ 5).$$

206  *Natural Generators of the Mathieu Groups*

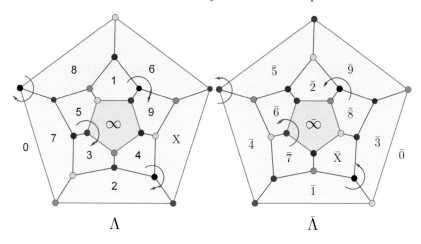

Figure 13.3  Generators for $M_{12} : 2$ on the faces of two dodecahedra

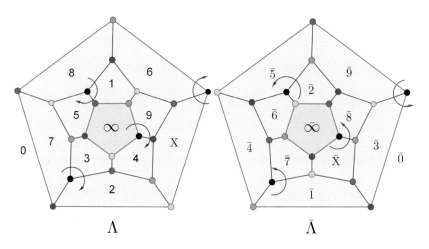

Figure 13.4  The alternative colouring of the vertices of the two dodecahedra

This element's action on $\bar{\Lambda}$ is obtained in precisely the same way except that we must now rotate *anticlockwise*, obtaining

$$s_1 \sim (\bar{\infty}\,\bar{3}\,\bar{5})(\bar{0}\,\bar{8}\,\bar{7})(\bar{1}\,\bar{6}\,\bar{9})(\bar{2}\,\bar{4}\,\bar{X}).$$

Such a permutation of the 12 faces of the dodecahedron is known as a *deep twist*. If instead of rotating the faces joined to a black vertex by an edge, we rotate the three faces with which it is incident, we obtain a *shallow twist*. Applying shallow

## 13.2 The Mathieu Group $M_{12}$

| Conjugation by 3 – cycle | Colouring 1 or 2 | deep/shallow twists | Action on $\Lambda$ and $\bar{\Lambda}$ |
|---|---|---|---|
| $(x\ y\ z)$ | 1 | deep | $(\infty\ 8\ X)(0\ 3\ 9)(1\ 4\ 7)(2\ 6\ 5)$ $(\bar{\infty}\ \bar{3}\ 5)(\bar{0}\ 8\ \bar{7})(\bar{1}\ \bar{6}\ 9)(\bar{2}\ \bar{4}\ \bar{X})$ |
| $(w\ z\ y)$ | 1 | shallow | $(\infty\ 3\ 5)((0\ 8\ 7)(1\ 6\ 9)(2\ 4\ X)$ $(\bar{\infty}\ \bar{6}\ \bar{7})(\bar{0}\ \bar{4}\ \bar{5})(\bar{1}\ \bar{3}\ \bar{X})(\bar{2}\ \bar{8}\ \bar{9})$ |
| $(w\ z\ x)$ | 2 | shallow | $(\infty\ 9\ 4)(0\ X\ 6)(1\ 5\ 8)(2\ 7\ 3)$ $(\bar{\infty}\ \bar{X}\ \bar{8})(\bar{0}\ \bar{9}\ \bar{3})(\bar{1}\ \bar{7}\ \bar{4})(\bar{2}\ \bar{5}\ \bar{6})$ |
| $(w\ x\ y)$ | 2 | deep | $(\infty\ 7\ 6)(0\ 5\ 4)(1\ X\ 3)(2\ 9\ 8)$ $(\bar{\infty}\ \bar{9}\ \bar{4})(\bar{0}\ \bar{X}\ \bar{6})(\bar{1}\ \bar{5}\ \bar{8})(\bar{2}\ \bar{7}\ \bar{3})$ |

Figure 13.5 Correspondence between the combinatorial and geometric approaches

twists consistently results in a different version of $M_{12}$. The other two copies are obtained by applying deep and shallow twists to the colouring displayed in Figure 13.4. In Figure 13.5 we show how these four geometric constructions correspond to the four combinatorial constructions of the previous section. One can observe in Figure 13.5 how the four copies of $M_{12}$ produced are cycled by the 'squaring' permutation $\tau$ of the previous section. As remarked above, it is the first of these constructions that yields our preferred version of $M_{12}$. Of course, all even permutations of the colours are permissible in any of the four diagrams in Figures 13.3 and 13.4; however, odd permutations interchange the two dodecahedra in each of the two figures. Thus, for instance, interchanging blue and green (which is equivalent to conjugating the 5-cycles of Table 13.3 by (4 5)) interchanges $\Lambda$ and $\bar{\Lambda}$ and so affords the outer automorphism of $M_{12}$. This results in

(blue $\leftrightarrow$ green) $\sim$
$(\infty\ \bar{9})(0\ \bar{7})(1\ \bar{5})(2\ \bar{X})(9\ \bar{0})(7\ \bar{\infty})(4\ \bar{3})(8\ \bar{6})(3\ \bar{8})(6\ \bar{4})(5\ \bar{2})(X\ \bar{1})$.

### 13.2.3 The Algebraic Approach

In Curtis (2007b) we explain in detail how the structure of $M_{12}$ requires these constructions to work, and we produce an element of order 4 that cycles the four resulting copies of the group. Here, though, we shall content ourselves by showing how permutations generating $M_{12}$ may be written down from an abstract consideration of the alternating group $A_5$.

For suppose $\Sigma \cong S_{12}$ and that $A \cong A_5$ is a subgroup of $\Sigma$ acting transitively on the 12 points permuted by $\Sigma$. Then the point stabilizer in $A$ is cyclic of order 5 and so the only non-trivial elements of $A$ that fix a point have order 5. In particular, a subgroup $H \cong A_4$ of $A$ acts regularly on the 12 points. As we saw in Section 13.1.2, the centralizer of the left regular representation is the right regular representation, and so the centralizer in $S_{12}$ of $H$ is $K$ another copy of the alternating group $A_4$. Now $K$ contains four cyclic subgroups of order 3 any one of which could be taken as our generator $s_1$. Its five images under conjugation by $A$ will generate a copy of $M_{12}$. These, of course, correspond to the four ways we have seen in the combinatorial and geometric approaches that a copy of $A_5$ acting transitively on 12 letters can be extended to a copy of the Mathieu group $M_{12}$.

# 14

# Symmetric Generation Using $M_{24}$

## 14.1 Introduction to Symmetric Generation

In Curtis (2007b) the familiar notation $H \star K$ for the *free product* of two groups $H$ and $K$ was modified to denote the free product of cyclic groups. Thus

$$m^{\star n} \cong \underbrace{C_m \star C_m \star \cdots \star C_m}_{n},$$

the free product of $n$ copies of the cyclic group $C_m$. Thus

$$m^{\star n} \cong \langle t_1, t_2, \ldots, t_n \mid t_i^m = 1 \text{ for } i \in \{1, \ldots, n\}\rangle,$$

a group generated by $n$ elements of order $m$ with no further relations between them. Thus we saw in Chapter 13 that $M_{24}$ is a homomorphic image of $2^{\star 7}$; and $M_{12}$ is a homomorphic image of $3^{\star 5}$. Moreover in both cases these generating sets are highly symmetric, so that if $\mathcal{T} = \{t_0, t_1, \ldots, t_6\}$ is such a set of seven involutions generating $M_{24}$ then,

$$N_{M_{24}}(\mathcal{T}) = \langle \pi \in M_{24} \mid \mathcal{T}^\pi = \mathcal{T}\rangle \cong L_2(7).$$

Similarly, if $\mathcal{S} = \{s_0, s_1, \ldots, s_4\}$ is such a set of five elements of order three generating $M_{12}$ then,

$$N_{M_{24}}(\mathcal{S}) = \langle \pi \in M_{24} \mid \mathcal{S}^\pi = \mathcal{S}\rangle \cong L_2(5).$$

That these two projective special linear groups $L_2(p)$, which are naturally permutation groups on $p + 1$ letters, are acting on 7 and 5 letters respectively is, of course, due to the exceptional isomorphisms identified by Galois, see Section 11.7.2:

$$L_2(7) \cong L_3(2) \text{ and } L_2(5) \cong A_5.$$

Within $M_{24}$ the way that our copy of $L_3(2)$ acts on $\mathcal{T}$ is by conjugation, but we can instead regard $\mathcal{T}$ as an abstract set of seven involutions with no further relations between them, in other words as a free product of cyclic groups as described above. Then a permutation $\pi$ of $L_3(2)$ acts as

$$t_i^\pi = \pi^{-1} t_i \pi = t_{(i)\pi},$$

where we are writing permutations acting on the right. This enables us to extend the free product $2^{\star 7}$ by $N \cong L_3(2)$ to form an infinite semi-direct product

$$2^{\star 7} : N = \{\pi w \mid \pi \in N, w \text{ a word in the } t_i\},$$

where

$$\pi u \cdot \sigma v = \pi \sigma u^\sigma v.$$

We can generalize this construction to semi-direct products of the form

$$P = m^{\star n} : N,$$

where $N$ possesses a *monomial* action on the $n$ *symmetric generators*. That is to say $N$ permutes the $n$ symmetric generators $t_i$ and raises them to a power co-prime to their order $m$. The subgroup $N$ is referred to as the *control group*, and $P$ is termed the *progenitor*. In this chapter we shall restrict our attention to the case when $N \leq S_n$ is a transitive permutation group of degree $n$; in fact, the examples we shall consider will all have $m = 2$ so that the symmetric generators are involutions. However, the more general case is considered briefly in Chapter 15 and the more general construction is dealt with in Curtis (2007b, Chapter 6).

In order to clarify how multiplication in these infinite progenitors is carried out, let us take the example

$$Q \cong 2^{\star 4} : S_4.$$

Then we have, for instance:

$$
\begin{aligned}
(1\ 2)t_2 t_4 \cdot (1\ 4)(2\ 3) t_1 t_2 &= (1\ 2)(1\ 4)(2\ 3)(t_2 t_4)^{(1\ 4)(2\ 3)} t_1 t_2 \\
&= (1\ 3\ 2\ 4) t_3 t_1 t_1 t_2 \\
&= (1\ 3\ 2\ 4) t_3 t_2.
\end{aligned}
$$

### 14.1.1 The Lemma

A given progenitor $P \cong 2^{\star n} : N$ will in general possess many finite homomorphic images; these are known as its *progeny*. Indeed, any group that is generated by a set of four involutions that possesses all the symmetries of the symmetric group $S_4$ is an image of the last considered progenitor $Q$. As is tabulated in Curtis (2007b, page 85), $L_2(23), L_3(5) : 2, A_9$ and indeed

## 14.1 Introduction to Symmetric Generation

Aut $M_{12} \cong M_{12} : 2$ and many others are all progeny of $Q$. The question arises: What should we factor out in order to obtain an interesting finite image?

Now every element of $P$ has the form $\pi w$ and so factoring out a normal subgroup is equivalent to mapping various such elements to the identity. Note that $\pi w$ is equivalent to $\pi = w^{-1}$, which is telling us how to write $\pi$ as a word in (the images of) the symmetric generators. In this way, outer automorphisms of the free product $2^{\star n}$ become inner automorphisms in the image; this explains how an image of a progenitor can be a simple group. The following simple lemma, which indicates which permutations can be written as words in a particular subset of $\mathcal{T}$, has proved useful in producing interesting images. In the following two sections we shall describe two such images in which the group $M_{24}$ serves as the control subgroup.

**Lemma 14.1** *If, in an image of the progenitor $P \cong 2^{\star n} : N$, the permutation $\pi \in N$ can be written as a word in the images of the symmetric generators $\{t_{k_1}, t_{k_2}, \ldots, t_{k_r}\}$ then*

$$\pi \in C_N(N_{k_1 k_2 \cdots k_r}),$$

*where $N_{k_1 k_2 \cdots k_r}$ denotes the stabilizer in $N$ of $k_i$ for all $i \in \{1, \ldots, r\}$.*

*Proof* If $\pi$ can be written as the image of a word in $\{t_{k_1}, t_{k_2}, \ldots, t_{k_r}\}$ then any permutation that fixes each of these generators must commute with $\pi$. Thus, conversely, $\pi$ must commute with every element of $N$ that fixes these symmetric generators, and thus lies in the centralizer in $N$ of $N_{k_1 k_2 \cdots k_r}$. $\square$

The lemma has proved remarkably effective even when $r = 2$, when we seek permutations that can be written in terms of just two of the symmetric generators. Indeed in Curtis (2007b, pages 170–171), we see how a relation of the form $\pi = t_i t_j t_i$ defines the Hall Janko group HJ and the exceptional Lie group $G_2(4) : 2$. However, before proceeding to the examples mentioned above, note the following.

**Note 14.2** It may happen that the stabilizer of two symmetric generators fixes several more elements of $\mathcal{T}$. Explicitly, we may have

$$\text{Fix}(N_{k_1 k_2}) = \{k_1, k_2, \ldots, k_r\},$$

so

$$N_{k_1 k_2} = N_{k_1, k_2, \ldots, k_r}.$$

So, instead of looking to write $\pi$ as a word in $t_{k_1}$ and $t_{k_2}$, it may be more fruitful to write it as a word in $\{t_{k_1}, t_{k_2}, \ldots, t_{k_r}\}$.

## 14.2 The Conway Group ·O

In our first example using $N \cong M_{24}$ we take the action of $M_{24}$ on tetrads of points in $\Omega$. Note that this is *not* a primitive action as the tetrad stabilizer is properly contained in the stabilizer of the sextet defined by that tetrad. Thus our progenitor will take the form

$$P \cong 2^{\star \binom{24}{4}} : M_{24},$$

and the symmetric generators will be

$$\mathcal{T} = \{t_T \mid T \subset \Omega, |T| = 4\}.$$

We shall be interested in two such symmetric generators $t_T$ and $t_U$, where $|T \cap U| = 2$ and $T \cup U$ is contained in an octad; that is to say $T \cup U$ is an $S_6$ in the classification of subsets of $\Omega$. An example of two such tetrads is given in Figure 14.1 where the pairing of the eight points of the octad defined by $T$ and $U$ is exhibited; thus $t_T = T_{ab}$ and $t_U = t_{ac}$. Once $T$ is chosen in $\binom{24}{4}$ ways, there are $\binom{4}{2}.5.\binom{4}{2} = 180$ ways of choosing $U$, and so $K$, the stabilizer in $M_{24}$ of $T$ and $U$ and hence of the pairing, has order

$$|K| = |\text{Stab}_{M_{24}}[T, U]| = \frac{|M_{24}|}{\binom{24}{4} \times 180} = 2^7.$$

$K$ has shape $2^4 : 2^3$, and consists of $L$, the elementary abelian group of order 16 fixing every point of the octad, extended by a group of order 8 every element of which flips an even number of the pairs $aa, bb, cc, dd$, see Figure 14.2.

According to Lemma 14.1, and to the Note 14.2 preceding it, the only elements of $M_{24}$ that can be written as words in

$$\mathcal{X} = \{t_{ab}, t_{ac}, t_{ad}, t_{bc}, t_{bd}, t_{cd}\}$$

without causing collapse lie in the centralizer in $M_{24}$ of $K$. But $K$ has orbits of sizes $2^4$ and 16 on $\Omega$, and so an element commuting with it will fix the octad $abcd$ and so lie in the octad stabilizing subgroup of shape $2^4 : A_8$. However,

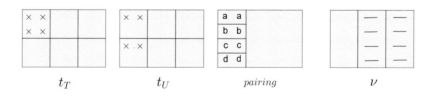

Figure 14.1 Symmetric generators for the Conway group ·O

## 14.2 The Conway Group ·O

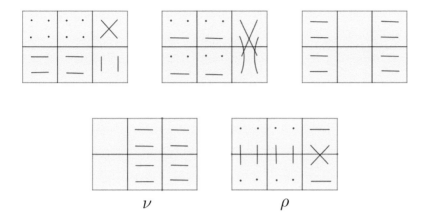

Figure 14.2 Elements of $K$, the central element $\nu$ and an element $\rho$ commuting with $\nu$

any element having non-trivial action on the octad acts non-trivially as in $L_4(2)$ on the 15 involutions of $L$ and so the centralizer of $K$ must lie in $L$. In fact, as can readily be checked using the MOG, see Figure 14.2, $K$ has orbits of lengths $1 + 2 + 2 + 2 + 8$ on these 15 involutions, and so the centralizer has order 2; it is the centre of $K$ that is generated by the element $\nu$ in Figure 14.2. We now seek the shortest possible word in the six elements of $X$ that could be set equal to $\nu$ without causing collapse. That is to say

$$\nu = w(t_{ab}, t_{ac}, t_{ad}, t_{bc}, t_{bd}, t_{cd})$$

with $w$ a word of shortest possible length. But $l(w) = 1$ is impossible as this would imply that $\nu$ commutes with the stabilizer in $M_{24}$ of $T$, which it does not. If $l(w) = 2$ then $\nu = t_{ab}t_{cd}$ or $\nu = t_{ab}t_{ac}$. The first case would require that $\nu$ lies in the centralizer of the stabilizer in $M_{24}$ of the two tetrads $ab$ and $cd$, which it does not. For instance, the element $[1001\omega\bar{\omega}]$ in the notation of Figure 8.4 fixes these two tetrads but does not commute with $\nu$. The second length 2 possibility is $\nu = t_{ab}t_{ac}$, but similarly we should then have $\nu = t_{ab}t_{ac} = t_{ac}t_{ad} = t_{ad}t_{ab}$ and multiplying these three relations together we get $\nu = 1$, a contradiction. So the minimum length for $w$ is 3, and $\nu = xyz$ with $x, y, z \in X$. But $\nu$ commutes with each of the elements of $X$ and so $xy = vz$ has order two and $x$ and $y$ commute with one another, and similarly all three elements $x, y$ and $z$ commute with one another. The elements $x, y, z$ must therefore be distinct and, if two of them are complementary within the octad, then without loss of generality we must have $\nu = t_{ab}t_{cd}t_{ac}$. But the element $[111\,100]$ in the hexacode notation of

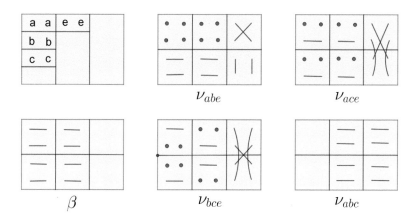

Figure 14.3 Some elements $v_{xyz}$

Figure 8.4 acts as $(a\,b)(c\,d)$ on the pairs and commutes with $v$, which implies that $t_{ac} = t_{bd}$, a contradiction. So we have just two possibilities

(i) $v = t_{ab}t_{ac}t_{bc}$, or
(ii) $v = t_{ab}t_{ac}t_{ad}$.

In order to see that case (i) fails we introduce a further pair $ee$ as in Figure 14.3 such that $aabbccee$ is an umbral octad $U_8$ as in the Todd–Conway diagram in Figure 7.12. Every triple of pairs of $aa, bb, cc, ee$ determines an element analogous to $v$ and the four resulting elements are given in Figure 14.3 where our familiar $v$ is now labelled $v_{abc}$ as

$$v_{abc} = t_{ab}t_{ac}t_{bc}.$$

Then

$$\begin{aligned}
1 &= t_{ab}^2 t_{ac}^2 t_{ae}^2 = t_{ab}t_{ac}.t_{ac}t_{ae}.t_{ae}t_{ab} \\
&= v_{abc}t_{bc}v_{ace}t_{ce}v_{abe}t_{be} \\
&= v_{abc}v_{ace}v_{abe}t_{bc}t_{ce}t_{be} \\
&= v_{abc}v_{ace}v_{abe}v_{bce}t_{be}t_{be} \\
&= \beta, \text{ say,}
\end{aligned}$$

where $\beta$ is the involution shown in Figure 14.3. Thus the relation in (i) causes the progenitor to collapse. We are thus left with relation (ii) and assert that

## 14.2 The Conway Group ·O

**Theorem 14.3**

$$G = \frac{2\star\binom{24}{4} : M_{24}}{t_{ab}t_{ac}t_{ad} = \nu} \cong \cdot O.$$

We may take this as the definition of the group ·O, as indeed is done in Curtis (2007b) where the definition is then used to construct the Leech lattice. However, since we have already constructed the lattice in this book and are familiar with the Conway elements that generate ·O, we prefer here to first demonstrate that the relation of Theorem 14.3 does indeed hold in ·O. For convenience of notation we recall the labelled tetrads $A, B, \ldots, G$ of Figure 4.2 as they occur as subsets of $\Lambda_1$, the first octad of the MOG; so we have $t_A = t_{cd}, t_{A'} = t_{ab}, \ldots$. Thus the relation now becomes $t_{A'}t_{B'}t_{C'} = \nu$.

**Lemma 14.4** *Let $\xi_T$ denote the Conway element of Section 11.1 where $T$ is a tetrad of points of $\Omega$. Then setting $t_T = -\xi_T$ we have*

$$t_{A'}t_{B'}t_{C'} = t_{ab}t_{ac}t_{ad} = \nu.$$

*Proof* We shall let the given product of Conway elements act on certain vectors of form $8\nu_i$ and show that it has the same effect as the permutation $\nu$. We know that these generators commute with one another and so centralizing the product in $M_{24}$ is a subgroup fixing the pair $aa$, fusing the three pairs $bb, cc, dd$ and transitive on the remaining 16 points. So we need only apply our product to three vectors in the Leech lattice. With the standard MOG arrangement these may be taken to be $8\nu_\infty, 8\nu_3$ and $8\nu_{11}$. In Figure 14.4 the action of our word in the $t_T$ on these three vectors of the Leech lattice is exhibited explicitly. Note that for each $t_T$ we must apply the corresponding Conway element $\xi_T$ and then

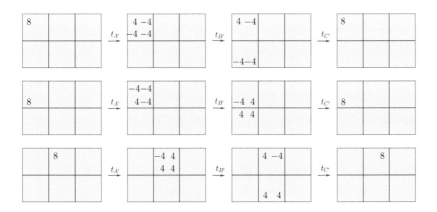

Figure 14.4 Verification that the relation of Theorem 14.3 holds in ·O

negate. This demonstrates that this word has the same effect on the 24 vectors $\{v_i \mid i \in \Omega\}$ as the permutation $v$ and, since these form a basis for the vector space, we have equality. □

Lemma 14.4 shows that ·O is a homomorphic image of the group defined by the symmetric presentation in Theorem 14.3, but as yet it could define an infinite group. Before proving that it does in fact define ·O, we explore the implications of the relation further. We claim

**Lemma 14.5** *The involutions $t_A$ and $t_{A'}$ commute with one another, and so $t_A t_{A'}$ is itself an involution.*

*Proof* We know that if $|T| = |U| = 4$, $|T \cap U| = 2$ and $T \cup U \subset O$, an octad of the Steiner system, then $t_T$ and $t_U$ commute. So $t_A$ commutes with $t_{B'}, t_{C'}$ and $v$, and thus it commutes with $t_{A'}$. □

We now show that the involution $t_T t_U$ where $T \cup U = O \in C_8$ is independent of the partition of the octad $O$ into two disjoint tetrads chosen.

**Lemma 14.6** *Suppose $X \cup X' = O = \Lambda_1$ is any other partition of $\Lambda_1$ into two disjoint tetrads. Then $t_A t_{A'} = t_X t_{X'}$, and so we may write $t_A t_{A'} = \epsilon_O$, an involution that depends only on the octad $O$ and thus commutes with the octad stabilizing subgroup of $M_{24}$.*

*Proof* The element $t_A t_{A'}$ certainly commutes with the maximal subgroup $H$ of the octad stabilizing subgroup of shape $2^4 : (A_4 \times A_4).2$ that preserves a partition of the octad into the two tetrads $A$ and $A'$. But

$$t_A t_{A'} = t_A t_B^2 t_{A'} = t_{cd} t_{bd} \cdot t_{bd} t_{ab} = t_{ad} v t_{bc} v = t_{C'} t_C,$$

and the element $\rho$ of Figure 14.2 preserves the partition of octad $O$ as $C : C'$ but does not lie in the copy of $2^4 : (A_4 \times A_4) \cdot 2$ preserving the partition $A : A'$. Thus

$$\langle H, \rho \rangle \cong 2^4 : A_8$$

and $\epsilon_O = t_A t_{A'}$ is independent of the partition into two tetrads. □

**Corollary** *If $O$ and $O'$ are two octads with $|O \cap O'| = 4$ then*

$$\epsilon_O \cdot \epsilon_{O'} = \epsilon_{O+O'},$$

*where $O + O'$ denotes the symmetric difference of $O$ and $O'$.*

*Proof* We write $O \setminus O' = T_1$, $O' \setminus O = T_2$ and $O \cap O' = T_3$, then $T_1, T_2, T_3$ are three tetrads of a sextet and

$$\epsilon_O \epsilon_{O'} = t_{T_1} t_{T_3} t_{T_3} t_{T_2} = t_{T_1} t_{T_2} = \epsilon_{O+O'}. \qquad \square$$

## 14.2 The Conway Group ·O

In order to prove Theorem 14.3 we shall need to perform a coset enumeration, see Section 15.4, over a subgroup; however, the index of $M_{24}$ in ·O is too large and so we seek to establish a larger subgroup. Hence

**Lemma 14.7** *The subgroup of G generated by the 759 involutions*

$$E = \langle \epsilon_O \mid O \in C_8 \rangle$$

*is an elementary abelian subgroup of order $2^{12}$, which is a copy of the binary Golay code.*

*Proof* Let $U: V: W$ be a trio of octads, and let $\{T_1, T_2, \ldots, T_6\}$ be a sextet refining this trio such that $U = T_1 \cup T_2$, $V = T_3 \cup T_4$ and $W = T_5 \cup T_6$. Then $\epsilon_U = \epsilon_{T_1+T_2}$ commutes with $\epsilon_{T_1+T_3}$ and $\epsilon_{T_1+T_4}$. But

$$\epsilon_{T_1+T_3} \cdot \epsilon_{T_1+T_4} = \epsilon_{T_3+T_4} = \epsilon_V,$$

and so, if $R$ and $S$ are two disjoint octads, $\epsilon_R$ commutes with $\epsilon_S$. Thus, for our trio $U: V: W$ we have that $\zeta = \epsilon_U \epsilon_V \epsilon_W$ is an involution that is preserved by the whole of the whole of the maximal trio group. But we have

$$\zeta = \epsilon_U \epsilon_V \epsilon_W = t_{T_1} t_{T_2} \cdot t_{T_3} t_{T_4} \cdot t_{T_5} t_{T_6} = t_{T_1} t_{T_3} \cdot t_{T_2} t_{T_4} \cdot t_{T_5} t_{T_6}$$

since the $t_{T_i}$ commute with one another. So $\zeta$ is preserved by elements of the sextet group preserving our refinement that do not preserve the trio; it is thus preserved by the whole of $M_{24}$ and we may rename $\zeta$ as $\epsilon_\Omega$; it is the product $\epsilon_{O_1} \epsilon_{O_2} \epsilon_{O_3}$ for $O_1 : O_2 : O_3$ any trio. Moreover, if $O$ is any octad then $\epsilon_O \epsilon_\Omega = \epsilon_{O+\Omega}$, an involution, and so $\epsilon_\Omega$ commutes with all the $\epsilon_O$.

We have seen that if $O$ and $U$ are octads such that $|O \cap U| = 0$ or 4 then $\epsilon_O$ and $\epsilon_U$ commute. Suppose now that $O, U \in C_8$ and $|O \cap U| = 2$. In Figure 14.5 we see two such octads and, by inserting the square of a 'dummy' octad type element whose octad intersects both $O$ and $U$ in 4 points, we see that this element is expressible as the product of two octad type involutions in two different ways. But each of these expressions is preserved by a subgroup of $M_{12}$ of shape $S_6$ fixing a partition into two hexads. Together these two copies of $S_6$ generate $M_{12}$, and so the element $\epsilon_O \epsilon_U$ depends only on the dodecad $O + U$ and we may legitimately write it as

$$\epsilon_O \epsilon_U = \epsilon_V \epsilon_W = \epsilon_{O+U} = \epsilon_U \epsilon_O = \epsilon_D,$$

where dodecad $D = O + U$ and $V$ and $W$ are any two octads such that $V + W = D$. This implies that the number of such dodecad type elements is $|M_{24}|/|M_{12}| = 2576$ and that octad type elements that intersect in two points commute with one another. So $E$ is certainly elementary abelian. It remains to show that it has the desired order.

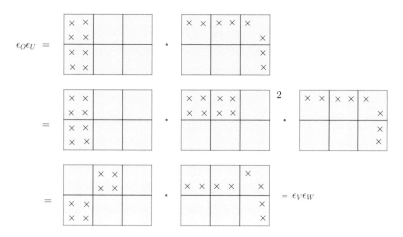

Figure 14.5 Octads demonstrating that reflection on a dodecad is well defined

We first claim that $\epsilon_D \epsilon_\Omega = \epsilon_{D+\Omega}$. To see this let $\Lambda_1 : \Lambda_2 : \Lambda_3$ be the MOG trio, so that $\epsilon_O = \epsilon_{\Lambda_1}$ in Figure 14.5. Then

$$\epsilon_D \epsilon_\Omega = \epsilon_{\Lambda_1} \epsilon_U \epsilon_{\Lambda_1} \epsilon_{\Lambda_2} \epsilon_{\Lambda_3} = \epsilon_{\Lambda_2} \epsilon_{U+\Lambda_3} = \epsilon_{D+\Omega},$$

as required.

It remains to show that multiplying a dodecad type element $\epsilon_D$ by an octad type generator $\epsilon_O$ results in one of the elements we have already seen. Now $|D \cap O| = 6, 4$ or $2$. But if $|D \cap O| = 6$ then $\epsilon_D \epsilon_O = \epsilon_{D+O}$ as we have already seen, and if $|D \cap O| = 2$ then $|(D+\Omega) \cap O| = 6$ and so $\epsilon_D \epsilon_\Omega \epsilon_O = \epsilon_{D+\Omega} \epsilon_O = \epsilon_{D+O+\Omega}$. So $\epsilon_D \epsilon_O = \epsilon_{D+O+\Omega} \epsilon_\Omega$. Finally, if $|D \cap O| = 4$, then we may choose $V$ to be an octad containing $D \cap O$ and two further points of $D$, then

$$\epsilon_D \epsilon_V \epsilon_V \epsilon_O = \epsilon_{D+V} \epsilon_{V+O} = \epsilon_{D+O},$$

where $V + O$ and $D + O$ are two octads intersecting in two points. Thus the group $E$ consists of the identity element, 759 octad type involutions $\epsilon_O$, 759 elements $\epsilon_{O+\Omega}$, 2576 dodecad type elements $\epsilon_D$ and the element $\epsilon_\Omega$. We have

$$1 + 759 + 2576 + 759 + 1 = 2^{12}$$

and we have confirmed that $E$ is essentially a copy of the binary Golay code. □

### 14.2.1 The Double-Coset Enumerator

The way in which groups are defined by means of a symmetric presentation lends itself to enumeration of the double cosets of form $NwN$, where $N$ is the control subgroup. The author devised a modification of the Todd–Coxeter single-coset enumeration for groups defined in this manner, and he carried out a number of hand enumerations. John Bray implemented this approach using the algebra package MAGMA, see Bosma and Cannon (1994), Bray and Curtis (2004) and Curtis (2007b, page 66), and extended it so as to enumerate double cosets of the form $HwN$; this improvement is particularly effective when $N \le H$.

The general form for a symmetrically generated group $G$ with involutory generators is as

$$G = \frac{2^{\star n} : N}{\pi_1 w_1, \pi_2 w_2, \ldots},$$

where $2^{\star n}$ denotes a free product of $n$ copies of $C_2$, the cyclic group of order 2; $N$ is the control subgroup, a permutation group of degree $n$; $\pi_i$ is a permutation in $N$ and $w_i$ is a word in the $n$ symmetric generators. Since every element of this group can be written as $\pi w$ for some permutation $\pi \in N$ and $w$ some word in the symmetric generators, the double cosets take the form

$$N\pi w N = NwN,$$

since $\pi \in N$. The double coset $N t_{k_1} t_{k_2} \cdots t_{k_r} N$ will be denoted by the sequence of subscripts $[k_1, k_2, \ldots, k_r]$. The enumerator finds all such double cosets and calculates how many single cosets each of them contains, hence it obtains the index of $N$ in $G$. The additional relators $\pi_i w_i$ are fed into the enumerator in the form $\langle ss, \pi \rangle$ where $ss$ is a sequence of subscripts $[l_1, l_2, \ldots, l_s]$ such that $\pi = t_{l_1} t_{l_2} \cdots t_{l_s}$. We may now use the enumerator to confirm mechanically Lemma 14.7, which we have just proved by hand. Thus

**Theorem 14.8**

$$\frac{2^{\star 759} : M_{24}}{\epsilon_O \epsilon_U \epsilon_{U+V} = 1} \cong 2^{12} : M_{24},$$

*where $O$ and $U$ are octads intersecting in four points, and $O + U$ denotes their symmetric difference.*

*Proof* We must first obtain $M_{24}$ as a permutation group of degree 759 acting on the octads of the Steiner system. We do this by asking for the **coset action** of $M_{24}$ on the cosets of the stabilizer of the octad labelled $[A', A', O]$ in the notation of Section 4.2, and we label this group **nn**. The function that maps permutations in $M_{24}$ of degree 24 to permutations of degree 759 acting on the octads is denoted by **f** and **kk** denotes the kernel of the mapping, which in the cases we shall consider will be trivial. Taking this octad as $O$, we let $U$ be the octad $[B', B', O]$ that intersects $O$ in four points, which is to say it lies in the 280-orbit of the stabilizer of $O$. The sum of these two octads $O + U$ will thus be $[C, C, O]$.

```
> m24:=sub<Sym(24)|
> (1,2,3,4,5,6,7,8,9,10,11,12,13,14,15,16,17,18,19,20,21,22,23),
> (24,23)(3,19)(6,15)(9,5)(11,1)(4,20)(16,14)(13,21)>;
> oct:=Stabilizer(m24,{24,23,3,19,11,1,4,20});
> f,nn,kk:=CosetAction(m24,oct);
> oo:=Orbits(Stabilizer(nn,1));
> [#oo[i]:i in [1..#oo]];
[ 1, 30, 280, 448 ]
> e3:=f(m24!(3,6,9)(19,15,5)(4,16,13)(20,14,21)(18,8,12)(10,17,7));
> e2:=f(m24!(24,9)(3,6)(23,5)(19,15)(11,13)(4,16)(1,21)(20,14));
> RR:=[<[1,1^e3,1^(e2*e3^2)],Id(nn)>];
> CT:=DCEnum(nn,RR,[nn]:Print:=5,Grain:=100);
Index: 4096 === Rank: 5 === Edges: 13 === Status: Early closed
=== Time: 0.31
> CT[7];
[ 1, 759, 2576, 759, 1 ]
```

The three octads $O, U$ and $O + U$ are all contained in the first four columns of the MOG; O is the first two rows, U is the 1st and 3rd, and so $O + U$ is the 2nd and 3rd. So the single additional relator becomes

$$\langle [t_O, t_U, t_{O+U}], \mathrm{Id}(nn)] \rangle = \langle [1, 1^{e3}, 1^{e2.e3^2}], \mathrm{Id}(nn) \rangle.$$

The enumerator confirms that $M_{24}$ has index $2^{12}$ in the group defined by this symmetric presentation. It returns a great deal of information about the action including representatives for each of the double cosets. Here we have simply printed out the lengths of the five double cosets which, of course, correspond to the empty set, the 759 octads, the 2576 dodecads, the 759 16-ads and the whole set $\Omega$. The additional relation holds in the affine group $2^{12} : M_{24}$ and so this group is a homomorphic image of the group defined by the symmetric presentation of Theorem 14.8. The enumeration confirms that it is an isomorphism and so, in particular, the subgroup generated by the symmetric generators is elementary abelian.

*Proof of Theorem 14.3* We now have a much larger group $H \cong 2^{12} : M_{24}$ that we can use in our enumeration of double cosets of the form $HwN$. We first need $M_{24}$ acting on the $\binom{24}{4}$ tetrads and so let **tetab** denote the tetrad consisting of the two pairs *aabb* of Figure 14.1, and take $M_{24}$ acting on the cosets of its stabilizer. The function **ff** now maps permutations of degree 24 onto actions on this set of $\binom{24}{4}$ points. The permutation **e3** of $M_{24}$ that fixes the columns and

## 14.2 The Conway Group ·O

the top row of the MOG while rotating the other three rows downwards acts as $(a)(b\ c\ d)$ on the pairs and so maps the tetrad $aabb$ to $aacc$ to $aadd$, the three tetrads we require in the additional relation of Theorem 14.3. Now $t_{ab}t_{cd} \in H$ and so we need the tetrad $ccdd$. The element **e2** in the MAGMA code below acts as $(a\ d)(b\ c)$ on the pairs and so maps $aabb$ to $ccdd$ as required. We may now define $H$ by extending $nnn \cong M_{24}$ by $t_{ab}t_{cd} = \epsilon_O$, where here $O$ is the first brick of the MOG. This element is represented in the coset enumerator as $\langle [1, 1^{e2}], \mathrm{Id}(nnn) \rangle$.

```
> m24:=sub<Sym(24)|
> (1,2,3,4,5,6,7,8,9,10,11,12,13,14,15,16,17,18,19,20,21,22,23),
> (24,23)(3,19)(6,15)(9,5)(11,1)(4,20)(16,14)(13,21)>;
> tetab:={24,23,3,19};
> stab:=Stabilizer(m24,tetab);
> ff,nnn,kk:=CosetAction(m24,stab);
> e3:=ff(m24!(3,6,9)(19,15,5)(4,16,13)(20,14,21)(18,8,12)(10,17,7));
> e2:=ff(m24!(24,9)(3,6)(23,5)(19,15)(11,13)(4,16)(1,21)(20,14));
> stabcd:=Stabilizer(nnn,[1,1^e3]);
> #stabcd;
128
> cabcd:=Centralizer(nnn,stabcd);
> #cabcd;
2
> RR:=[<[1,1^e3,1^(e3^2)],cabcd.1>];
> HH:=[*nnn,<[1,1^e2],Id(nnn)>*];
> CTdotto:=DCEnum(nnn,RR,HH:Print:=5,Grain:=100);
Index: 8292375 === Rank: 19 === Edges: 1043 === Status: Early closed
=== Time: 31.64
> Factorisation(8292375);
[ <3, 6>, <5, 3>, <7, 1>, <13, 1> ]
```

The stabilizer in $M_{24}$ of $t_{ab}$ and $t_{ac}$, which is the group of order $2^7$ fixing all four pairs $aa, bb, cc$ and $dd$ that we have examined before, is labelled **stabcd**. Its centralizer in $M_{24}$ is its centre **cabcd** $= \langle v \rangle$, and so $v = $ **cabcd.1**. The subgroup $H$, written **HH** in the code, is defined by placing generators for it between asterisks. The enumerator informs us that $H$ has index $8\,292\,375 = 3^6.5^3.7.13$ in $G$ and so

$$|G| = 3^6.5^3.7.13.2^{12}.|M_{24}| = 2^{22}.3^9.5^4.7^2.11.13.23 = |\cdot O|.$$

Since we know that ·O is an image of $G$, we have demonstrated isomorphism. □

Note that the enumerator also tells us that $M_{24}$ has 19 orbits on the cosets of $H$. Since $H$ is the stabilizer in ·O of a cross this is telling us that $M_{24}$ has 19 orbits on the crosses. The lengths of these 19 orbits are returned by the program.

## 14.3 The Janko Group $J_4$

This section is based on joint work of S. Bolt, J. N. Bray and the author; see Bolt (2002), Bray (1997).

The first sporadic finite simple group to be discovered after the celebrated Mathieu groups was the Janko group $J_1$, see Janko (1965, 1966). This was in 1965 and Zvonimir Janko went on to discover three further finite simple groups, culminating in the magnificent $J_4$, see Janko (1976). This group was investigated by many people, for instance Lempken (1978) and Benson (1980), the second of whom used MOG techniques extensively in his approach. It is a delightful fact that Janko thus discovered the first sporadic finite simple group to be found since the 19th century and the last.

The lowest degree of an irreducible complex representation of $J_4$ is in 1333 dimensions, but Norton (1980) discovered that it can be represented in 112 dimensions over $\mathbb{Z}_2$, the integers modulo 2. Generators for the group as $112 \times 112$ matrices over $\mathbb{Z}_2$ were constructed by Norton, Parker and Thackray. Details of the construction can be found in Benson (1980) and in Norton (1980).

So far we have investigated progenitors in which $M_{24}$ acts on a set of involutions corresponding to octads, and have used its action on tetrads of points to define and construct the largest Conway group $\cdot O$. We shall now investigate its action on trios and on sextets, diagrams of which are given in Figures 12.5 and 12.4. The five ways in which two trios can intersect one another and the manner in which these intersections correspond to the suborbits are shown in Figure 14.6, and the four ways in which two sextets can intersect and the correspondence with suborbits is shown in Figure 14.7.

$$\begin{pmatrix} 8 & 0 & 0 \\ 0 & 8 & 0 \\ 0 & 0 & 8 \end{pmatrix} \begin{pmatrix} 8 & 0 & 0 \\ 0 & 4 & 4 \\ 0 & 4 & 4 \end{pmatrix} \begin{pmatrix} 0 & 4 & 4 \\ 4 & 0 & 4 \\ 4 & 4 & 0 \end{pmatrix} \begin{pmatrix} 0 & 4 & 4 \\ 4 & 2 & 2 \\ 4 & 2 & 2 \end{pmatrix} \begin{pmatrix} 4 & 2 & 2 \\ 2 & 4 & 2 \\ 2 & 2 & 4 \end{pmatrix}$$
$$\begin{matrix} I_0 & I_1 & I_2 & I_3 & I_4 \\ (1) & (42) & (56) & (1008) & (2688) \end{matrix}$$

Figure 14.6 The intersection matrices for two trios

Consider, then, the progenitor

$$P_J = 2^{\star 3795} : M_{24},$$

in which the 3795 symmetric generators correspond to the trios. The generators for $J_4$ given in Parker et al. (1999) may be used to verify that the group is indeed a homomorphic image of $P_J$ and, furthermore, that the relations which we are about to deduce also hold. So there is no question of the resulting symmetric presentation collapsing. However, it is our intention to show that two short

$$J_0 = \begin{pmatrix} 4 & 0 & 0 & 0 & 0 & 0 \\ 0 & 4 & 0 & 0 & 0 & 0 \\ 0 & 0 & 4 & 0 & 0 & 0 \\ 0 & 0 & 0 & 4 & 0 & 0 \\ 0 & 0 & 0 & 0 & 4 & 0 \\ 0 & 0 & 0 & 0 & 0 & 4 \end{pmatrix} \quad J_1 = \begin{pmatrix} 2 & 2 & 0 & 0 & 0 & 0 \\ 2 & 2 & 0 & 0 & 0 & 0 \\ 0 & 0 & 2 & 2 & 0 & 0 \\ 0 & 0 & 2 & 2 & 0 & 0 \\ 0 & 0 & 0 & 0 & 2 & 2 \\ 0 & 0 & 0 & 0 & 2 & 2 \end{pmatrix}$$
$$(1) \qquad\qquad\qquad (90)$$

$$J_2 = \begin{pmatrix} 3 & 1 & 0 & 0 & 0 & 0 \\ 1 & 3 & 0 & 0 & 0 & 0 \\ 0 & 0 & 1 & 1 & 1 & 1 \\ 0 & 0 & 1 & 1 & 1 & 1 \\ 0 & 0 & 1 & 1 & 1 & 1 \\ 0 & 0 & 1 & 1 & 1 & 1 \end{pmatrix} \quad J_3 = \begin{pmatrix} 2 & 0 & 0 & 0 & 1 & 1 \\ 0 & 2 & 0 & 0 & 1 & 1 \\ 0 & 0 & 2 & 0 & 1 & 1 \\ 0 & 0 & 0 & 2 & 1 & 1 \\ 1 & 1 & 1 & 1 & 0 & 0 \\ 1 & 1 & 1 & 1 & 0 & 0 \end{pmatrix}$$
$$(240) \qquad\qquad\qquad (1440)$$

Figure 14.7 The intersection matrices for two sextets

relations that can be deduced using Lemma 14.1 without reference to the group are sufficient to define the group. Indeed, Bray has shown that one of these two relations is essentially redundant, as will be explained at the end of this section.

### 14.3.1 Deducing Suitable Relations by Which to Factor $P_J$

If two trios have an octad in common, then they must intersect as in $I_1$ of Figure 14.6. The stabilizer of two trios related in this manner fixes a third trio. Explicitly, if the two trios are $E = \{U, V, W\}$, the MOG trio, and $A = \{U, X, Y\}$ of Figure 14.8 then the third fixed trio is

$$C = \{U, (V + X), (W + X)\}$$

where, for instance, $V + X$ denotes the symmetric difference of the two octads $V$ and $X$; we denote this trio by $E + A$ a 'sum' which is only defined if $E$ and $A$ intersect as in $I_1$. The stabilizer of these three trios must fix the sextet four tetrads of which are $V \cap X$, $V \cap Y$, $W \cap X$ and $W \cap Y$. Indeed, if we label the tetrads of this sextet 1, 2, 3, 4, 5, 6, then these three trios can be taken as 12.34.56, 12.35.46 and 12.36.45, and the subgroup of the sextet stabilizer that fixes them all can be seen to have shape $2^6 : (S_3 \times V_4)$, acting as $\langle (1\,2), (3\,4)(5\,6), (3\,5)(4\,6) \rangle$ on the six tetrads of the fixed sextet. This subgroup has trivial centralizer in $M_{24}$ and so Lemma 14.1 says that the only element of $M_{24}$ that can be written in terms of them without causing collapse is the identity. We thus set the product of all three to be the identity, which is to say

$$t_E t_A t_{E+A} = 1,$$

where $E$ and $A$ are two trios intersecting as in $I_1$ of Figure 14.6.

We now investigate two trios whose intersection matrix is as in $I_3$, an example of which is the top two trios in Figure 14.8 which are labelled $A$ and $B$. Note that the first brick of the MOG is the unique octad of trio $A$ to intersect the three octads of $B$ as $4.4.0$, and that the third brick of the MOG is the only octad of trio $B$ to intersect the three octads of $A$ as $4.4.0$. Thus the stabilizer of both $A$ and $B$ must fix the MOG trio which is labelled $E$ in Figure 14.8. But now $A$ and $E$ intersect as in $I_1$ and so their sum, namely $A + E = C$ of Figure 14.8, must also be fixed; and similarly $B + E = D$ is fixed. Thus $st_{AB}$, the stabilizer of $A$ and $B$, fixes the five trios $A, B, C, D$ and $E$. We now apply Lemma 14.1 to this set of symmetric generators. The stabilizer in $M_{24}$ of $A$ and $D$ has order

$$|st_{AB}| = |\text{Stabilizer}(M_{24}, [A, B])| = \frac{|M_{24}|}{3795 \times 1008} = 64.$$

The centralizer in $M_{24}$ of $st_{AB}$ is its centre of order 4 whose non-trivial elements are exhibited in Figure 14.9 as $v_1, v_2$ and $v_3$. The top row of Figure 14.9 gives generators for $st_{AB}$, which by the above fixes $A, B, C, D$ and $E$. So Lemma 14.1 tells us that the only non-trivial elements of $M_{24}$ that can be written in terms of $t_A, t_B, t_C, t_D$ and $t_E$ without causing the progenitor to collapse are the $v_i$ for $i = 1, 2, 3$. However, the normalizer in $M_{24}$ of $st_{AB}$ is a subgroup of order $2^9$

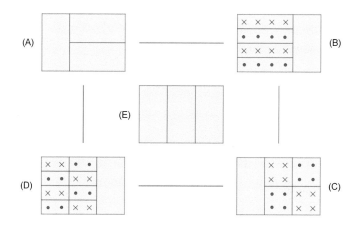

Figure 14.8 The five trios yielding the defining relation for the Janko group $J_4$

## 14.3 The Janko Group $J_4$

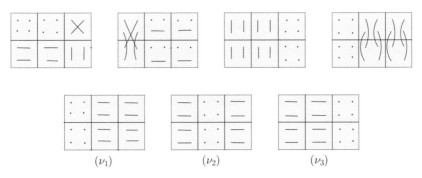

Figure 14.9 Generators for the subgroup of $M_{24}$ fixing two trios that intersect one another as in matrix $I_3$

consisting of $\mathrm{st}_{AB}$ extended by a group isomorphic to the dihedral group $D_8$ which acts as $\langle (A\,B\,C\,D), (B\,D)\rangle$ on the five fixed trios, and interchanges $v_1$ and $v_3$. Thus, if any one of the three $v_i$ can be written as a word in the $t_A, t_B, t_C, t_D$ and $t_E$ then $v_2 = v_1 \cdot v_3$ can be. We seek the shortest possible word $w$ such that $v_2 = w(t_A, t_B, t_C, t_D, t_E)$. Note that, since $t_A t_C t_E = t_B t_D t_E = 1$, if $w$ involves $t_E$ then we can immediately shorten it. Now $l(w) = 1$ is impossible as it would imply that $M_{24}$ contains involutions commuting with the trio stabilizer, which it does not. If $l(w) = 2$ then without loss of generality $v_2 = t_A t_C = t_E$, a contradiction; or $v_2 = t_A t_B = t_B t_A = t_B t_C$, which implies $t_A = t_C$, again a contradiction. If $v_2 = t_X t_Y t_Z$ for $X, Y, Z \in \{A, B, C, D\}$ then $t_X t_Y = v_2 t_Z$, an involution since $v_2$ fixes all five trios. Thus $t_X$ and $t_Y$ commute with one another, and similarly they both commute with $t_Z$. Thus $X, Y$ and $Z$ must be distinct; but now two of them must be diagonally opposite one another in Figure 14.8 and so multiply to $t_E$. So we have $l(w) \geq 4$. By the symmetries of Figure 14.8 we may assume $t_A t_B t_X t_Y = v_2$; but $X \neq B$ and $X \neq D$, or we can shorten $w$. So we have $X = A$ or $X = C$; but if it is the latter then we may replace $t_C$ by $t_A t_E$ and so we may assume that $t_A t_B t_A t_Y = v_2$ with $Y = B$ or $Y = D$. The former case leads to collapse and so we are left with the relation

$$t_A t_B t_A t_D = v_2.$$

The remainder of this section is devoted to proving the following

**Theorem 14.9** *We have*

$$G = \frac{2^{\star 3795} : M_{24}}{t_T t_U t_{T+U} = 1, t_A t_B t_A t_D = v_2} \cong J_4,$$

*where $T$ and $U$ are two trios having intersection matrix as in $I_1$, and $A, B, D$*

and $v_2$ are as in Figure 14.8. $J_4$ denotes the largest of the four sporadic simple groups discovered by Janko.

As mentioned earlier, we can use the 112-dimensional representation over $\mathbb{Z}_2$ to find matrices satisfying this presentation, and so it remains to show that the group defined in Theorem 14.9 has the right order.

### 14.3.2 A Maximal Subgroup of $G$ Containing $N \cong M_{24}$

Unfortunately, as in the case of $\cdot O$, subgroups isomorphic to $M_{24}$ are much too small to afford an enumeration of their cosets, and so we need to deduce that the presentation implies the existence of a supergroup of considerably larger order. In order to do so we recall the relationship between trios and sextets.

Recall that, as in Section 4.2, the six tetrads of a sextet $X = \{a, b, c, d, e, f\}$ may be grouped together in pairs such as $ab \cdot cd \cdot ef$ so as to form a trio in 15 ways, the *coarsenings* of the sextet, denoted by $\text{coa}(X)$. Thus a given trio is a coarsening of $1771 \times 15/3795 = 7$ sextets, the *refinements* of the trio. The seven refinements of the MOG trio correspond to the Points in Figure 4.1 and so they are labelled $\{A, B, \ldots, G\}$, and they form the seven non-trivial elements of a 3-dimensional subgroup of the cocode of the binary Golay code which is acted on by the copy of $L_3(2)$ in the trio stabilizer. Our first move is to investigate the subgroup of $G$ generated by the 15 symmetric generators corresponding to the 15 trios that coarsen a given sextet $D$. That is,

$$H_D = \langle t_T : T \in \text{coa}(D) \rangle.$$

The first additional relation of Theorem 14.9 says that

$$t_{ab \cdot cd \cdot ef} t_{ab \cdot ce \cdot df} t_{ab \cdot cf \cdot de} = 1. \qquad (14.1)$$

Now the sextet group fixing $D$, which is isomorphic to $2^6 : 3 \cdot S_6$, permutes these 15 synthemes, see Section 9.12, and its subgroup of shape $2^6 : 3$ fixes all the tetrads and so acts trivially on the synthemes. We shall use Relation 14.1 to prove

**Lemma 14.10**

$$H_D \cong 2^5 \cong 2^{1+4},$$

*an elementary abelian 2-group containing a special involution which is fixed by all permutations of* $M_{24}$ *preserving* $D$.

*Proof* The symmetric group $S_6$ acts isomorphically on the 15 synthemes and the 15 unordered pairs of the six points, with the two actions being interchanged by the outer automorphism of $S_6$. Since we find it easier to work with pairs

## 14.3 The Janko Group $J_4$

than with synthemes, we shall notionally apply the outer automorphism, and so the 15 symmetric generators under consideration coarsening the fixed sextet $D$ will (for now only) become $\mathcal{U} = \{u_{ij} \mid 1 \leq i \leq j \leq 6\}$, and Relation 14.1 will become

$$u_{12}u_{34}u_{56} = 1.$$

Thus $u_{ij}$ and $u_{kl}$ commute if $|\{i,j,k,l\}| = 4$. Then

$$u_{12}^{u_{13}} = u_{13}u_{12}u_{13} = u_{13}u_{45}u_{36}u_{13} = u_{26}u_{36}u_{25}u_{46} = u_{26}u_{14}u_{46} = u_{35}u_{46} = u_{12},$$

and so $u_{12}$ commutes with $u_{13}$ and $\langle \mathcal{U} \rangle$ is an elementary abelian 2-group. Now let $z_{abc} = u_{ab}u_{bc}u_{ca}$, where the commutativity of $\langle \mathcal{U} \rangle$ means that the order does not matter. We then have

$$z_{123}z_{124} = u_{12}u_{23}u_{13}u_{12}u_{24}u_{14} = u_{23}u_{13}u_{24}u_{14} = u_{23}u_{56}u_{14} = u_{14}u_{14} = 1,$$

and so $z_{abc} = z_{abd} = z_{bde} = z_{def}$ and $z = z_{abc}$ is independent of the choice of $a, b$ and $c$. It is now clear that the subgroup $H_D = \langle \mathcal{U} \rangle$ consists of 15 elements $u_{ab}$, 15 elements $zu_{ab}$, $z$ and the identity. □

Reverting to the original notation, in which synthemes correspond to trios coarsening the fixed sextet $D$, we see that

$$t_{ab \cdot cd \cdot ef} t_{ad \cdot cf \cdot eb} t_{af \cdot bc \cdot ed} = s_D,$$

where $s_D$ (the element $z$ above) is an involution centralized by the whole of the sextet group stabilizing $D$. There are thus just 1771 of these *sextet elements* or *sextet involutions* under the action of $M_{24}$. We deduce

**Corollary** *If $T$ and $U$ are two trios having intersection matrix as in $I_2$ then $t_T$ and $t_U$ commute with one another.*

*Proof* If sextet $D = abcdef$ then the trios $ab \cdot cd \cdot ef$ and $ad \cdot cf \cdot eb$ have intersection matrix $I_2$, and we have seen that $t_{ab \cdot cd \cdot ef}$ commutes with $t_{ad \cdot cf \cdot eb}$. □

### The Seven Sextets Refining a Trio

We have seen that a given trio $T$ possesses seven refinements that are denoted by $\text{ref}(T)$, and each of these seven sextets, $D$ say, gives rise to a sextet-type involution of $G$ denoted by $s_D$. We define

$$K_T = \langle s_X \mid X \in \text{ref}(T) \rangle$$

in the case when $T$ is the MOG trio we thus have

$$K_T = \langle s_A, s_B, s_C, s_D, s_E, s_F, s_G \rangle,$$

where $A, B, \ldots, G$ are the sextets defined by the tetrads in Figure 4.1. In $C^*$, the cocode of the binary Golay code, sextets such as $A, B, C$ form a 2-dimensional subspace with $A + B + C = 0$; this is an even line of sextets, see Figure 7.1. We now use the relations of Theorem 14.9 to prove

**Lemma 14.11** *If $A, B$ and $C$ are three sextets refining a trio such that $A + B + C = 0$ then $s_A s_B s_C = 1$.*

*Proof* The subgroup $L$ of the trio group fixing a line of refinements has index 7 and shape $2^6 : (S_3 \times S_4)$. There are $3 \times 3 \times 2 \times 2 = 36$ octads of the form $[0, X \text{ or } X', X \text{ or } X']$, where $X$ is one of the tetrads $A, B$ or $C$, using the notation of Section 4.2. So there are 18 trios that contain one of the three bricks of the MOG and two of these 36 octads. We shall denote the complements within the octad of the tetrads $A, B, C$ by $A', B'$ and $C'$ respectively. Then these 18 trios will be denoted by symbols $X_i^\epsilon$, such as

$$A_1^+ = \{[0', 0, 0], [0, A, A], [0, A', A']\} = \begin{array}{|cc|cc|cc|} \hline x & x & o & o & o & o \\ x & x & o & o & o & o \\ \hline x & x & . & . & . & . \\ x & x & . & . & . & . \\ \hline \end{array},$$

$$A_1^- = \{[0', 0, 0], [0, A, A'], [0, A', A]\} = \begin{array}{|cc|cc|cc|} \hline x & x & o & o & . & . \\ x & x & o & o & . & . \\ \hline x & x & . & . & o & o \\ x & x & . & . & o & o \\ \hline \end{array},$$

where $X$ is one of $A, B, C$; $i \in \{1, 2, 3\}$ denoting which brick of the MOG is included in the trio; and $\epsilon = +$ or $-$. The $+$ sign signifies that the tetrad $X$ is repeated in the two non-empty bricks, and the $-$ sign signifies that one of them is complemented. We thus obtain a progenitor $P_L$ of form

$$P_L = 2^{\star 18} : (2^6 : (S_3 \times S_4)),$$

which inherits several relations from the two additional relations of Theorem 14.9:

(i) $A_1^+ A_1^- = t_E$,
(ii) $A_1^+ B_1^+ C_1^+ = 1$,
(iii) $A_1^+ A_2^+ = A_2^+ A_1^+$ (derived from $A_1^+ A_2^+ A_3^- = s_A$),
(iv) $A_1^+ B_3^+ A_1^+ B_3^- = v_2$, of Theorem 14.9,

together, of course, with all their conjugates under the group $L$. Note first that these four relations imply that $(X_i^\pm Y_j^\pm)^2 = v_k t_E$ for $X, Y \in \{A, B, C\}$,

### 14.3 The Janko Group $J_4$

$X \neq Y$, and $i, j, k$ distinct. We are now in a position to perfrom the necessary calculation:

$$\begin{aligned}
s_A s_B &= A_1^+ A_2^+ A_3^- B_3^- B_2^- B_1^- = A_1^+ A_2^+ C_3^+ B_2^- B_1^- \\
&= A_1^+ A_2^+ B_2^+ C_3^+ B_1^- v_1 = A_1^+ C_2^+ C_3^+ B_1^- v_1 \\
&= A_1^+ C_2^+ B_1^+ C_3^+ v_2 v_1 = A_1^+ B_1^- C_2^+ C_3^+ v_3 v_2 v_1 \\
&= C_1^- C_2^+ C_3^+ = s_C,
\end{aligned}$$

as required. □

We thus see that our group $G$ contains 1771 involutions that correspond to the sextets in the Golay cocode, and that these involutions satisfy the relation that $s_A s_B s_C = 1$ whenever $A, B$ and $C$ are three sextets which form an even line in the cocode – which is to say that $A + B + C = 0$ and any two of them have intersection matrix as in $J_1$, see Figure 7.1. We wish to show that these 1771 involutions generate an elementary abelian group of order $2^{11}$. To do this we prove

**Theorem 14.12**

$$\frac{2^{\star 1771} : M_{24}}{s_A s_B s_C = 1} = 2^{11} : M_{24},$$

*where $\{A, B, C\}$ forms an even line of sextets in the cocode of the binary Golay code.*

Note that the right-hand side is certainly an image of the left-hand side as this relation holds in the Golay cocode, which consists of the 1771 sextets, 276 duads and the zero vector. So verification will be achieved by simply showing that $M_{24}$ has the correct index in the image.

A manual proof of Theorem 14.12, which is crucial to verifying Theorem 14.9, appears in Curtis (2007b, page 193). However, our double-coset enumerator is ideal for confirming this kind of result and so we include appropriate MAGMA code.

#### 14.3.3 Verification of Theorem 14.12 Using MAGMA

Firstly, we must obtain the action of $M_{24}$ on sextets. We let **sextAgp** denote the stabilizer of the sextet $A$ as defined in Figure 4.1. We then take the **CosetAction** on the cosets of **sextAgp**, labelling as **fs** the function which maps elements of $M_{24}$ as permutations on 24 points to permutations on the 1771 sextets; and letting **gs** denote the image group of degree 1771. We then let **e3** denote the image under **fs** of the familiar element that cycles the 2nd, 3rd and 4th rows of the MOG whilst fixing the columns; it will act as $(A\ B\ C)$ on the sextets of the canonical even line. Since the sextet *A* is labelled 1 in the permutation action of degree 1771, the relation becomes

$$[1, 1^{e3}, 1^{e3^2}] = \text{the identity of } \mathbf{gs}.$$

```
> sextAgp:=sub<m24|Stabilizer(m24,{24,23,3,19}),
> Stabilizer(m24,{6,15,9,5})>;
> Index(m24,sextAgp);
1771
> fs,gs,ks:=CosetAction(m24,sextAgp);
> e3:=fs(m24!(3,6,9)(19,15,5)(4,16,13)(20,14,21)(18,8,12)(10,17,7));
> RR:=[<[1,1^e3,1^(e3^2)],Id(gs)>];
> CT:=DCEnum(gs,RR,[gs]:Print:=0, Grain:=100);
> CT[1];CT[7];
2048
[ 1, 1771, 276 ]
```

This relation is fed into **RR** and the enumerator is asked to calculate all double cosets of the form $N\pi N$ where $N = \mathbf{gs}$. We are told that there are three double cosets of lengths 1, 1771 and 276, respectively; thus **gs** has index (1 + 1771 + 276) = 2048 = $2^{11}$ in the group defined by our symmetric presentation as required. □

### 14.3.4 Using MAGMA to Verify Theorem 14.9

Having identified the subgroup $H \cong 2^{11} : M_{24}$, being the even part of the Golay cocode extended by $M_{24}$, we are in a position to use the double-coset enumerator to find all cosets of the form $HwN$, where $N = \mathbf{gt} \cong M_{24}$ acting on the 3795 trios. We define the trios labelled *E*, *A* and *B* in Figure 14.8 as **trE**, **trA** and **trB**. In order to facilitate feeding the two relations into MAGMA, we next define the three involutions shown in Figure 14.10; involutions **e2a** and **e2b** map *E*, the MOG trio, to trios *A* and *B* of Figure 14.8 respectively. The subgroup **stAB** of order 64 mentioned earlier is obtained as the intersection of **stA** and **stB**, the stabilizers of trios *A* and *B* respectively. We verify that the centralizer of this subgroup **cstAB** is its centre of order 4, and in order to identify which

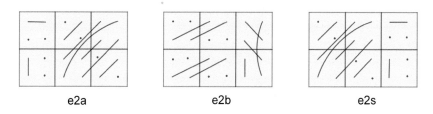

Figure 14.10 Involutions used to obtain the $J_4$ relations

## 14.3 The Janko Group $J_4$

of the three non-trivial elements in this fourgroup is our $v_1$ we extend **stAB** to its normalizer **nstAB**. Then **nu** = $v_1$ is the generator of the centre of **nstAB**.

```
> trE:={{24,23,3,6,9,19,15,5},{11,4,16,13,1,20,14,21},
> {22,18,8,12,2,10,17,7}};
> trA:={{24,3,6,9,23,19,15,5},{11,1,22,2,4,20,18,10},
> {16,14,8,17,13,21,12,7}};
> trB:={{24,23,11,1,6,15,16,14},{3,19,4,20,9,5,13,21},
> {22,18,8,12,2,10,17,7}};
> e2a:=m24!(24,23)(6,9)(1,4)(16,22)(14,18)(2,13)(10,21)(12,17);
> e2b:=m24!(3,11)(1,19)(9,16)(5,14)(10,22)(17,18)(2,7)(8,12);
> e2s:=m24!(23,3)(6,11)(1,9)(4,15)(5,20)(13,14)(2,22)(8,12);
> trE^e2a eq trA;
true
> trE^e2b eq trB;
true
> stA:=Stabilizer(m24,trA);
> stB:=Stabilizer(m24,trB);
> stAB:=stA meet stB;
> #stAB;
64
> cstAB:=Centralizer(m24,stAB);
> #cstAB;
4
> nstAB:=Normalizer(m24,stAB);
> nu:=Centre(nstAB).1;
```

We are now in a position to take the **CosetAction** of $M_{24}$ on the 3795 trios by considering the action on the cosets of the stabilizer of the MOG trio $E$. We let **ft** denote the function from permutations on 24 letters to permutations on 3795 letters, and we denote the image of degree 3795 by **gt**. The MOG trio is then labelled 1 and its image under **ft(e2A)** is trio A.

```
> stE:=Stabilizer(m24,trE);
> ft,gt,kt:=CosetAction(m24,stE);
> Degree(gt);
3795
> Fix(Stabilizer(gt,[1,1^ft(e2a)]));
{ 1, 1775, 3044 }
> Fix(Stabilizer(gt,[1^ft(e2a),1^ft(e2b)]));
{ 1, 1775, 2433, 2876, 3044 }
> Fix(Stabilizer(gt,[1,1^ft(e2a*e2s)]));
{ 1, 2757, 3321 }
```

The fix of the stabilizer of trios $E$ and $A$ is thus

$$\{E, A, E + A = C\} = \{1, 3044, 1775\}.$$

Similarly the fix of the stabilizer of trios $A$ and $B$ is

$$\{A, B, C, D, E\} = \{3044, 2876, 1775, 2433, 1\}.$$

The two relations of **RR** then become

$$t_1 t_{3044} t_{1775} = 1, \text{ and } t_{3044} t_{2876} t_{3044} t_{2433} = v_1.$$

Finally we need to produce a sextet type involution which, together with **gt** $\cong M_{24}$ will generate $H \cong 2^{11} : M_{24}$. The trios $E$ and its image under **e2a** × **e2s** are both coarsenings of the same sextet, with intersection matrix of type $I_2$ in Figure 14.6, namely the one labelled $A$ in Figure 4.1. As in Lemma 14.10, the product of the corresponding trio elements and the trio element corresponding to third syntheme they determine results in the sextet type element $s_A$. The fix of the stabilizer of these two trios consists of $\{1, 2757, 3321\}$ and so we adjoin

$$s_A = t_1 t_{2757} t_{3321}$$

to **gt** to obtain $H$. We are now in a position to apply the double-coset enumerator (DCE):

```
> RR:=[<[3044,2876,3044,2433],ft(nu)>,<[1,3044,1775],Id(gt)>];
> HH:=[*gt,<[1,2757,3321],Id(gt)>*];
> CT:=DCEnum(gt,RR,HH:Print:=1,Grain:=100);
Index: 173067389 === Rank: 20 === Edges: 3121 ===
Status: Early closed === Time:34.172
> Factorisation(173067389);
[ <11, 2>, <29, 1>, <31, 1>, <37, 1>, <43, 1> ]
```

The DCE confirms that $H$ does indeed have index $173\,067\,389$ in $G$ which is the index of subgroups $2^{11} : M_{24}$ in $J_4$. Since we are able to verify, using the 112-dimensional representation, that $J_4$ is indeed an image of $G$ this proves that Theorem 14.9 holds.

### 14.3.5 The (Virtual) Redundancy of One of the Two Relations

The word *virtual* is used here in the sense that the following corollary holds:

**Corollary** (J. N. Bray)  *Omitting the first relation in Theorem 14.9 we have*

$$G = \frac{2^{\star 3795} : M_{24}}{t_A t_B t_A t_D = v_2} \cong J_4 \times 2.$$

*Proof* First note that $G$ possesses a subgroup with index 2, namely the set of all elements of form $\pi w$ where $\pi \in N$ and $w$ is a word in the symmetric generators of even length. So $G$ cannot be simple. We shall show that the element $t_T t_U t_{T+U}$, which is set equal to the identity in Theorem 14.9, is an

## 14.3 The Janko Group $J_4$

element of order 2 in the centre of $G$. Thus factoring it out (i.e., setting it equal to 1) yields the group $J_4$. In order to simplify the notation, in this section only, we shall denote the symmetric generator $t_T$, for $T$ a trio, by $T$. Thus the relation and its conjugates become

$$ABAD = BCBA = CDCB = DADC$$
$$= ADAB = BABC = CBCD = DCDA = v_2.$$

Inverting these expressions for the involution $v_2$, we obtain

$$DABA = ABCB = BCDC = CDAD$$
$$= BADA = CBAB = DCBC = ADCD = v_2.$$

So we have $C = BABv_2$, and $D = ABAv_2$; thus $C, D \in \langle A, B, v_2 \rangle$. Moreover,

$$v_2 = DADC = ABAAABABABv_2^3 = (AB)^4 v_2$$

since $v_2$ fixes all these trios. Thus $(AB)^4 = 1$. Furthermore, $AC = v_2 ABAB$ and so $CA = v_2 BABA = AC$ and we see that $A$ and $C$ commute. Similarly $B$ and $D$ commute. Indeed, any two symmetric generators whose associated trios have an octad in common commute; and so, in particular, $E$ commutes with $A, B, C$ and $D$. So $ACE$ is an involution independent of the ordering of $A, C$ and $E$. However,

$$ACE = ABABEv_2 \text{ and } BDE = BABAEv_2 = ACE,$$

and so the element $ACE = z$, say, is fixed by any permutation of $M_{24}$ that fixes the set $\{A, C, E\}$ and by any permutation of $M_{24}$ that fixes the set $\{B, D, E\}$. It is readily verified that

$$\langle \text{Stab}_{M_{24}}(\{A, C, E\}), \text{Stab}_{M_{24}}(\{B, D, E\}) \rangle = N \cong M_{24},$$

and so $z$ commutes with $N$ and with $t_E$ and thus $z$ lies in the centre of $G$. Theorem 14.9 tells us that factoring out the subgroup $\langle z \rangle$ gives the Janko group $J_4$ and so $G \cong J_4 \times 2$ as required. □

# 15

# The Thompson Chain of Subgroups of $Co_1$

## 15.1 Introduction

As has been mentioned earlier, Conway often referred to ·O, the largest group named after him, as '$M_{24}$ writ large'. After all, $M_{24}$ is most naturally regarded as a permutation group on 24 letters, and ·O is a group of orthogonal transformations of $\mathbb{R}^{24}$, 24-dimensional Euclidean space. Moreover, $M_{24}$ is used to define the Leech lattice $\Lambda$, which has ·O as its group of symmetries, thus giving the sequence

$$M_{24} \mapsto \Lambda \mapsto {\cdot}O.$$

This is an example of a highly successful way of finding new groups: use a familiar group $H$ to construct some combinatorial object – a graph, a code, a block design or a lattice – and then observe that the new object possesses far more symmetries than those inherited from $H$. There is little doubt that the best way of studying ·O is through its action on $\Lambda$, and we would argue that an understanding of this action is facilitated by use of the Miracle Octad Generator.

However, we conclude this book by presenting a theoretical approach to defining the simple group $Co_1$, the perfect group ·O factored by its centre of order 2, but an approach that nonetheless has its roots in the MOG trio. This approach produces a nested chain of subgroups of $Co_1$ the terms of which give rise to a fascinating collection of maximal subgroups. Explicitly, the large Conway simple group $Co_1$ contains a copy of the alternating group $A_9$ and thus contains a nested sequence $A_3 \leq A_4 \leq \cdots \leq A_9$. Shortly after $Co_1$ was discovered, J. G. Thompson recognized that the normalizer of each of the groups in this sequence (apart from that of $A_8$) is maximal in $Co_1$ and the resulting collection of subgroups

$$3{\cdot}\text{Suz} : 2,\ (A_4 \times G_2(4)) : 2,\ (A_5 \times HJ) : 2,\ (A_6 \times U_3(3)) : 2,$$
$$(A_7 \times L_2(7)) : 2,\ A_8 \times S_4,\ A_9 \times S_3$$

is now known as the *Thompson chain*, where Suz denotes the Suzuki simple group, which acts on the *complex Leech lattice*, see Wilson (1983); HJ denotes the Hall–Janko group that was discovered by Zvonimir Janko (1969), and constructed by Marshall Hall Jr. and David Wales, see Hall and Wales (1969).

Remarkably, we can start at the other end in the sense that if we consider $U_3(3)$ in a certain way we obtain a construction that produces each of the groups

$$U_3(3) : 2, \ HJ : 2, \ G_2(4) : 2, \ 3\dot{}Suz : 2, \ 2 \times Co_1$$

spontaneously. Indeed, a presentation containing a parameter $n$ is given which, for $n = 3, 4, 5, 6, 7$, defines each of the above groups; $n$ appears just twice in the presentation.

Specifically, we associate with each directed edge $ij$ of $K_n$, the complete graph on $n$ vertices, an element $t_{ij}$ of order 7 in some group $G$, where $t_{ji} = t_{ij}^{-1}$. We insist that $G$ possesses automorphisms corresponding to the symmetric group permuting the $n$ vertices of our $K_n$, and in addition an automorphism $z$ that commutes with this symmetric group and squares each of the $t_{ij}$ by conjugation. If we now factor by a relation that ensures that a triangle generates $U_3(3)$, then a $K_4$ generates HJ, a $K_5$ generates $G_2(4)$, a $K_6$ generates $3\dot{}$Suz and a $K_7$ generates $Co_1$. What happens for $n \geq 8$ is explained fully in the text.

So this is not simply a sequence of nested subgroups in a larger group, but a finite family of closely related perfect groups.[1]

## 15.2 The Progenitor

Thus, in the notation of Section 14.1, we are seeking images of the progenitor

$$P_n = 7^{\star\binom{n}{2}} : (3 \times S_n).$$

The action of the control subgroup on the free product of cyclic groups is clear in this instance, but more generally we may use representation theory to construct suitable progenitors. This is discussed in more detail in Curtis (1996, 2007b, pages 277 and 287), Bray and Curtis (2003) where progenitors of shape

$$7^{\star(15+15)} : 3\dot{}S_7 \ \text{and} \ 5^{\star(176+176)} : 2\dot{}HS : 2$$

---

[1] The construction described in this chapter was kindly presented at the meeting *Finite Simple Groups: Thirty Years of the Atlas and Beyond*, see Curtis (2017), by Kay Magaard (1962–2018) on behalf of the author who was unwell at the time.

are factored by a single short relation to obtain respectively the Held group, see Held (1969), and the Harada–Norton sporadic simple groups, see Norton (1975), Harada (1976).

As things stand the progenitor $P_n$ only involves a genuine free product for $n > 2$ and we have $P_2 \cong 7:6$, a Frobenius group of order 42; however, we shall describe a *doubling up* process below that leads to a more interesting interpretation of the group defined by a single edge.

Before proceeding we ask what can be said about the (infinite) progenitor $P_n$.

**Lemma 15.1** *Let* $P_n \cong 7^{\star\binom{n}{2}} : (3 \times S_n)$ *as above. Then for* $n \geq 5$,

$$P'_n \cong 7^{\star\binom{n}{2}} : A_n$$

*and so* $|P_n : P'_n| = 6$. *Moreover* $P''_n = P'_n$ *and* $P'_n$ *is perfect*.

*Proof* Denote by $z$ the element of order 3 that commutes with $S_n$ and squares each of the symmetric generators. Then

$$[t_{ij}, z] = t_{ji} t_{ij}^z = t_{ji} t_{ij}^2 = t_{ij} \in P'_n \text{ for all } i, j.$$

If $N \cong 3 \times S_n$ for $n \geq 5$ then $N' \cong A_n$ and the first result follows. Moreover, $N'' = N' \cong A_n$ is contained in $P''$, and $[t_{12}, (1\ 2)(3\ 4)] = t_{21} t_{21} = t_{12}^5 \in P''_n$. Since $N \cong A_n$ acts doubly transitively on the $n$ vertices of the underlying graph, we see that $t_{ij} \in P''_n$ for all $i, j$ and $P''_n = P'_n$. □

**Corollary** *The derived group of any homomorphic image of* $P_n$ *is perfect and has index dividing 6*.

*Proof* Let $K$ be any normal subgroup of $P_n$. Then

$$(P_n/K)' = P'_n K/K \text{ and } (P_n/K)'' = P''_n K/K = P'_n K/K = (P_n/K)'.$$

Furthermore,

$$|P_n/K : P'_n K/K| = |P_n : P'_n K| \mid |P_n : P'_n| = 6.$$ □

In order to get started we need an image of $P_3$, and so we need to say what the image of a triangle of symmetric generators is to be.

## 15.3 Possible Images of a Triangle

We seek images of the progenitor

$$P_3 = 7^{\star 3} : 3 \times S_3.$$

## 15.3 Possible Images of a Triangle

So we seek a group $G_3$ generated by three elements of order 7, which we shall label $t_{12}, t_{23}$ and $t_{31}$, and which possesses automorphisms (inner or outer) consisting of:

(i) a copy of the symmetric group $S_3$ such that

$$(1\ 2\ 3) : t_{12} \to t_{23} \to t_{31} \to t_{12}$$
$$(1\ 2) : t_{12} \to t_{21} = t_{12}^{-1},\ t_{23} \to t_{13} = t_{31}^{-1},\ t_{31} \to t_{32} = t_{23}^{-1},$$

mapping by conjugation,

(ii) an element $z$ that commutes with this copy of $S_3$ and is such that

$$t_{ij}^z \to t_{ij}^2 \text{ for all } ij.$$

Since for values of $n$ greater than 4 the derived group $P'_n$ is perfect we shall assume, factoring out a maximal normal subgroup if necessary, that the image of $P'_3$ is simple, and a consideration of character tables reveals that there are several candidates for $G_3$. It is readily shown that both the linear group $L_2(8) : 3$ and the symmetric group $S_7$ are images of $P_3$; however, extending these cases to $P_4$, corresponding to a complete graph on 4 vertices, leads to collapse. This leads us to the unitary group $U_3(3)$.

### 15.3.1 The Unitary Group $U_3(3)$

$GF_9$, the field of order 9, is taken to be $\{0, \pm 1, \pm i, \pm 1 \pm i\ |\ i^2 = -1 = 1+1\}$ and we let $\alpha$ denote the field automorphism that interchanges $i$ and $-i$. We let

$$A := \begin{pmatrix} 1+i & i & i \\ i & 1+i & i \\ i & i & 1+i \end{pmatrix} \text{ and } Z := \begin{pmatrix} 0 & 1 & 0 \\ 0 & 0 & 1 \\ 1 & 0 & 0 \end{pmatrix},$$

see Section 11.7.2. Matrix $A = I + iJ$, where $J$ denotes the all 1s matrix, and so $J^2 = 0$ over $\mathbb{Z}_3$. Then $A^3 = I + i^3 J^3 = I$; so $A$ has order 3 and $A^2 = I - iJ - J^2 = I - iJ$, the 'complex' conjugate of $A$. Thus conjugation by $\alpha$ inverts $A$; moreover $Z$, which corresponds to a rotation of the three coordinates, commutes with $A$. Thus

$$\langle Z, A, \alpha \rangle \cong 3 \times S_3.$$

As symmetric generators we take $ts := [t_{12}, t_{23}, t_{31}] =$

$$\left[ \begin{pmatrix} i & 1 & -1-i \\ 1 & 1-i & i \\ -1-i & i & -i \end{pmatrix}, \begin{pmatrix} 1 & -1 & -1-i \\ -1+i & -1+i & 0 \\ 1 & -1 & 1+i \end{pmatrix}, \begin{pmatrix} 1 & -1+i & 1 \\ -1 & -1+i & -1 \\ -1-i & 0 & 1+i \end{pmatrix} \right].$$

We find that

$$t_{12}t_{23}^4 t_{12} t_{23}^3 t_{12}^3 = A^2 Z = (3\ 2\ 1)Z. \qquad (15.1)$$

We now use the procedure **act3(ts,uu)** below, written in MAGMA, to confirm that $A$, $Z$ and $\alpha$ act on these three symmetric generators by conjugation in the required manner, and that this relation holds.

```
> procedure act3(ts,uu);
> for i in [1..3] do
> for j in [1..3] do
> for k in [1..6] do
> if ts[i]^uu eq ts[j]^k then
> i,j,k;
> end if;
> end for;end for;end for;
> end procedure;
```

which yields

```
> act3(ts,Z);
1 1 2
2 2 2
3 3 2
> act3(ts,A);
1 2 1
2 3 1
3 1 1
```

and

```
> act3(ts,ts[1]*ts[2]^4*ts[1]*ts[2]^3*ts[1]^3);
1 3 2
2 1 2
3 2 2
> act3(ts,A^2*Z);
1 3 2
2 1 2
3 2 2
```

which is to say

$$Z : t_{12} \mapsto t_{12}^2, t_{23} \mapsto t_{23}^2, t_{31} \mapsto t_{31}^2;\quad A : t_{12} \mapsto t_{23} \mapsto t_{31} \mapsto t_{12}.$$

### 15.3 Possible Images of a Triangle

The element $\alpha$, that is complex conjugation, inverts the symmetric matrix $t_{12}$; moreover we soon check that $\bar{t}_{23} = (\bar{t}_{31})^t = t_{31}^{-1} = t_{13}$ and so $\alpha$ acts as the permutation (1 2) on the subscripts, as required. The final piece of MAGMA code confirms that $t_{12}t_{23}^4 t_{12}t_{23}^3 t_{12}^3$ and $A^2Z$ conjugate the generating set $\{t_{12}, t_{23}, t_{31}\}$ in the same manner, and so one is a central element times the other. Since $U_3(3)$ has trivial centre they are thus equal. So, with $x = Z\alpha$ of order 6 and $y = A$, we find that the following is a presentation for $U_3(3) : 2$.

$$\langle x, y, t \mid x^6 = y^3 = y^x y = t^7 = t^x t^2 = t(t^y)^4 t(t^y)^3 t^3 x^2 y = 1 \rangle \cong U_3(3) : 2,$$

as can be readily verified in MAGMA, noting that $\langle x, y \rangle$ visibly has order 18:

```
> gg<x,y,t>:=Group<x,y,t|x^6=y^3=y^x*y=t^7=t^x*t^2=
> t*(t^y)^4*t*(t^y)^3*t^3*x^2*y=1>;
> Index(gg,sub<gg|x,y>);
672
> 672*18;
12096
> Index(gg,sub<gg|t,t^y>);
2
```

That is to say,

$$\frac{7^{\star 3} : (3 \times S_3)}{R = 1} \cong U_3(3) : 2,$$

where $R = 1$ is the above relation of length 5 in the symmetric generators.

**Note** There are other simpler relations that hold in $U_3(3) : 2$ and define the group as a factor of the progenitor; for instance, the relators $[t_{21}, t_{31}] t_{23}^2$ and $((1\ 2\ 3)zt_{12})^7$ are sufficient. However the additional relation that we have given earlier is the shortest *single* relation to suffice. We see that all even permutations of our $S_3$ and the element $z$ that squares the symmetric generators induce *inner* automorphisms of our group $U_3(3)$, which is itself generated by the three symmetric generators. So a single relation that defines the group from the progenitor must express an even permutation times a non-trivial power of $z$ as a word in those symmetric generators. Thus any feasible single defining relation must take the form

$$\pi z = w(t_{12}, t_{13}, t_{23}),$$

where $\pi$ is a non-trivial even permutation, and $w$ is a word in the symmetric generators.

### 15.3.2 Important Identities

In the context of $U_3(3) : 2$, and in what follows when we consider complete graphs on more than three vertices, we shall define

$$a_{ij} = t_{ki}t_{jk} \text{ for distinct } i, j, k,$$

which will turn out later to be independent of the subscript $k$ and thus be well defined for $n \geq 3$. We have $a_{ji} = t_{kj}t_{ik} = a_{ij}^{-1}$. For matrices in $U_3(3)$ we have

$$a_{12} = t_{31}t_{23} = \begin{pmatrix} -1+i & 1+i & 0 \\ 1+i & -1+i & 0 \\ 0 & 0 & i \end{pmatrix} \text{ and } d := a_{12}^2 = \begin{pmatrix} 0 & -1 & 0 \\ -1 & 0 & 0 \\ 0 & 0 & -1 \end{pmatrix}.$$

The element we have called $d$ has order 2; it visibly commutes with $A \sim (1\,2\,3)$ and with $\alpha \sim (1\,2)$ and is thus independent of the order in which $i, j, k$ appear in its definition. Moreover, it clearly inverts $Z$ and so

$$\langle A, \alpha, Z, d \rangle \cong S_3 \times S_3.$$

It is natural to ask what is generated by a single 'edge' of our triangle, $t_{12}$ say, together with the element $d$. We find that $dt^3$ has order 3, and that $[d, t^3]$ has order 4, which is essentially the classical presentation of the simple group $L_2(7)$. In fact we have

```
> g<t,d>:=Group<t,d|t^7=d^2=(d*t^3)^3=(d,t^3)^4=1>;
> Order(g);
168
```

when the permutation representation

$$\begin{aligned} d &\sim (\infty\,0)(1\,6)(2\,3)(4\,5) \equiv u \mapsto -\frac{1}{u}; \\ t &\sim (0\,5\,3\,1\,6\,4\,2) \equiv u \mapsto u - 2 \end{aligned}$$

verifies the assertion. The element $Z$ that squares $t_{12}$ is already in this simple group, and the element $\alpha$ that inverts $t_{12}$ and commutes with $Z$ and $d$ provides the outer automorphism. If we let $s_{12} = t_{12}^d$, and more generally $s_{ij} = t_{ij}^d$, then we see that

$$s_{ij}^z = t_{ij}^{dz} = t_{ij}^{z^2 d} = (t_{ij}^4)^d = (t_{ij}^d)^4 = s_{ij}^4,$$

so $z$ squares $t_{ij}$ and fourth powers $s_{ij}$. In the notation of symmetric generation what we are seeing is

$$\frac{7^{\star 2} : (S_3 \times 2)}{\left[\begin{pmatrix} 0 & 1 \\ 1 & 0 \end{pmatrix} t\right]^3} \cong L_2(7) : 2,$$

## 15.3 Possible Images of a Triangle

where the two symmetric generators of order 7 are $t' = t^3$ and $s' = (t')^d$, say; the action of $Z$ may be denoted by $\begin{pmatrix} 2 & 0 \\ 0 & 4 \end{pmatrix}$, meaning that $Z$ squares $t'$ and fourth powers $s'$; $d \sim \begin{pmatrix} 0 & 1 \\ 1 & 0 \end{pmatrix}$ interchange the two symmetric generators; and $\alpha \sim \begin{pmatrix} -1 & 0 \\ 0 & -1 \end{pmatrix}$ inverts them both. With $x = Z\alpha$ we have

```
> g<t,x,d>:=Group<t,x,d|t^7=x^6=d^2=(x*d)^2=t^x*t^2=(t*d)^3=1>;
> #g;
336
```

as claimed; indeed, a manual coset enumeration over the Frobenius subgroup $\langle t, x \rangle \cong 7 : 6$ will reveal eight cosets with $t \sim (0\ 1\ 2\ 3\ 4\ 5\ 6)$, $x \sim (1\ 5\ 4\ 6\ 2\ 3)$ and $d \sim (\infty\ 0)(1\ 6)(2\ 3)(4\ 5)$.

A relation that only makes sense when the underlying complete graph is on at least four vertices is

$$t_{ki}t_{jk} = t_{li}t_{jl} \text{ for distinct } i, j, k, l.$$

We shall show that this holds in the case $n = 4$, when it will be clear that $a_{ij}$ is indeed well defined for all $n$. Moreover, since $d = a_{ij}^2$ commutes with all permutations of $i, j, k$ as we have already seen, we can then deduce that in the case of a complete graph on $n$ letters $d$ will commute with $S_n$. Using $d$ we shall later 'double up' to a control subgroup of shape $S_3 \times S_n$.

Let us now define

$$T_{ijk} = \langle t_{ij}, t_{jk}, t_{ki} \rangle \cong U_3(3) \text{ for all distinct } i, j, k,$$

the subgroup generated by the triangle on vertices $i, j$ and $k$. Then, if the configuration does not collapse to the identity,[2] we define

$$E_{12} = T_{123} \cap T_{124} \geq \langle t_{12}, d \rangle = \langle t_{12}, s_{12} \rangle \cong L_2(7),$$

the subgroup generated by the edge 12, where $s_{12} = t_{12}^d$ and more generally $s_{ij} = t_{ij}^d$. Since $T_{123} \neq T_{124}$, as we shall confirm when we investigate the case $n = 4$ below, and since $L_2(7)$ is maximal in $U_3(3)$, we must have

$$E_{12} \cong L_2(7) \cong E_{ij} = \langle t_{ij}, d \rangle = \langle t_{ij}, s_{ij} \rangle \text{ for all distinct } i, j.$$

We can verify directly within $U_3(3)$ that, corresponding to each vertex, we have a subgroup

$$V_1 = E_{12} \cap E_{13} \cong S_4 \cong V_i = E_{ij} \cap E_{ik} \text{ for distinct } i, j, k.$$

---

[2] In fact our presentation does collapse if we take a complete graph on $n$ vertices for $n > 7$, as we shall see below. However, for $n \leq 7$ we obtain distinct nested groups.

However, note that

**Lemma 15.2** $E_{12} \cap E_{13} \cong S_4$, or the group collapses.

*Proof* We have

$$d = (t_{12}t_{31})^2 \Rightarrow t_{21}dt_{12} = t_{31}dt_{13} \in E_{12} \cap E_{13}.$$

Thus the involution $d^{t_{12}} = d^{t_{13}} \in E_{12} \cap E_{13}$; moreover our element $z$ that squares all the $t_{ij}$ and fourth powers all the $s_{ij} = t_{ij}^d$ also lies in $E_{ij}$ for all $i, j$. We have

$$d^{t_{12}} = \begin{pmatrix} 1 & 1-i & -1 \\ 1+i & 0 & 1+i \\ -1 & 1-i & 1 \end{pmatrix}; \quad zd^{t_{12}} = \begin{pmatrix} 1+i & 0 & 1+i \\ -1 & 1-i & 1 \\ 1 & 1-i & -1 \end{pmatrix},$$

which have orders 2 and 4, respectively. So $\langle d^{t_{12}}, z \rangle$ satisfies the presentation $\langle x, y \mid x^2 = y^3 = (xy)^4 \rangle$ that defines the symmetric group $S_4$, and since it contains elements of order 4, it is $S_4$. The maximality of $S_4$ in $L_2(7)$ and the fact that $E_{12} \neq E_{13}$ shows that $E_{12} \cap E_{13} \cong S_4$. □

## 15.4 Extension to $K_n$ for Higher $n$

We now wish to extend this construction to general $n$. As generators for the control subgroup $3 \times S_n$ we shall take

$$x = (1\ 2\ \cdots\ n) \text{ and } y = (1\ 2)z; \tag{15.2}$$

$t$ will as usual represent $t_{12}$, the directed edge from vertex 1 to vertex 2. Visibly $x$ will have order $n$, $y$ will have order 6, and $xy^3$ will have order $n - 1$. The element $y^2 = z^2$ must commute with $x$ and we may now factor by sufficient commutators to define the group. In fact, it is readily verified that

$$\langle x, y \mid x^n = y^6 = [x, y^2] = (xy^3)^{n-1} = [x, y]^3 = [x^2, y]^2 = 1 \rangle \cong 3 \times S_n$$

for $n = 4, 5, 6, 7$. In fact the last two relations are redundant for $n = 4$, and the final relation is redundant for $n = 5$, but they all hold for all $n > 3$.

```
>    for i in [4..7] do
for>    g<x,y>:=Group<x,y|x^i=y^6=(x,y^2)=(x*y^3)^(i-1)=
for>    (x,y)^3=(x^2,y)^2=1>;
for>    i,#g;
for> end for;
```

## 15.4 Extension to $K_n$ for Higher $n$

4  72
5  360
6  2160
7  15120

Note that in MAGMA commutators are denoted by round brackets rather than the more customary square brackets. We must now ensure that this control subgroup acts on the symmetric generators in the required manner. Certainly $z$ must square $t = t_{12}$ by conjugation, and (1 2) must invert it; so we shall have $t^y = t^{-2}$. Now

$$x[y,x] = x(x^{-1})^y x = (1\ 2\ \cdots\ n)(n\ \cdots\ 3\ 1\ 2)(1\ 2\ \cdots\ n) = (3\ 4\ \cdots\ n),$$

and

$$(y^3)^{x^2} = (1\ 2)^{(1\ 2\ \cdots\ n)^2} = (3\ 4).$$

Thus $\langle x[y,x], (y^3)^{x^2}\rangle \cong S_{n-2}$ commuting with $t = t_{12}$. Factoring by the three relators

(i) $t^y t^2$, (ii) $[t, x[y,x]]$ and (iii) $[t, (y^3)^{x^2}]$

is thus sufficient to define the progenitor $P_n$ for $n = 4, 5, 6, 7$ and we have:

$$P_n \cong 7^{\star\binom{n}{2}} : (3 \times S_n) \cong$$

$$\left\langle x, y, t \;\middle|\; \begin{array}{l} x^n = y^6 = [x, y^2] = (xy^3)^{(n-1)} = [x, y]^3 = [x^2, y]^2 = \\ t^7 = t^y t^2 = [t, x[y,x]] = [t, (y^3)^{x^2}] = 1 \end{array} \right\rangle. \quad (15.3)$$

It remains to factor out our additional relation that ensures that every triangle generates (a quotient of) $U_3(3)$. We have

$$[x, y] = x^{-1} x^y = (n\ \cdots\ 2\ 1)(2\ 1\ 3\ \cdots\ n) = (1\ 2\ 3),$$

and the relator found in Section 15.3.1 above was $t_{12} t_{23}^4 t_{12} t_{23}^3 t_{12}^3 z^2 (1\ 2\ 3)$. So we factor by

$$R = t(t^x)^4 t(t^x)^3 t^3 y^2 [x, y].$$

We have seen that the symmetric generators $t_{ij}$ and even permutations of our symmetric group $S_n$ lie in the derived group of $P_n$, and so their images lie in the derived group of any quotient group of $P_n$. Setting $t = [x, y] = 1$ in the relator $R$ shows that $y^2$ lies in the derived group of $P_n/\langle R\rangle$. Thus $P_n/\langle R\rangle$ has a perfect derived group to index 2.

We shall define

$$K_n = P_n/\langle R^{P_n}\rangle \cong \frac{7^{\star\binom{n}{2}} : (3 \times S_n)}{R = 1},$$

where we know that $K_3 \cong U_3(3) : 2$.

Now $xy = (2\ 3\ \cdots n)z$ and $y^x = (2\ 3)z$ and so $\langle xy, y^x \rangle \cong 3 \times S_{n-1}$; thus

$\langle xy, y^x, t^x = t_{23} \rangle$ maps to (an image of) $K_{n-1}$ as a subgroup of $K_n$.

So in each case we perform a coset enumeration of $K_n$ over this subgroup. The algorithm employed by MAGMA is based on that devised by J. A. Todd and H. S. M. Coxeter, both of whom were students of the eminent geometer H. F. Baker.

```
> for i in [4..7] do
for>    g<x,y,t>:=Group<x,y,t|x^i=y^6=(x,y^2)=(x*y^3)^(i-1)=
for>    (x,y)^3=(x^2,y)^2=
for>    t^7=t^y*t^2=(t,x*(y,x))=(t,(y^3)^(x^2))=
for>    t*(t^x)^4*t*(t^x)^3*t^3*y^2*(x,y)=1>;
for>    h:=sub<g|t^x,x*y,y^x>;
for>    i,Index(g,h:Hard:=true,CosetLimit:=20000000);
for> end for;
4 100
5 416
6 5346
7 3091200
```

The reader familiar with the groups in question will recognize that:

- the index of $U_3(3)$ in HJ is 100;
- the index of HJ in $G_2(4)$ is 416;
- the index of $G_2(4)$ in $3\!\cdot\!\text{Suz}$ is 5346; and
- the index of $3\!\cdot\!\text{Suz}$ in $\text{Co}_1$ is 3 091 200.

Of course our groups $K_n$ are not simple, but have perfect subgroups to index 2. We assert that $K_4 \cong \text{HJ} : 2$; $K_5 \cong G_2(4) : 2$; $K_6 \cong 3\!\cdot\!\text{Suz} : 2$. However, the Conway group $\text{Co}_1$ does not have an outer automorphism and so we assert that $K_7 \cong 2 \times \text{Co}_1$. Indeed, we may factor out an additional relation that removes the direct factor of order 2:

```
>    g<x,y,t>:=Group<x,y,t|x^7=y^6=(x,y^2)=(x*y^3)^6=
>    (x,y^3)^3=(x^2,y^3)^2=
>    t^7=t^y*t^2=(t,x*(y,x))=(t,(y^3)^(x^2))=
>    t*(t^x)^4*t*(t^x)^3*t^3*y^2*(x,y)=(x*y^3*t)^23=1>;
>    h:=sub<g|t^x,x*y,y^x>;
>    Index(g,h:Hard:=true,CosetLimit:=20000000);
1545600
>
```

## 15.5 The Progenitor $7^{\star\binom{n}{2}} : S_n$

The element $z$, which squares each of the symmetric generators $t_{ij}$ by conjugation, was instrumental in identifying $U_3(3)$ as a suitable triangle group, thus serving as a seed group for the sequence that follows. In Curtis (2016) it was the progenitor given in the presentation 15.3 that was used in our development, sometimes 'doubled up' to

$$7^{\star\binom{n}{2}+\binom{n}{2}} : (S_3 \times S_n).$$

However, it turns out that we can dispense with its services in the control subgroup, although it will emerge spontaneously in the images. So we consider the simpler progenitor

$$P_n = 7^{\star\binom{n}{2}} : S_n.$$

Note that an analogous argument to Lemma 15.1 shows that, for $n \geq 5$,

$$|P_n : P'_n| = 2 \text{ and } P''_n = P'_n,$$

and so any image of $P_n$, $n \geq 5$, contains a perfect subgroup to index at most 2.

A suitable presentation for $S_n$ in terms of $x \sim (1\ 2\ 3\ \cdots\ n)$ and $y \sim (1\ 2)$ is given by

$$S_n \cong \langle x, y \mid x^n = y^2 = (xy)^{n-1} = [x, y]^3 = [x^2, y]^2 = 1 \rangle,$$

which holds for $n = 4, 5, 6, 7$. As in Section 15.4, only the first three and the first four relations are needed for $n = 4$ and 5 respectively, but they all hold for $n > 3$. As before, we take $t \sim t_{12}$ and so $t$ is inverted by $y$ and commutes with $x[y, x] \sim (3\ 4\ \cdots\ n)$ and $y^{x^2} \sim (3\ 4)$. Using the matrices of Section 15.3.1 we find that

$$t_{12}t_{23}^2 t_{12}^5 t_{31}^4 t_{12}^5 = [x, y] \sim (1\ 2\ 3), \tag{15.4}$$

and we know that

$$d = (t_{23}t_{12})^2 \text{ commutes with } S_3.$$

If $K_n$ is the group defined by these relations, then the subgroup $K_{n-1}$ is generated by $\langle t^x, y^x, xy \rangle$. We may thus perform the analogous coset enumerations to those in Section 15.4:

```
> for i in [4..7] do
for> g<x,y,t>:=Group<x,y,t|
for> x^i=y^2=(x*y)^(i-1)=(x,y)^3=(x^2,y)^2=t^7=
for> (t*y)^2=(t,x*(y,x))=(t,y^(x^2))=
for> t*(t^x)^2*t^5*(t^(x*y))^3*t^5*(y,x)=
```

```
for> ((t^x*t)^2,x) =1>;
for> h:=sub<g|t^x,y^x,x*y>;
for> i,Index(g,h:Hard:=true, CosetLimit:=20000000);
for> end for;
4 100
5 416
6 5346
7 3091200
```

## 15.6 Identification of the Groups in the Chain

Given permutation or matrix representations of the groups in the chain, one may readily obtain generators that satisfy the presentations given and thus demonstrate that the groups defined are as claimed. On the other hand, one may start with the presentation and use MAGMA to obtain a permutation representation of the group acting on cosets of the previous group in the chain. In the self-explanatory MAGMA code that follows $kn$ stands for the group $K_n$ defined by the presentation, $kns$ for the image of $K_n$ in $K_{n+1}$, and $knp$ for the permutation representation of $K_n$ acting on the cosets of $K_{n-1}$.

### 15.6.1 The Hall–Janko Group HJ

We have seen that $K_3$ has index 100 in $K_4$, and so we shall obtain $K_4$ as a permutation group of degree 100. We then ask for the lengths of the orbits of the stabilizer of a point and obtain $K_4$ as a rank 3 permutation group with suborbit lengths 1, 36 and 63.

```
> k4<x,y,t>:=Group<x,y,t|x^4=y^2=(x*y)^3=t^7=t*t^y=(t,y^(x^2))=
> t*(t^x)^2*t^5*(t^(x*y))^3*t^5*y*y^x=((t^x*t)^2,x)=1>;
> k3s:=sub<k4|y^x,x*y,t^x>;
> ct4:=CosetTable(k4,k3s);
> k4p:=CosetTableToPermutationGroup(k4,ct4);
> Degree(k4p);#k4p;
100
1209600
> orb4:=Orbits(Stabilizer(k4p,1));
> [#orb4[i]:i in [1..#orb4]];
[ 1, 36, 63 ]
>
```

In fact we may readily show that

$$K_4 = K_3 \cup K_3(1\ 2)K_3 \cup K_3 t_{12} K_3,$$

## 15.6 Identification of the Groups in the Chain

where $K_3 = \langle t_{23}, t_{24}, t_{34}, (2\ 3) \rangle$.

Having obtained a permutation representation for the group $K_4$, this is a convenient point at which to verify our claims about the elements $a_{ij}$ and $d$, namely: that $a_{ij} = t_{ki} t_{jk}$ is independent of $k$; that $d = (t_{ki} t_{jk})^2$ is independent of the choice of distinct $i, j, k$ and thus commutes with the automorphisms of $S_n$; moreover, that $d$ is an involution inverting $z$. We denote the images of $x, y$ and $t$ in this representation by $xp, yp$ and $tp$, and show that $t_{23} t_{12} = t_{43} t_{14}$, that is to say $t^x t = (t^{x^2})^{-1} (t^{x^3})^{-1}$, and so $a_{31}$ is well defined. We saw in $U_3(3)$ that $d := (t_{ki} t_{jk})^2$ is independent of the order of $i, j$ and $k$, and so $d = a_{ij}^2$ commutes with the transposition $(i\ j)$ and every 3-cycle involving $i$ and $j$ and thus with $S_n$.

```
> xp:=k4p.1;
> yp:=k4p.2;
> tp:=k4p.3;
> tp^xp*tp eq (tp^(xp^2))^-1*(tp^(xp^3))^-1;
true
> dp:=(tp^xp*tp)^2;
> Order(dp);
2
> yp^dp eq yp;
true
> dp2:=((tp^xp)^2*tp^2)^2;
> zp:=dp2*dp;
> tp^zp eq tp^2;
true
> #sub<k4p|zp,dp>;
6
> [Order((zp,xp)), Order((zp,yp)),Order((dp,xp)),Order((dp,yp))];
[ 1, 1, 1, 1 ]
>
```

We define $d2$ to be $(t_{23}^2 t_{12}^2)^2$ and $z$ to be $d2.d$, when we see that $\langle z, d \rangle \cong S_3$, commuting with the $S_4$ of the control group. Since $z$ squares $t = t_{12}$ and commutes with our $S_4$, it must square each of the symmetric generators. In this way the element $z$ that we recently abandoned appears spontaneously in each term of the chain.

### 15.6.2 The Exceptional Lie Group $G_2(4)$

Taking $n = 5$ and repeating the exercise, we obtain $K_5$ as a rank 3 permutation group with suborbit lengths 1, 100 and 315.

```
> k5<x,y,t>:=Group<x,y,t|x^5=y^2=(x*y)^4=(x,y)^3=(x^2,y)^2=t^7=
> t^y*t=(t,x*(y,x))=t*(t^x)^2*t^5*(t^(x*y))^3*t^5*y*y^x=
> ((t^x*t)^2,x)=1>;
> k4s:=sub<k5|y^x,x*y,t^x>;
```

```
> ct5:=CosetTable(k5,k4s);
> k5p:=CosetTableToPermutationGroup(k5,ct5);
> Degree(k5p);
416
> xp:=k5p.1;yp:=k5p.2;tp:=k5p.3;
> orb5:=Orbits(Stabilizer(k5p,1));
> [#orb5[i]:i in [1..#orb5]];
[ 1, 100, 315 ]
>
```

Again we have

$$K_5 = K_4 \cup K_4(1\ 2)K_4 \cup K_4 t_{12} K_4,$$

where $K_4 = \langle t_{23}, (2\ 3\ 4\ 5), (2\ 3) \rangle$. This exceptional Lie group, see Section 11.7.3, may be thought of as the automorphism group of the $\mathbb{F}_4$ octonions or Cayley algebra, see the ATLAS (Conway et al., 1985, page 97). More revealingly from the point of view of the Thompson Chain, it is also the automorphism group of the quaternionic Leech lattice, see Wilson (2009a, page 118).

### 15.6.3 The Triple Cover of the Suzuki Simple Group Suz

Unlike the previous cases, the group $K_6'$ is not simple but is the triple cover of the Suzuki group. Indeed $3 \cdot \text{Suz} : 2$ contains a subgroup $(3 \times G_2(4)) : 2$ and so $K_5 \cong G_2(4) : 2$ is not a maximal subgroup. Thus our permutation group will not be primitive but will have blocks of size 3.

```
> k6<x,y,t>:=Group<x,y,t|x^6=y^2=(x*y)^5=(x,y)^3=(x^2,y)^2=t^7=
> t^y*t=(t,x*(y,x))=
> t*(t^x)^2*t^5*(t^(x*y))^3*t^5*y*y^x=((t^x*t)^2,x)=1>;
> k5s:=sub<k6|y^x,x*y,t^x>;
> ct6:=CosetTable(k6,k5s);
> k6p:=CosetTableToPermutationGroup(k6,ct6);
> Degree(k6p);
5346
> orb6:=Orbits(Stabilizer(k6p,1));
> [#orb6[i]:i in [1..#orb6]];
[ 1, 2, 416, 832, 4095 ]
```

Moreover

```
> xp:=k6p.1;yp:=k6p.2;tp:=k6p.3;
> [#(1^k5sp),#((1^(tp*xp*yp)^13)^k5sp),#((1^yp)^k5sp),
> #((1^(yp*(tp*xp*yp)^13))^k5sp),#((1^tp)^k5sp)];
[ 1, 2, 416, 832, 4095 ]
```

shows that the five double cosets of form $K_5 \pi w K_5$ are given by

$$K_6 = K_5 \cup K_5(t_{12}(2\,3\,4\,5\,6))^{13}K_5 \cup K_5(1\,2)K_5 \cup \\ K_5(1\,2)(t_{12}(2\,3\,4\,5\,6))^{13}K_5 \cup K_5 t_{12} K_5,$$

where the double cosets have lengths as in the MAGMA code and $K_5 = \langle t_{23}, (2\,3\,4\,5\,6), (2\,3) \rangle$ as usual.

The centre of the perfect group 3˙Suz may be generated by $(t_{12}(2\,3\,4\,5\,6))^{13} = (txy)^{13}$ in terms of our generators, and so we may perform a coset enumeration over $(3 \times G_2(4)) : 2$:

```
> k5sx3:=sub<k6|k5s,(t*x*y)^13>;
> ct6b:=CosetTable(k6,k5sx3);
> k6pq:=CosetTableToPermutationGroup(k6,ct6b);
> Degree(k6pq);
1782
> orb6q:=Orbits(Stabilizer(k6pq,1));
> [#orb6q[i]:i in [1..#orb6q]];
[ 1, 416, 1365 ]
```

The action on the blocks of size 3 is the familiar rank 3 permutation action of degree 1782 of the Suzuki simple group with suborbit lengths 1, 416 and 1365, see the ATLAS (Conway et al., 1985, page 128). It is the automorphism group of the complex Leech lattice, as is explained in Wilson (2009a, page 229).

### 15.6.4 The Conway Simple Group $Co_1$

As mentioned earlier, the group $K_7$ is a direct product of shape $2 \times Co_1$; the centre of this group may be generated by $((2\,3\,4\,5\,6\,7)t_{12})^{23}$. Factoring this out we obtain a perfect group that may be represented as permutations of degree 1 545 600 over $K_6 \cong 3\,\dot{}\,Suz : 2$.

```
> k7<x,y,t>:=Group<x,y,t|x^7=y^2=(x*y)^6=(x,y)^3=(x^2,y)^2=t^7=
> t^y*t=(t,x*(y,x))=t*(t^x)^2*t^5*(t^(x*y))^3*t^5*y*y^x=
> ((t^x*t)^2,x)=(x*y*t)^23=1>;
> k6s:=sub<k7|y^x,x*y,t^x>;
> Index(k7,k6s:Hard:=true, CosetLimit:=20000000);
1545600
> ct7:=CosetTable(k7,k6s);
> k7p:=CosetTableToPermutationGroup(k7,ct7);
> orb7:=Orbits(Stabilizer(k7p,1));
> [#orb7[i]:i in [1..#orb7]];
[ 1, 5346, 22880, 405405, 1111968 ]
> IsSimple(k7p);
true
```

As we see this permutation action has rank 5 with suborbit lengths as given. The simplicity of $K_7'$ has been confirmed computationally above; however, it is straightforward to prove that a perfect group acting primitively in this manner is simple.

The results may be summarized in the following table:

| $n$ | $K_n$ |
|---|---|
| 3 | $U_3(3) : 2$ |
| 4 | $HJ : 2$ |
| 5 | $G_2(4) : 2$ |
| 6 | $3\,\dot{}\,Suz : 2$ |
| 7 | $2 \times Co_1$ |

In each case, all even permutations in the symmetric group $S_n$ of the control subgroup are contained in the perfect subgroup of index 2 in the above table. Odd permutations lie outside this subgroup so, if we wish to obtain presentations of these perfect (simple except in the case of $3\,\dot{}\,Suz$) groups then we must take as control subgroup $N_n \cong A_n$. However, care must be taken here to ensure that the subgroup over which we are performing the coset enumeration is indeed the group we claim; in particular, we must ensure that the triangle groups are still copies of $U_3(3)$.

## 15.7 The Real Normed Division Algebras

Formally we construct the field of complex numbers $\mathbb{C}$ from the reals $\mathbb{R}$ by taking $\mathbb{R}[x]$, the ring of all finite polynomials over the indeterminate $x$ with coefficients in $\mathbb{R}$, and factoring out the principal ideal $(x^2+1)$. In this way we have extended the reals by a symbol $i$, where $i^2 = -1$ and we have $\mathbb{C} = \{a + bi \mid a, b \in \mathbb{R}\}$, where

$$(a + bi)(c + di) = (ac - bd) + (ad + bc)i \text{ for } a, b, c, d \in \mathbb{R}.$$

If $\alpha = a+bi \in \mathbb{C}$ then $\bar{\alpha} = a-bi$ is the *complex conjugate* of $\alpha$, and $\alpha\bar{\alpha} = a^2+b^2$. Then $N(\alpha)$, the *norm* of $\alpha$, is defined to be

$$N(\alpha) = \alpha\bar{\alpha} = a^2 + b^2,$$

the square of the length of $\alpha$ in the complex plane. We readily obtain the familiar multiplicative property that

$$N(\alpha\beta) = N(\alpha)N(\beta) \text{ for } \alpha, \beta \in \mathbb{C}.$$

## 15.7 The Real Normed Division Algebras

For some years William Rowan Hamilton and his friend John Thomas Graves had discussed ways of extending the 2-dimensional algebra $\mathbb{C}$ to a 3-dimensional real algebra, but it was on 16th October 1843 that Hamilton realized that what they were looking for was the 4-dimensional algebra, now known as the *quaternions*. In his exuberance at making this discovery, Hamilton carved the defining relations for this new algebra, which is denoted by $\mathbb{H}$ in his honour, on the Broom Bridge in Cabra, Dublin, see Halberstam and Ingram (1967). We have

$$\mathbb{H} = \{a + bi + cj + dk \mid a, b, c, d \in \mathbb{R}, \ i^2 = j^2 = k^2 = ijk = -1\},$$

from which we may deduce that $ij = k$ but that $ji = -k$ and so $\mathbb{H}$ is *not* commutative, although it is associative. If $\alpha = a + bi + cj + dk$ then, as with $\mathbb{C}$, $a$ is known as the *real* part of $\alpha$ and $bi + cj + dk$ is the *imaginary* part. Furthermore, $\bar{\alpha} = a - bi - cj - dk$ is the *conjugate* of $\alpha$ and, in a directly analogous manner to $\mathbb{C}$, we define the *quaternion norm* $N$ by[3]

$$N(\alpha) = \alpha\bar{\alpha} = a^2 + b^2 + c^2 + d^2.$$

The multiplicative property for norms also holds:

$$N(\alpha\beta) = N(\alpha)N(\beta) \text{ for } \alpha, \beta \in \mathbb{H},$$

thus demonstrating the famous *four squares theorem* that the product of two sums of four squares is itself the sum of four squares. Note that if $0 \neq \alpha \in \mathbb{H}$ then

$$\alpha^{-1} = \frac{\bar{\alpha}}{N(\alpha)} \qquad (15.5)$$

and so every non-zero element of $\mathbb{H}$ has an inverse and $\mathbb{H}$ is a *division algebra*. In further correspondence between Graves and Hamilton it is clear that the former had found an 8-dimensional extension to $\mathbb{H}$, see Graves (1844) and Rice (2004), now known as the *octonions* $\mathbb{O}$. Sir Arthur Cayley had discovered this structure independently and it was for many years known as the *Cayley Numbers*; however, it is now accepted that the original discovery was due to Graves. We have

$$\mathbb{O} = \left\{ a_\infty + a_0 i_0 + \cdots + a_6 i_6 \ \middle| \ \begin{array}{l} a_\infty, a_j \in \mathbb{R}, \ i_j^2 = -1 = i_{j+1} i_{j+2} i_{j+4} \\ \text{for } j \in \{0, 1, \ldots, 6\} \end{array} \right\},$$

where all subscripts are to be read modulo 7. Note that, for instance,

$$\langle i_1, i_2, i_4 \rangle \cong \mathbb{H};$$

---

[3] Many authors take the norm of $\alpha$ to be the square root of $N(\alpha)$ as defined here.

in fact, any two octonions that do not commute with one another generate a copy of $\mathbb{H}$, and thus generate an associative subsystem. However, $\mathbb{O}$ is *not* associative:

$$i_0(i_1 i_2) = i_0 i_4 = i_5; \quad (i_0 i_1) i_2 = i_3 i_2 = -i_5.$$

The real and imaginary parts, the conjugate and the norm of an octonion are defined in a precisely analogous manner to those in $\mathbb{C}$ and $\mathbb{H}$ and, remarkably, the norm is still multiplicative. This latter fact gives rise to the *eight squares theorem*, see Dickson (1919). Equation 15.5 shows that every non-zero element of $\mathbb{O}$ possesses an inverse and so $\mathbb{O}$ is a further normed division algebra.

In fact *Hurwitz' Theorem*, see Hurwitz (1923), states that up to isomorphism $\mathbb{R}$, $\mathbb{C}$, $\mathbb{H}$ and $\mathbb{O}$, of dimensions 1, 2, 4 and 8 respectively, are the only finite dimensional, normed division algebras over $\mathbb{R}$.

The *Cayley–Dickson construction* that was introduced by Leonard Eugene Dickson in 1919 provides a uniform method of producing each term of this sequence from the previous one, see Dickson (1919). To the algebra $A$ that is required to possess a conjugation function denoted by $\alpha \mapsto \bar{\alpha}$, we adjoin a square root of $-1$ denoted by $e$. Then a multiplication is defined on the set $\{\alpha + \beta e \mid \alpha, \beta \in A\}$ by

$$(\alpha + \beta e)(\gamma + \delta e) = (\alpha \gamma - \bar{\delta} \beta) + (\delta \alpha + \beta \bar{\gamma}) \text{ for } \alpha, \beta, \gamma, \delta \in A, \quad (15.6)$$

and addition is simply term by term:

$$(\alpha + \beta e) + (\gamma + \delta e) = (\alpha + \gamma) + (\beta + \delta) e.$$

Note that the order of the terms in Equation 15.6 is crucial as $\mathbb{H}$ is not commutative.

### 15.7.1 Connection with the Thompson Chain

As is made clear in Chapters 10 and 11, the Conway group ·O, the double cover of $Co_1$, is the group of symmetries of the Leech lattice, a 24-dimensional lattice over $\mathbb{R}$. In his ground-breaking paper, Jacques Tits (1980a) produced three more presentations of the Leech lattice: as a 12-dimensional lattice over $\mathbb{C}$; as a 6-dimensional lattice over $\mathbb{H}$; and as a 3-dimensional lattice also over $\mathbb{H}$. The groups of symmetries of these versions of the Leech lattice are respectively the double covers of: $3^{\cdot}$Suz, see Wilson (1983); $G_2(4)$, see Wilson (1982); and HJ, see Tits (1980b).

Given the classification of finite-dimensional real normed division algebras, the reader may be surprised that the final construction is over $\mathbb{H}$ rather than over

○, and the same thought occurred to Wilson who, in (2009b) and (2009a, page 230), constructed a 3-dimensional *octonionic* version of the Leech lattice.

Thus we see that there is an extraordinary correspondence between the groups of the chain that we have constructed here and the Hurwitz division algebras. When regarded as subgroups of $Co_1$ these groups have traditionally been referred to as the *Suzuki chain*. However, this chapter makes it clear that the chain does not end with 3·Suz but includes $Co_1$ itself, and so I have chosen to refer to it as the *Thompson chain* as it was John Thompson who recognized that, with one exception, the normalizers of the groups in the chain were maximal subgroups of $Co_1$. However, given the correspondence described in this section, it would perhaps be better to refer to it as the *Leech chain*.

## 15.8 The Group $K_n$ for $n > 7$

As usual, in presenting $K_8$ we take $x \sim (1\,2\,3\,4\,5\,6\,7\,8)$ and $y \sim (1\,2)$, but we need one further commutator relation, namely $[x^3, y]^2$, to define the symmetric group $S_8$:

```
> n8<x,y>:=Group<x,y|x^8=y^2=(x*y)^7=(x,y)^3=
> (x^2,y)^2=(x^3,y)^2=1>;
> #n8;
40320
```

We shall be insisting that every complete 7-graph generates a copy of $Co_1$ and so must factor by a relator equivalent to $((2\,3\,4\,5\,6\,7)t_{12})^{23}$. In fact we find that $x[y, x] = (3\,4\,5\,6\,7\,8)$ and so the relation $(x[y, x]t^x)^{23}$ will suffice. Everything else stays as before.

```
> k8<x,y,t>:=Group<x,y,t|x^8=y^2=(x*y)^7=(x,y)^3=
> (x^2,y)^2=(x^3,y)^2=t^7=t^y*t=(t,x^(y,x))=(t,y^(x^2))=
> t*(t^x)^2*t^5*(t^(x*y))^3*t^5*(y,x)=((t^x*t)^2,x)=
> (x*(y,x)*t^x)^23=1>;
> k7s:=sub<k8|t^x,x*y,y^x>;
> Index(k8,k7s:Hard:=true, CosetLimit:=20000000);
1
```

This means that **k8** defines the same group as **k7** or a proper homomorphic image of it. But the Conway group $Co_1$ is simple and so its only proper (non-isomorphic) homomorphic image is the trivial group. So **k8** ≅ $Co_1$ or **k8** = 1. In **k8** the transposition (1 2) commutes with the subgroup

$$\langle t_{ij} \mid i \neq j,\ i, j \in \{3, 4, \ldots, 8\}\rangle,$$

which in **k7** ≅ Co$_1$ is a copy of 3˙Suz whose centralizer in Co$_1$ is its centre of order 3. Thus **k8** cannot be a copy of Co$_1$ and must therefore collapse to the trivial group. So, if we wish to consider values of $n$ greater than 7 we must restrict the control subgroup to even permutations.

## 15.9 Doubling Up

In the constructions of the Held group and of its sequel on the Harada–Norton group, mentioned briefly in Section 15.2, it proved beneficial to 'double up' the number of symmetric generators. Thus in the Held case, see Curtis (1996), instead of considering a progenitor of form $7^{\star 15} : 3\,\mathrm{A}_7$ in which the central element of order 3 in the control subgroup squares all the symmetric generators, we took a progenitor of shape $7^{\star(15+15)} : 3\,\mathrm{S}_7$ in which the same 'central' 3 squares one set of 15 symmetric generators and fourth powers the other 15. Similarly in the Harada–Norton case, see Bray and Curtis (2003), we doubled up from a progenitor of shape $5^{\star 176} : 2\,\mathrm{HS}$, where 2˙HS denotes the double cover of the Higman–Sims sporadic simple group, to $5^{\star(176+176)} : 2\,\mathrm{HS} : 2$. Note that in each case the control subgroup acts non-permutation identically on the two sets of symmetric generators: when S$_7$ acts on 15 + 15 points, the stabilizer of a point has suborbits of lengths $(1 + 14) + (7 + 8)$; and when HS : 2 acts on 176 + 176 points, the stabilizer of a point has suborbits of lengths $(1 + 175) + (50 + 126)$. In both these examples a single additional relation of the form $(\pi t)^3 = 1$ for $\pi \in N$ and $t$ a symmetric generator was sufficient to define the group in question.

The case for doubling up is far less compelling here as generically A$_n$ has a unique action on the $\binom{n}{2}$ unordered pairs. Nonetheless, the procedure does have a pay-off in that (i) a single additional relation – the same, of course, for all $n$ – is still sufficient to define the image group, and (ii) the group $E_{ij} = \langle t_{ij}, s_{ij} \rangle$, corresponding to each (directed) edge of the complete graph is the image of a subprogenitor.

Accordingly in Curtis (2016), we investigated the progenitor

$$7^{\star(\binom{n}{2}+\binom{n}{2})} : \mathrm{S}_3 \times \mathrm{S}_n,$$

where we have adjoined an element $d = (t_{ij} t_{ki})^2$ that inverts $z$ and commutes with our S$_n$. Presentations are obtained in terms of $u = (1\ 2\ \ldots\ n)$ and $v = (1\ 2)z^2$, so $d$ commutes with $u$ and inverts $v$.

## 15.10 The Case $N \cong A_9$

To proceed beyond $n = 7$, we have seen that we must restrict our control subgroup to the alternating group $A_n$ and, since $A_{10} \geq S_8$ we may go no further than $n = 9$. So we consider the progenitor

$$P_9 = 7^{\star 36} : A_9.$$

As generators for the group $N \cong A_9$, we take $x \sim (1\ 2\ 3\ 4\ 5\ 6\ 7\ 8\ 9)$ and $y \sim (1\ 2\ 3)$. Then $y^x y = (1\ 2)(3\ 4)$ and $y^{x^3} = (4\ 5\ 6)$; moreover, $(xy^{-1})^7 = (x^2 y)^4 = 1$. This enables us to write down a suitable presentation for $A_9$:

$$\langle x, y \mid x^9 = y^3 = (xy^{-1})^7 = (y^x y)^2 = (x^2 y)^4 = [y^{x^3}, y] = 1 \rangle \cong A_9$$

as may be verified using MAGMA by

```
> a9<x,y>:=
> Group<x,y|x^9=y^3=(y^x*y)^2=(y^(x^3),y)=(x*y^-1)^7=(x^2*y)^4=1>;
> #a9;
181440
```

Our generator $t$ will as usual stand for $t_{12}$ and we must identify a subgroup of $\langle x, y \rangle$ that normalizes $\langle t \rangle$ and has index $\binom{9}{2} = 36$ in $N$. We see from the above that $y^x y = (1\ 2)(3\ 4)$ inverts $t$ and $xy^{-1} = (3\ 4\ 5\ 6\ 7\ 8\ 9)$ commutes with $t$. Together these two elements visibly generate $S_7$. We must now factor by the additional relations

$$t(t^y)^2 t^5 (t^{y^2})^4 t^5 = y \quad \text{and} \quad [(t^y t)^2, x] = 1.$$

Finally, in a delightful flourish, $Co_1$ ties together what has come before in the chain: any complete 6-graph in our complete 9-graph generates a copy of $KA_6 \cong 3\text{·}Suz$; the central elements in this group are precisely the 3-cycles of the $A_9$ that fix every vertex of its associated 6-graph. A non-trivial central element of the $KA_6$ on the vertices $\{1, 2, 3, 4, 5, 6\}$ is given by $((2\ 3\ 4\ 5\ 6)t_{12})^{13}$, which must be put equal to the 3-cycle $(7\ 8\ 9)$ or its inverse. This is neatly accomplished by requiring the element $(2\ 3\ 4\ 5\ 6)(7\ 8\ 9)t_{12}$ to have order 13. In our presentation below we achieve this by observing that $xyx^2y^2 = (1\ 5\ 8)(3\ 6\ 9\ 4\ 7)$, and so requiring this element times $t_{23}$ to have order 13 will suffice. We then claim that

$$\begin{aligned}
\langle x, y, t \mid\ & x^9 = y^3 = (xy^{-1})^7 = (y^x y)^2 = (x^2 y)^4 \\
&= [y^{x^3}, y] = t^7 = t^{y^x y} t = [t, xy^{-1}] = [(t^x t)^2, x] \\
&= t(t^y)^2 t^5 (t^{y^2})^4 t^5 y^2 = (xyx^2 y^2 t^x)^{13} = 1 \rangle \\
\cong\ & Co_1.
\end{aligned} \qquad (15.7)$$

We shall, of course, perform our coset enumeration over a copy of $3\dot{\,}Suz:2$ and choose the $KA_6$ on vertices $\{1, 2, \ldots, 6\}$. In order to include an odd permutation of these six vertices, we use the fact that $(y^x y)^{x^{-2}} = (1\ 2)(8\ 9)$.

```
> co1<x,y,t>:=Group<x,y,t|
> x^9=y^3=(y^x*y)^2=(y^(x^3),y)=(x*y^-1)^7=(x^2*y)^4=
> t*(t^y)^2*t^5*(t^(y^2))^4*t^5*y^2=(x*y*x^2*y^2*t^x)^13=
> t^7=t^(y^x*y)*t=(t,x*y^-1)=((t^x*t)^2,x)=1>;
> suz:=sub<co1|t,y,y^x,y^(x^2),y^(x^3),x^2*y^x*y*x^-2>;
> ct9:=CosetTable(co1,suz:Hard:=true, CosetLimit:=20000000);
> co1p:=CosetTableToPermutationGroup(co1,ct9);
> Degree(co1p);
1545600
```

Note that $xy^{-1} = (3\ 4\ 5\ 6\ 7\ 8\ 9), y^x y = (1\ 2)(3\ 4)$ and $t^{x^2} = t_{34}$ and so $\langle xy^{-1}, y^x y, t^{x^2} \rangle$ generates an image of $P_7 = 7^{\star 21} : S_7$ that we have seen above, subject to the additional relations, generates a homomorphic image of $Co_1$. This cannot be trivial as is demonstrated by the foregoing enumeration over **suz**, so it must be $Co_1$. However,

```
> co1sub:=sub<co1|x*y^-1,y^x*y,t^(x^2)>;
> Index(co1,co1sub:Hard:=true, CosetLimit:=20000000);
1
```

So our complete 9-graph generates the same group as a 7-graph, namely $Co_1$.

In the above code MAGMA calculates the coset table for $Co_1$ acting on the cosets of $3\dot{\,}Suz:2$, and thus obtains its permutation action of degree 1 545 600:

```
> co1p:=CosetTableToPermutationGroup(co1,ct9);
> xp:=co1p.1;yp:=co1p.2;tp:=co1p.3;
> d1p:=(tp^xp*tp)^2;d2p:=((tp^xp)^2*tp^2)^2;
> zp:=d1p*d2p;
> Order(zp);
3
> Order((zp,xp));Order((zp,yp));
1
1
> #sub<co1p|d1p,d2p>;
6
> E12:=sub<co1p|d1p,tp>;
> #E12;
168
```

```
> T123:=sub<co1p|yp,tp>;
> #T123;
6048
> L12:=sub<co1p|d1p,tp>;L23:=sub<co1p|d1p,tp^yp>;
> V2:=(L12 meet L23);
> #V2;
24
```

This enables us to verify computationally the claims made in the text. For instance, we see above that

$$\langle (t_{23}t_{12})^2, (t_{23}^2 t_{12}^2)^2 \rangle = \langle z, d \rangle \cong S_3,$$

commuting with our copy of $A_9$. The edge group $E_{12}$ generated by $t_{12}$ and $d$ does indeed have the order of $L_2(7)$; and the triangle group $T_{123}$ has the order of $U_3(3)$. Lastly we verify that the intersection of edge groups meeting at a vertex has the order of $S_4$.

The lowest degree permutation representation of $Co_1$ is on 98 280 letters, being the index of $Co_2$, the stabilizer of a type 2 vector (modulo, of course, the central involution). This representation is obtained at the end of the chapter. It should be emphasized that the reader who wishes to work computationally with $Co_1$ can download this permutation representation of degree 98 280 directly from the MAGMA library, which makes a wealth of such information readily available. The package is remarkably efficient at working with such permutation groups, and with those of much larger degree (such as $Co_1$ acting on 1 545 600 letters). Indeed, a recent development will obtain a complete list of maximal subgroups from such a permutation representation.

## 15.11 Properties of the Complete 9-Graph

What has been described in this final chapter is a complete graph on nine vertices in which:

I. Each **vertex** represents a copy of the symmetric group $S_4$.
II. Each **edge** represents a copy of the linear group $L_3(2)$.
III. Each **triangle** represents a copy of the unitary group $U_3(3)$.
IV. Each **complete 4-graph** represents a copy of HJ, the Hall–Janko sporadic simple group.
V. Each **complete 5-graph** represents a copy of the exceptional Lie group $G_2(4)$.

**VI.** Each **complete 6-graph** represents a copy of the 3˙Suz, the triple cover of the Suzuki sporadic simple group in which the central elements of order 3 act as 3-cycles on the remaining three vertices.

**VII.** Every **complete 7-graph** represents the same copy of $Co_1$, the largest Conway sporadic simple group.

**VIII, IX.** Every **complete 8-graph and the complete 9-graph** represent this copy of $Co_1$.

**X.** Any other subgraph generates the same group as that generated by the smallest complete graph in which it is contained; thus, for example, two disjoint edges generate HJ.

**XI.** For $2 \leq n \leq 6$ odd permutations on $n$ vertices furnish the outer automorphism of the relevant group, leading to $L_3(2):2$, $U_3(3):2$, HJ : 2, $G_2(4):2$ and 3˙Suz : 2.

The progenitor

$$7^{\star 36} : A_9$$

contains subprogenitors of shape

$$7^{\star \binom{k}{2}} : ((A_k \times A_{9-k}) : 2),$$

for $3 \leq k \leq 6$, where the 'outer' elements of the control subgroup act oddly on the two alternating groups $A_k$ and $A_{9-k}$. We then have

$$\begin{aligned}
7^{\star 36} : A_9 &\mapsto Co_1 \\
7^{\star 15} : ((A_6 \times A_3) : 2) &\mapsto 3\text{˙Suz} : 2 \\
7^{\star 10} : ((A_5 \times A_4) : 2) &\mapsto (G_2(4) \times A_4) : 2 \\
7^{\star 6} : ((A_4 \times A_5) : 2) &\mapsto (HJ \times A_5) : 2 \\
7^{\star 3} : ((A_3 \times A_6) : 2) &\mapsto (U_3(3) \times A_6) : 2 \\
[7^{\star 2} : (S_3 \times (A_2 \times A_7) : 2) &\mapsto (L_2(7) \times A_7) : 2],
\end{aligned}$$

where each of the image groups on the right-hand side is a maximal subgroup of $Co_1$, as listed in the ATLAS (Conway et al., 1985, page 183). Note that the final, bracketed term has a modified control subgroup in that it refers to the doubling up process obtained by adjoining the element $d = (t_{ki}t_{jk})^2$, which inverts $z$ and commutes with all the permutations of the vertices, see Section 15.9. The alternating group $A_2$ is, of course, the trivial group.

## 15.12 Connection with the Leech Lattice

We conclude this book, and tie the current chapter in with the general theme, by showing how the construction of the groups in the Thompson chain comes

## 15.12 Connection with the Leech Lattice 259

| 8 | 7 |
|---|---|
| 1 | 3 |
| 2 | 6 |
| 4 | 5 |

| 16 | 15 | 24 | 23 |
|----|----|----|----|
| 9  | 11 | 17 | 19 |
| 10 | 14 | 18 | 22 |
| 12 | 13 | 20 | 21 |

Figure 15.1 The brick labelling of $\Omega$

directly from the symmetric group permuting the three bricks of the MOG. We do this by constructing the group $3 \times A_9$ acting as $24 \times 24$ matrices on the Leech lattice, and then producing an element of order 7 as one of the 36 edges of the complete 9-graph. The central element of order 3, which corresponds to our element $z$, cycles the three bricks. A word of caution though: It is the covering group $\cdot O \cong 2 \cdot Co_1$ that has the 24-dimensional representation, not the simple group $Co_1$, and so we must in a sense work modulo $\langle -I_{24} \rangle$. Indeed, the alternating group $A_9$ lifts in $\cdot O$ to $2 \cdot A_9$ and, as can be seen in the ATLAS (Conway et al., 1985, page 37), elements of cycle shape $1^5 \cdot 2^2$ in $A_9$ lift in the double cover to elements of order 4 squaring to $-I_{24}$. Elements of $A_9$ of cycle shape $1 \cdot 2^4$ lift to involutions.

In $M_{24}$ the centralizer of $z$, which cycles the three bricks of the MOG, is a group of shape $3 \times L_2(7)$ where the $L_2(7)$ acts identically on the three bricks. Accordingly we label the points of the first brick with the eight elements of the projective line $P_1(7)$ with 0 replaced by 7 and $\infty$ by 8; the other two bricks are labelled by adding 8 and then 16 to these labels, see Figure 15.1.

In Figure 15.2 we exhibit elements of $\cdot O$ that have the same action in each of the three bricks. Thus $a$, which corresponds to $x \mapsto x + 1$, and $b$, which corresponds to $x \mapsto -1/x$, together generate $L_2(7)$; $c$ and $e$ also lie in $L_2(7)$ and are useful in the sequel. The element $d$ (which is not to be confused with the $d$ in the first half of this chapter) denotes negation on the dodecad indicated as having the same four points in each of the three bricks. The images of these sign changes under conjugation by our $L_2(7)$ generate an elementary abelian group of order $2^3$, resulting in a visible group of shape $2^3 : L_2(7)$. In the table of Figure 15.2 we indicate how these elements embed in the alternating group $A_9$.

We shall work over the Galois field $\mathbb{Z}_3$, the integers modulo 3, when the rows of the matrices in $\cdot O$ that correspond to 4-vectors in the Leech lattice, will have norm 1 (64 being congruent to 1 modulo 3); so no further normalization will be required. Using MAGMA to construct $a, b, c, d$ and $e$ as $24 \times 24$ matrices is quite straightforward:

```
> z3:=GaloisField(3);
> gl24:=GeneralLinearGroup(24,z3);
> procedure mat(gl24,pi,~xx);
procedure> xx8:=PermutationMatrix(z3,pi);
procedure> xx:=gl24!DirectSum(DirectSum(xx8,xx8),xx8);
procedure> end procedure;
> mat(gl24,Sym(8)!(1,2,3,4,5,6,7),~aa);
> mat(gl24,Sym(8)!(7,8)(1,6)(2,3)(4,5),~bb);
> mat(gl24,Sym(8)!(1,2,4)(3,6,5),~cc);
> mat(gl24,Sym(8)!(8,5)(7,4)(1,2)(3,6),~ee);
> #sub<gl24|aa,bb>;
168
> #sub<gl24|aa,bb,cc,ee>;
168
> dd8:=DiagonalMatrix(z3,8,[-1,-1,1,-1,1,1,-1,1]);
> dd:=gl24!DirectSum(DirectSum(dd8,dd8),dd8);
> #sub<gl24|aa,bb,dd>;
2688
```

Note that the subgroup $\langle a, b, d \rangle$ contains the central involution $-I_{24}$.

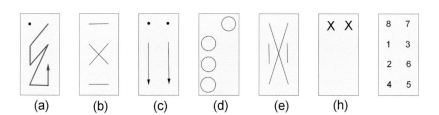

| Label | Action in $L_2(7)$ | Action in $A_9$ |
|---|---|---|
| (a) | (1 2 3 4 5 6 7) | (1 2 3 4 5 6 7) |
| (b) | (8 7)(1 6)(2 3)(4 5) | (8 7)(1 6)(2 3)(4 5) |
| (c) | (1 2 4)(3 6 5) | (1 2 4)(3 6 5) |
| (d) |  | (8 7)(1 3)(2 6)(4 5) |
| (e) | (8 5)(7 4)(1 2)(3 6) | (8 5)(7 4)(1 2)(3 6) |
| (h) |  | (1 4 2)(3 6 5) |
| (j) |  | (8 9)(1 6)(2 5)(3 4) |

Figure 15.2 Elements of $\cdot O$ commuting with $z$

## 15.12 Connection with the Leech Lattice

The element $h$ is defined to be

$$h := \xi_T \epsilon_{T+L},$$

where $\xi_T$ is the Conway element of Section 11.1, $T$ is the first column of the MOG, $\epsilon_X$ denotes negation on the coordinates corresponding to the subset $X \subset \Omega$ and $L$ denotes the top row of the MOG, which is congruent modulo $\mathscr{C}$ to the sextet consisting of the columns of the MOG. So the element $h$ is completely determined by $L$ and, if for $i \in \Omega$ we let $\gamma(i)$ denote the column of the MOG containing $i$, we have

$$h : x_i \mapsto \begin{cases} x_i - \frac{1}{2}\sum_{j \in \gamma(i)} x_j = x_i + \sum_{j \in \gamma(i)} x_j \ (\text{modulo } 3) \text{ for } i \notin L, \\ -x_i - \sum_{j \in \gamma(i)} x_j \ (\text{modulo } 3) \text{ for } i \in L. \end{cases}$$

Thus the image of $x_i$ under the action of $h$ only depends on whether $i$ is in $L$ or not. In particular $h$ must commute with elements of $M_{24}$ that fix the top row of the MOG, such as our $b$ and $c$. Using MAGMA we construct the action of $h$ on the first brick by coercing a string of length 64 into $GL_8(3)$ and then duplicating this $8 \times 8$ matrix in the other two bricks:

```
> z3:=GaloisField(3);
> gl8:=GeneralLinearGroup(8,z3);
> hh8:=gl8![-1,1,0,1,0,0,0,-1,
> 1,-1,0,1,0,0,0,-1,
> 0,0,-1,0,1,1,-1,0,
> 1,1,0,-1,0,0,0,-1,
> 0,0,1,0,-1,1,-1,0,
> 0,0,1,0,1,-1,-1,0,
> 0,0,1,0,1,1,1,0,
> 1,1,0,1,0,0,0,1];
> Order(hh8);
3
> hh:=gl24!DirectSum(DirectSum(hh8,hh8),hh8);
> Order(hh);
3
> Order(hh*aa);
5
> #sub<gl24|aa,bb,cc,dd,ee,hh>;
40320
```

We readily check that $h$ has order 3 and, since it commutes with $b$ and $c$ it must correspond in $A_9$ to the element shown in Figure 15.2 or its inverse. Checking the order of $ah$ confirms that it is indeed the element shown; so we now have the group $2^{\cdot}A_8$ generated by rather familiar elements. In order to complete the $2^{\cdot}A_9$ we first note that any element commuting with the brick group

$$B = \left\langle \begin{bmatrix} 0 & I_8 & 0 \\ 0 & 0 & I_8 \\ I_8 & 0 & 0 \end{bmatrix}, \begin{bmatrix} I_8 & 0 & 0 \\ 0 & 0 & I_8 \\ 0 & I_8 & 0 \end{bmatrix} \right\rangle \cong S_3,$$

which is defined in MAGMA by

```
> gl24:=GeneralLinearGroup(24,z3);
> mr3:=MatrixRing(z3,3);mr8:=MatrixRing(z3,8);
> bk3:=gl24!TensorProduct(mr3![0,1,0,0,0,1,1,0,0],mr8!Id(gl8));
> bk2:=gl24!TensorProduct(mr3![1,0,0,0,0,1,0,1,0],mr8!Id(gl8));
```

must have form

$$\begin{bmatrix} X & Y & Y \\ Y & X & Y \\ Y & Y & X \end{bmatrix} = \begin{bmatrix} 1 & 0 & 0 \\ 0 & 1 & 0 \\ 0 & 0 & 1 \end{bmatrix} \otimes X + \begin{bmatrix} 0 & 1 & 1 \\ 1 & 0 & 1 \\ 1 & 1 & 0 \end{bmatrix} \otimes Y,$$

where $\otimes$ denotes the tensor product of matrices. We seek such an element $j$ corresponding to $(8\,9)(1\,6)(2\,5)(3\,4)$. This will be a genuine involution; it will invert $a$ and commute with $c$ and so both $X$ and $Y$ must have the form

$$X, Y \sim \begin{bmatrix} u & v & u & v & v & w & u & z \\ v & u & v & v & w & u & u & z \\ u & v & v & w & u & u & v & z \\ v & v & w & u & u & v & u & z \\ v & w & u & u & v & u & v & z \\ w & u & u & v & u & v & v & z \\ u & u & v & u & v & v & w & z \\ \hline y & y & y & y & y & y & y & x \end{bmatrix}.$$

The condition that $j$ inverts $h$ means that

$$hh8 \times X \times hh8 = X \quad \text{and} \quad hh8 \times Y \times hh8 = Y,$$

or, equivalently, since $hh8$ is an orthogonal matrix

$$hh8 \times X = X \times hh8^t; \qquad hh8 \times Y = Y \times hh8^t. \tag{15.8}$$

The set of candidates for $X$ and $Y$ is closed under addition and under scalar multiplication and so forms a vector space. Equating the $i$8th and $8j$th entries in (15.8) we find that

$$u = z = y, \quad v = x + y, \quad w = -x,$$

and so this space has dimension 2, and there are just nine candidates for $X$ and $Y$.

## 15.12 Connection with the Leech Lattice

From Figure 15.2 we see that $bj \sim (2\ 4)(3\ 5)(7\ 9\ 8)$, which must have order 12 in $2^{\cdot}A_9$, and $dj \sim (1\ 4\ 2)(3\ 6\ 5)(7\ 9\ 8)$ which must have order 3 or 6 and commute with $c$ and $h$. Multiplying by $-I_{24}$, if necessary, we may assume that $dj$ has order 3. We may deduce the $X$ and $Y$ we need manually from Figure 10.1 using the fact that the rows of any element of $\cdot O$ form the twenty-four 4-vectors in a cross (in our case, read modulo 3). Or we may use MAGMA to obtain

$$X = \left(\begin{array}{cccccc|c} 2 & 0 & 2 & 0 & 0 & 2 & 2 \\ 0 & 2 & 0 & 0 & 2 & 2 & 2 \\ 2 & 0 & 0 & 2 & 2 & 2 & 0 & 2 \\ 0 & 0 & 2 & 2 & 2 & 0 & 2 & 2 \\ 0 & 2 & 2 & 2 & 0 & 2 & 0 & 2 \\ 2 & 2 & 2 & 0 & 2 & 0 & 0 & 2 \\ 2 & 2 & 0 & 2 & 0 & 0 & 2 & 2 \\ 2 & 2 & 2 & 2 & 2 & 2 & 1 \end{array}\right); \quad Y = \left(\begin{array}{cccccc|c} 2 & 2 & 2 & 2 & 2 & 0 & 2 & 2 \\ 2 & 2 & 2 & 2 & 0 & 2 & 2 & 2 \\ 2 & 2 & 2 & 0 & 2 & 2 & 2 & 2 \\ 2 & 2 & 0 & 2 & 2 & 2 & 2 & 2 \\ 2 & 0 & 2 & 2 & 2 & 2 & 2 & 2 \\ 0 & 2 & 2 & 2 & 2 & 2 & 2 & 2 \\ 2 & 2 & 2 & 2 & 2 & 2 & 0 & 2 \\ 2 & 2 & 2 & 2 & 2 & 2 & 2 & 0 \end{array}\right).$$

Thus the 8th and 1st rows of $j \sim (8\ 9)(1\ 6)(2\ 5)(3\ 4)$ correspond respectively to the 4-vectors

$$\begin{array}{|cc|cc|cc|} \hline 1 & 2 & 0 & 2 & 0 & 2 \\ 2 & 2 & 2 & 2 & 2 & 2 \\ \hline 2 & 2 & 2 & 2 & 2 & 2 \\ 2 & 2 & 2 & 2 & 2 & 2 \\ \hline \end{array} \equiv \text{mod } \mathbb{Z}_3 \;-\; \begin{array}{|cc|cc|cc|} \hline 5 & 1 & -3 & 1 & -3 & 1 \\ 1 & 1 & 1 & 1 & 1 & 1 \\ \hline 1 & 1 & 1 & 1 & 1 & 1 \\ 1 & 1 & 1 & 1 & 1 & 1 \\ \hline \end{array} \in \Lambda,$$

$$\begin{array}{|cc|cc|cc|} \hline 2 & 2 & 2 & 2 & 2 & 2 \\ 2 & 2 & 2 & 2 & 2 & 2 \\ \hline 0 & 2 & 2 & 0 & 2 & 0 \\ 0 & 0 & 2 & 2 & 2 & 2 \\ \hline \end{array} \equiv \text{mod } \mathbb{Z}_3 \;-\; \begin{array}{|cc|cc|cc|} \hline 1 & 1 & 1 & 1 & 1 & 1 \\ 1 & 1 & 1 & 1 & 1 & 1 \\ \hline -3 & 1 & 1 & -3 & 1 & -3 \\ -3 & -3 & 1 & 1 & 1 & 1 \\ \hline \end{array} \in \Lambda.$$

So, as in Figure 10.1, $j$ is a *triad type* element of $\cdot O$, see Remark in Section 11.4. We may check using MAGMA that $\langle a, b, c, d, h, j \rangle$ has order 362 880 and so these matrices generate the required copy of the double cover $2^{\cdot}A_9$.

It remains to find a symmetric generator of order 7; we thus seek $t_{89}$ that must commute with $a$, $c$ and $h$ and be inverted by $j$. It must also be squared by the element $z = \mathbf{bk3}$ that cycles the three bricks of the MOG. Suppose that

$$t_{89} = \begin{bmatrix} T_{11} & T_{12} & T_{13} \\ T_{21} & T_{22} & T_{23} \\ T_{31} & T_{32} & T_{33} \end{bmatrix}.$$

Then each of the constituent $8 \times 8$ matrices $T_{ij}$ must commute with $(1\,2\,3\,4\,5\,6\,7)$

and (1 2 4)(3 6 5) and so have form

$$T_{ij} \sim \begin{bmatrix} u & v & v & w & v & w & w & z \\ w & u & v & v & w & v & w & z \\ w & w & u & v & v & w & v & z \\ v & w & w & u & v & v & w & z \\ w & v & w & w & u & v & v & z \\ v & w & v & w & w & u & v & z \\ v & v & w & v & w & w & u & z \\ y & y & y & y & y & y & y & x \end{bmatrix}.$$

Similarly, we must have that $hh8 \cdot T_{ij} = T_{ij} \cdot hh8$, so we are seeking matrices that commute with an 8-dimensional representation of $\langle a, c, h \rangle \cong 2 \cdot A_7$. Clearly this set of matrices is closed under addition, scalar multiplication and matrix multiplication and so forms an algebra over $\mathbb{Z}_3$. Equating coefficients as for $j$ we find that

$$u = x, v = -w = -z = y,$$

and so there are just nine candidates for $T_{ij}$ and they satisfy

$$T_{ij} = xI_8 + yK,$$

where

$$K = \begin{bmatrix} 0 & 1 & 1 & 2 & 1 & 2 & 2 & 2 \\ 2 & 0 & 1 & 1 & 2 & 1 & 2 & 2 \\ 2 & 2 & 0 & 1 & 1 & 2 & 1 & 2 \\ 1 & 2 & 2 & 0 & 1 & 1 & 2 & 2 \\ 2 & 1 & 2 & 2 & 0 & 1 & 1 & 2 \\ 1 & 2 & 1 & 2 & 2 & 0 & 1 & 2 \\ 1 & 1 & 2 & 1 & 2 & 2 & 0 & 2 \\ 1 & 1 & 1 & 1 & 1 & 1 & 1 & 0 \end{bmatrix}.$$

We have $K^2 = -I_8$ and so this set of nine matrices corresponds to the elements of the field $GF_9$ in which $K$ is playing the role of $i$; moreover, we see that $K^t = -K$ and transposition corresponds to the field automorphism interchanging $i$ and $-i$. Each of the $T_{ij}$ must be one of these nine matrices, which we exhibit in Table 15.1. If we let $U \sim 1 + i$ then the correspondence is given by

$$\begin{bmatrix} 0 & 1 & -1 & i & -i & 1+i & 1-i & -1+i & -1-i \\ 0 & 1 & -1 & -U^2 & U^2 & U & U^t & -U^t & -U \end{bmatrix}.$$

## 15.12 Connection with the Leech Lattice

| No. | Matrix<br>Field Label | No. | Matrix<br>Field Label |
|---|---|---|---|
| 1 | $\begin{bmatrix} 0 & 0 & 0 & 0 & 0 & 0 & 0 & 0 \\ 0 & 0 & 0 & 0 & 0 & 0 & 0 & 0 \\ 0 & 0 & 0 & 0 & 0 & 0 & 0 & 0 \\ 0 & 0 & 0 & 0 & 0 & 0 & 0 & 0 \\ 0 & 0 & 0 & 0 & 0 & 0 & 0 & 0 \\ 0 & 0 & 0 & 0 & 0 & 0 & 0 & 0 \\ 0 & 0 & 0 & 0 & 0 & 0 & 0 & 0 \\ 0 & 0 & 0 & 0 & 0 & 0 & 0 & 0 \end{bmatrix}$<br>$[0 \sim 0]$ | | |
| 2 | $\begin{bmatrix} 1 & 0 & 0 & 0 & 0 & 0 & 0 & 0 \\ 0 & 1 & 0 & 0 & 0 & 0 & 0 & 0 \\ 0 & 0 & 1 & 0 & 0 & 0 & 0 & 0 \\ 0 & 0 & 0 & 1 & 0 & 0 & 0 & 0 \\ 0 & 0 & 0 & 0 & 1 & 0 & 0 & 0 \\ 0 & 0 & 0 & 0 & 0 & 1 & 0 & 0 \\ 0 & 0 & 0 & 0 & 0 & 0 & 1 & 0 \\ 0 & 0 & 0 & 0 & 0 & 0 & 0 & 1 \end{bmatrix}$<br>$[1 \sim 1]$ | 3 | $\begin{bmatrix} 2 & 0 & 0 & 0 & 0 & 0 & 0 & 0 \\ 0 & 2 & 0 & 0 & 0 & 0 & 0 & 0 \\ 0 & 0 & 2 & 0 & 0 & 0 & 0 & 0 \\ 0 & 0 & 0 & 2 & 0 & 0 & 0 & 0 \\ 0 & 0 & 0 & 0 & 2 & 0 & 0 & 0 \\ 0 & 0 & 0 & 0 & 0 & 2 & 0 & 0 \\ 0 & 0 & 0 & 0 & 0 & 0 & 2 & 0 \\ 0 & 0 & 0 & 0 & 0 & 0 & 0 & 2 \end{bmatrix}$<br>$[-1 \sim -1]$ |
| 4 | $\begin{bmatrix} 0 & 1 & 1 & 2 & 1 & 2 & 2 & 2 \\ 2 & 0 & 1 & 1 & 2 & 1 & 2 & 2 \\ 2 & 2 & 0 & 1 & 1 & 2 & 1 & 2 \\ 1 & 2 & 2 & 0 & 1 & 1 & 2 & 2 \\ 2 & 1 & 2 & 2 & 0 & 1 & 1 & 2 \\ 1 & 2 & 1 & 2 & 2 & 0 & 1 & 2 \\ 1 & 1 & 2 & 1 & 2 & 2 & 0 & 2 \\ 1 & 1 & 1 & 1 & 1 & 1 & 1 & 0 \end{bmatrix}$<br>$[i \sim -U^2]$ | 5 | $\begin{bmatrix} 0 & 2 & 2 & 1 & 2 & 1 & 1 & 1 \\ 1 & 0 & 2 & 2 & 1 & 2 & 1 & 1 \\ 1 & 1 & 0 & 2 & 2 & 1 & 2 & 1 \\ 2 & 1 & 1 & 0 & 2 & 2 & 1 & 1 \\ 1 & 2 & 1 & 1 & 0 & 2 & 2 & 1 \\ 2 & 1 & 2 & 1 & 1 & 0 & 2 & 1 \\ 2 & 2 & 1 & 2 & 1 & 1 & 0 & 1 \\ 2 & 2 & 2 & 2 & 2 & 2 & 2 & 0 \end{bmatrix}$<br>$[-i \sim U^2]$ |
| 6 | $\begin{bmatrix} 1 & 1 & 1 & 2 & 1 & 2 & 2 & 2 \\ 2 & 1 & 1 & 1 & 2 & 1 & 2 & 2 \\ 2 & 2 & 1 & 1 & 1 & 2 & 1 & 2 \\ 1 & 2 & 2 & 1 & 1 & 1 & 2 & 2 \\ 2 & 1 & 2 & 2 & 1 & 1 & 1 & 2 \\ 1 & 2 & 1 & 2 & 2 & 1 & 1 & 2 \\ 1 & 1 & 2 & 1 & 2 & 2 & 1 & 2 \\ 1 & 1 & 1 & 1 & 1 & 1 & 1 & 1 \end{bmatrix}$<br>$[1+i \sim U]$ | 7 | $\begin{bmatrix} 1 & 2 & 2 & 1 & 2 & 1 & 1 & 1 \\ 1 & 1 & 2 & 2 & 1 & 2 & 1 & 1 \\ 1 & 1 & 1 & 2 & 2 & 1 & 2 & 1 \\ 2 & 1 & 1 & 1 & 2 & 2 & 1 & 1 \\ 1 & 2 & 1 & 1 & 1 & 2 & 2 & 1 \\ 2 & 1 & 2 & 1 & 1 & 1 & 2 & 1 \\ 2 & 2 & 1 & 2 & 1 & 1 & 1 & 1 \\ 2 & 2 & 2 & 2 & 2 & 2 & 2 & 1 \end{bmatrix}$<br>$[1-i \sim U^t]$ |
| 8 | $\begin{bmatrix} 2 & 1 & 1 & 2 & 1 & 2 & 2 & 2 \\ 2 & 2 & 1 & 1 & 2 & 1 & 2 & 2 \\ 2 & 2 & 2 & 1 & 1 & 2 & 1 & 2 \\ 1 & 2 & 2 & 2 & 1 & 1 & 2 & 2 \\ 2 & 1 & 2 & 2 & 2 & 1 & 1 & 2 \\ 1 & 2 & 1 & 2 & 2 & 2 & 1 & 2 \\ 1 & 1 & 2 & 1 & 2 & 2 & 2 & 2 \\ 1 & 1 & 1 & 1 & 1 & 1 & 1 & 2 \end{bmatrix}$<br>$[-1+i \sim -U^t]$ | 9 | $\begin{bmatrix} 2 & 2 & 2 & 1 & 2 & 1 & 1 & 1 \\ 1 & 2 & 2 & 2 & 1 & 2 & 1 & 1 \\ 1 & 1 & 2 & 2 & 2 & 1 & 2 & 1 \\ 2 & 1 & 1 & 2 & 2 & 2 & 1 & 1 \\ 1 & 2 & 1 & 1 & 2 & 2 & 2 & 1 \\ 2 & 1 & 2 & 1 & 1 & 2 & 2 & 1 \\ 2 & 2 & 1 & 2 & 1 & 1 & 2 & 1 \\ 2 & 2 & 2 & 2 & 2 & 2 & 2 & 2 \end{bmatrix}$<br>$[-1-i \sim -U]$ |

Table 15.1 Candidates for the $T_{ij}$ forming $t_{89}$, stored in **tt89ss**

Now $t_{89}$ is a $24 \times 24$ orthogonal matrix over $\mathbb{Z}_3$ and so, if $R = (r_{ij})$ is the corresponding $3 \times 3$ matrix over $GF_9$, then we have

$$t_{89}^{-1} = t_{89}^t = \begin{bmatrix} T_{11}^t & T_{21}^t & T_{31}^t \\ T_{12}^t & T_{22}^t & T_{32}^t \\ T_{13}^t & T_{23}^t & T_{33}^t \end{bmatrix} \rightarrow \begin{bmatrix} \bar{r}_{11} & \bar{r}_{21} & \bar{r}_{31} \\ \bar{r}_{12} & \bar{r}_{22} & \bar{r}_{32} \\ \bar{r}_{13} & \bar{r}_{23} & \bar{r}_{33} \end{bmatrix} = \bar{R}^t = R^{-1}.$$

Thus $R$ is a unitary matrix and the problem involving $24 \times 24$ matrices reduces to one involving the corresponding $3 \times 3$ unitary matrices in $U_3(3)$.

The matrix $t_{89}$ that we seek must now satisfy a number of conditions:

(i) Each of the constituent $8 \times 8$ matrices $T_{ij}$ must be one of the nine shown in Figure 15.1.
(ii) The corresponding $3 \times 3$ matrix over $GF_9$ must be unitary and have order 7.
(iii) Its rows must correspond to a set of mutually orthogonal 4-vectors forming a cross, as in Table 10.1.
(iv) It must be inverted by our element $j$.
(v) It must satisfy the additional relation that defines $Co_1$.

In Section A we use MAGMA to find the required element $t_{89}$, which in the code is labelled **ww**. However, we now show how our knowledge of the elements of $\cdot O$ and the Leech vectors enables us to proceed by hand. From (iii) we see that, if such an element commutes with our permutation $a \in M_{24}$ of order 7, then three of these rows must be 4-vectors fixed by $a$. In our case these will visibly be the 8th, 16th and 24th rows. We work with respect to the standard Leech vectors written over $\mathbb{Z}$, where all the coordinates are even or all are odd.

In the *odd case* we see that a matrix corresponding to $\pm 1$ would be impossible as it would have seven odd entries congruent to 0 modulo 3 and thus have too great a norm. The other correspondences are

$$i \sim \begin{array}{|cc|} \hline -3 & 1 \\ 1 & 1 \\ 1 & 1 \\ 1 & 1 \\ \hline \end{array}; \quad 1+i \sim \begin{array}{|cc|} \hline 1 & 1 \\ 1 & 1 \\ 1 & 1 \\ 1 & 1 \\ \hline \end{array}; \quad -1+i \sim \begin{array}{|cc|} \hline 5 & 1 \\ 1 & 1 \\ 1 & 1 \\ 1 & 1 \\ \hline \end{array}.$$

But the cross containing

$$\begin{array}{|cc|cc|cc|} \hline 5 & 1 & -3 & 1 & -3 & 1 \\ 1 & 1 & 1 & 1 & 1 & 1 \\ 1 & 1 & 1 & 1 & 1 & 1 \\ 1 & 1 & 1 & 1 & 1 & 1 \\ \hline -1+i & & i & & i & \\ \end{array} \text{ also contains } \begin{array}{|cc|cc|cc|} \hline 1 & -3 & 1 & -3 & 1 & 1 \\ 1 & 1 & 1 & 1 & 1 & -3 \\ 1 & 1 & 1 & 1 & 1 & -3 \\ 1 & 1 & 1 & 1 & 1 & -3 \\ \hline \end{array},$$

and no row of our nine candidates contains three zeros modulo 3. So this is impossible.

## 15.12 Connection with the Leech Lattice

In the *even case* we have

$$1 \sim \begin{array}{|cc|} \hline 4 & 0 \\ 0 & 0 \\ 0 & 0 \\ 0 & 0 \\ \hline \end{array} \; ; \quad -i \sim \begin{array}{|cc|} \hline -6 & 2 \\ 2 & 2 \\ 2 & 2 \\ 2 & 2 \\ \hline \end{array} \; ; \quad 1-i \sim \begin{array}{|cc|} \hline -2 & 2 \\ 2 & 2 \\ 2 & 2 \\ 2 & 2 \\ \hline \end{array} \; ; \quad -1-i \sim \begin{array}{|cc|} \hline 2 & 2 \\ 2 & 2 \\ 2 & 2 \\ 2 & 2 \\ \hline \end{array}.$$

The cross containing

$$\begin{array}{|cc|cc|cc|} \hline -6 & 2 & 0 & 0 & 0 & 0 \\ 2 & 2 & 0 & 0 & 0 & 0 \\ 2 & 2 & 0 & 0 & 0 & 0 \\ 2 & 2 & 0 & 0 & 0 & 0 \\ \hline \end{array}$$
$$\begin{array}{ccc} -i & 0 & 0 \end{array}$$

also contains

$$\begin{array}{|cc|cc|cc|} \hline 0 & 0 & 2 & 2 & 2 & 2 \\ 0 & 0 & 2 & 2 & 2 & 2 \\ 0 & 0 & 2 & 2 & 2 & 2 \\ 0 & 0 & 2 & 2 & 2 & 2 \\ \hline \end{array} \text{ and } \begin{array}{|cc|cc|cc|} \hline 0 & 0 & 2 & 2 & -2 & -2 \\ 0 & 0 & 2 & 2 & -2 & -2 \\ 0 & 0 & 2 & 2 & -2 & -2 \\ 0 & 0 & 2 & 2 & -2 & -2 \\ \hline \end{array}.$$
$$\begin{array}{ccc} 0 & -1-i & -1-i \end{array} \qquad \begin{array}{ccc} 0 & -1-i & 1+i \end{array}$$

And the cross containing

$$\begin{array}{|cc|cc|cc|} \hline 0 & 0 & -2 & 2 & -2 & 2 \\ 0 & 0 & 2 & 2 & 2 & 2 \\ 0 & 0 & 2 & 2 & 2 & 2 \\ 0 & 0 & 2 & 2 & 2 & 2 \\ \hline \end{array}$$
$$\begin{array}{ccc} 0 & 1-i & 1-i \end{array}$$

also contains

$$\begin{array}{|cc|cc|cc|} \hline 2 & 2 & 4 & 0 & -4 & 0 \\ 2 & 2 & 0 & 0 & 0 & 0 \\ 2 & 2 & 0 & 0 & 0 & 0 \\ 2 & 2 & 0 & 0 & 0 & 0 \\ \hline \end{array} \text{ and } \begin{array}{|cc|cc|cc|} \hline -2 & -2 & 4 & 0 & -4 & 0 \\ -2 & -2 & 0 & 0 & 0 & 0 \\ -2 & -2 & 0 & 0 & 0 & 0 \\ -2 & -2 & 0 & 0 & 0 & 0 \\ \hline \end{array}.$$
$$\begin{array}{ccc} -1-i & 1 & -1 \end{array} \qquad \begin{array}{ccc} 1+i & 1 & -1 \end{array}$$

Thus

$$\begin{bmatrix} i & 0 & 0 \\ 0 & 1+i & 1+i \\ 0 & 1+i & -1-i \end{bmatrix} \text{ and } \begin{bmatrix} 1+i & 1 & -1 \\ -1-i & 1 & -1 \\ 0 & -1+i & -1+i \end{bmatrix} \quad (15.9)$$

are viable matrices, where we may of course permute and negate rows and columns, whilst ensuring that the determinant remains 1. Now, over $GF_9$, we have the factorization

$$x^7 - 1 = (x-1)(x^3 - (1+i)x^2 + (1-i)x - 1)(x^3 - (1-i)x^2 + (1+i)x - 1)$$

and so a $3 \times 3$ matrix of order 7 must have trace $1 \pm i$, which further restricts the number of possibilities we need to consider. In fact, it is the second matrix in (15.9) which is the matrix we seek.

$$t_{89} = \begin{bmatrix} U & 1 & -1 \\ -U & 1 & -1 \\ 0 & -U^t & -U^t \end{bmatrix} \sim \begin{bmatrix} 1+i & 1 & -1 \\ -1-i & 1 & -1 \\ 0 & -1+i & -1+i \end{bmatrix} \left\{ \begin{array}{|cc|c|c|} \hline -2 & -2 & 4 & -4 \\ -2 & -2 & & \\ -2 & -2 & & \\ -2 & -2 & & \\ \hline 2 & 2 & 4 & -4 \\ 2 & 2 & & \\ 2 & 2 & & \\ 2 & 2 & & \\ \hline & & 2 & -2 & 2 & -2 \\ & & -2 & -2 & -2 & -2 \\ & & -2 & -2 & -2 & -2 \\ & & -2 & -2 & -2 & -2 \\ \hline \end{array} \right\} .$$

The three vectors in the final column correspond to the 8th, 16th and 24th rows of the matrix that are fixed by $a$, the permutation of order 7, and of course they lie in the same cross. Thus $t_{89}$ is a $\cdot$O element of *involutory type* in the sense of Figure 10.1, see Remark in Section 11.4.

The relation of Equation 15.1 is equivalent to

$$t_{89}t_{97}^4 t_{89} t_{97}^3 t_{89}^3 (7\ 8\ 9) Z^2 = 1.$$

In order to verify that this holds, we use the MAGMA intrinsic **BlockMatrix** to produce a $24 \times 24$ matrix **ww**, corresponding to our $t_{89}$, out of the nine $8 \times 8$ matrices from the sequence **tt89ss** displayed in Figure 15.1. We check that *ww* is inverted by $j$. We then construct the 3-cycle $(7\ 8\ 9) = (jb)^4$, which we label *yy* in order to obtain $t_{97}$. The element **bk3** of order 3 that cycles the bricks of the MOG is labelled *zz*. We check that it squares *ww* by conjugation. Note that $(jb)^4 a = (1\ 2\ 3\ 4\ 5\ 6\ 7\ 8\ 9)$.

```
> zz:=bk3;
> ww:=gl24!BlockMatrix(3,3,[tt89sb[i]:i in [6,2,3,9,2,3,1,8,8]]);
> Order(ww);
7
> jj*ww*jj eq ww^-1;
true
> yy:=(jj*bb)^4;
> Order(ww*(yy^-1*ww*yy)^4*ww*(yy^-1*ww*yy)^3*ww^3*yy*zz^2);
1
> zz^-1*ww*zz eq ww^2;
true
```

We are now in a position to verify the claims made in the text.

```
> t89:=ww;t97:=yy^-1*ww*yy;t78:=yy*ww*yy^-1;
> bk2a:=(t89*t78)^2;
> s89:=bk2a^-1*t89*bk2a;
```

## 15.12 Connection with the Leech Lattice

```
> #sub<gl24|t89,s89>;
168
> #sub<gl24|t89,t97,t78>;
6048
> t18:=aa^-1*t78*aa;
> t28:=aa^-1*t18*aa;
> t38:=aa^-1*t28*aa;
> t48:=aa^-1*t38*aa;
> #sub<gl24|t78,t89,t18>;
1209600
> #sub<gl24|t78,t89,t18,t28>;
503193600
> Index(sub<gl24|t78,t89,t97,t18,t28,t38,cyc3>,
> sub<gl24|t78,t89,t97,cyc3,t18,t28>);
5346
>
```

The element $bk2a = (t_{ij}t_{ki})^2$ corresponds to the element bodily interchanging the 2nd and 3rd bricks of the MOG. [In fact, it is the negative of this element here.] It conjugates the $t_{ij}$ into the $s_{ij}$ in the doubling up process in Section 15.9. We verify here that $\langle t_{89}, s_{89} \rangle \cong L_2(7)$. We also verify that the group generated by a triangle has the order of $U_3(3)$; that the group generated by a complete 4-graph has the order of HJ; and that the group generated by a complete 5-graph has the order of $G_2(4)$. Recall that this is all being carried out modulo the centre of ·O, so the order may be multiplied by a factor of 2. We further check that the index of the group generated by a complete 5-graph has index 5346 in the group generated by a complete 6-graph, consistent with our claim that it is 3˙Suz, the triple cover of the Suzuki simple group.

Finally, we must identify elements in the group we have constructed that satisfy the presentation (15.7). At this point a slight subtlety is introduced. Since there is no odd permutation in the symmetric group $S_9$ commuting with a 9-cycle, the group $A_9$ has two classes of elements of order 9. Perhaps surprisingly these two classes remain distinct in $Co_1$; indeed, each element of one class fixes nine 2-vectors (and their inverses), while the other fixes none. It turns out that the permutation (1 2 3 4 5 6 7 8 9) in the above control subgroup does not belong to the class of 9-elements that occur in our presentation. Instead we must take

$$x \sim (jb)^2 a = (1\ 2\ 3\ 4\ 5\ 6\ 7\ 9\ 8),$$
$$y \sim (jb)^2 = (9\ 8\ 7), \text{ and}$$
$$t = (yt_{89}y^2)^{-1} = t_{79}$$

when the presentation (15.7) holds.

### 15.12.1 A Subgroup Isomorphic to $Co_2$

In the permutation character of the action of $Co_1$ acting on the cosets of $Co_2$, which in $\Lambda$ correspond to the pairs consisting of a 2-vector and its negative, has rank 4 and is given by

$$\chi_{98\,280} = \chi_1 + \chi_{299} + \chi_{17\,250} + \chi_{80\,730}.$$

From this, we see that our element $xx$, which lies in ATLAS conjugacy class 9C, see Conway et al. (1985, page 184), fixes precisely nine 2-vectors (and their negatives). These are

| 0 | 0 | 2 | 2 | −2 | −2 | | 0 | 0 | 2 | 0 | −2 | 0 | | 0 | 0 | 0 | −2 | 0 | 2 |
|---|---|---|---|---|---|---|---|---|---|---|---|---|---|---|---|---|---|---|---|
| 0 | 0 | 2 | −2 | −2 | 2 | | 0 | 0 | 2 | 0 | −2 | 0 | | 0 | 0 | 0 | 2 | 0 | −2 |
| 0 | 0 | 0 | 0 | 0 | 0 | | 0 | 0 | 0 | −2 | 0 | 2 | | 0 | 0 | 0 | −2 | 0 | 2 |
| 0 | 0 | 0 | 0 | 0 | 0 | | 0 | 0 | 0 | −2 | 0 | 2 | | 0 | 0 | 0 | −2 | 0 | 2 |

together with their images under the action of the $S_3$ bodily permuting the three octads in the MOG trio. Reading modulo 3 and with the ordering of Figure 15.1, the first of these vectors becomes

$$v2 = (0, 0, 0, 0, 0, 0, 0, 0, 2, 0, 1, 0, 0, 0, 2, 2, 1, 0, 2, 0, 0, 0, 1, 1);$$

it is fixed by x and by the involution

$$e2 = tx^2tx^3t^{-1}x^{-2}t^{-2}x^{-1}t^3,$$

and together these two elements generate the Conway group $Co_2$:

```
> v2:=Vector(24,
> [z3!0,0,0,0,0,0,0,0,2,0,1,0,0,0,2,2,1,0,2,0,0,0,1,1]);
> xx:=(jj*bb)^2*aa;
> yy:=(jj*bb)^2;
> tt:=(yy*ww*yy^2)^-1;
> ee2:=tt*xx^2*tt*xx^3*tt^-1*xx^-2*tt^-2*xx^-1*tt^3;
> v2*xx eq v2;
true
> v2*ee2 eq v2;
true
> #sub<gl24|xx,ee2>;
42305421312000
```

## 15.12 Connection with the Leech Lattice

The elements $txt$ and $x^{yx^{-2}y}$ also fix $v2$ and their inclusion facilitates the coset enumeration giving the action of $Co_1$ on the cosets of $Co_2$, which is to say the permutation action on the set of 98 280 2-vectors and their negatives.

```
> co1<x,y,t>:=Group<x,y,t|
> x^9=y^3=(y^x*y)^2=(y^(x^3),y)=(x*y^-1)^7=(x^2*y)^4=
> t*(t^y)^2*t^5*(t^(y^2))^4*t^5*y^2=(x*y*x^2*y^2*t^x)^13=
> t^7=t^(y^x*y)*t=(t,x*y^-1)=((t^x*t)^2,x)=1>;
> suz:=sub<co1|t,y,y^x,y^(x^2),y^(x^3),x^2*y^x*y*x^-2>;
> co2:=sub<co1|x,t*x*t,x^(y^x^-2*y),t*x^2*t*x^3*t^-1*x^-2*t^-2*x^-1*t^3>;
> ct9:=CosetTable(co1,co2:Hard:=true, CosetLimit:=40000000);
> co1p:=CosetTableToPermutationGroup(co1,ct9);
> Degree(co1p);
98280
> xp:=co1p.1;yp:=co1p.2;tp:=co1p.3;
> [#Fix(xp),#Fix(yp*xp),#Fix(yp),#Fix(tp)];
[ 9, 0, 0, 0 ]
```

We see from the above that $x \sim (1\ 2\ 3\ 4\ 5\ 6\ 7\ 9\ 8)$ in $A_9$ fixes nine such pairs, whereas $yx \sim (1\ 2\ 3\ 4\ 5\ 6\ 7\ 8\ 9)$ fixes none. Neither $y$ nor $t$ fixes a pair.

This permutation representation may readily be used to verify all the claims made in the text. For instance, the following verifies that $U_3(3)$, HJ, $G_2(4)$ and $3\,\!\cdot\!Suz$ are generated by complete graphs on $n$-letters for $n = 3, 4, 5, 6$ respectively. Finally, it shows that the 3-cycle $y = (9\ 8\ 7)$ lies in the centre of the copy of $3\,\!\cdot\!Suz$ generated by $\{t_{ij} : i, j \in \{1, \ldots, 6\}\}$.

```
> #sub<co1p|tp,yp>;
6048
> #sub<co1p|[tp^(xp^i):i in [1..3]]>;
604800
> #sub<co1p|[tp^(xp^i):i in [1..4]]>;
251596800
> sz123456:=sub<co1p|[tp^(xp^i):i in [3..7]]>;
> #sz123456;
1345036492800
> #Centre(sz123456);
3
> yp in Centre(sz123456);
true
> #sub<co1p|xp,xp^(yp*xp^-2*yp)>;
1512
```

The last line indicates that $\langle x, x^{yx^{-2}y}\rangle \cong L_2(8) : 3$, which is the intersection of the copy of $Co_2$ fixing $v2$ with $\langle x, y \rangle \cong A_9$. Generators for its intersection with

⟨aa, bb, dd⟩ ≅ $2^4 : L_2(7)$, a copy of the alternating group $A_4$, are shown in Figure 15.3.

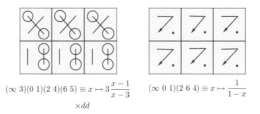

$(\infty\ 3)(0\ 1)(2\ 4)(6\ 5) \equiv x \mapsto 3\dfrac{x-1}{x-3}$
×dd

$(\infty\ 0\ 1)(2\ 6\ 4) \equiv x \mapsto \dfrac{1}{1-x}$

Figure 15.3 Generators for the intersection of $Co_2$ with $2^{12} : M_{24}$

# Appendix

## MAGMA Code for $7^{\star 36} : A_9 \mapsto Co_1$

For the convenience of the reader who has access to MAGMA, we append the code used in this chapter. This enables one to explore the subgroups associated with the edges of the complete 9-graph either as $24 \times 24$ matrices over $\mathbb{Z}_3$ or as permutations of degree 98 280.

```
z3:=GaloisField(3);
gl24:=GeneralLinearGroup(24,z3);
procedure mat(gl24,pi,~xx);
   xx8:=PermutationMatrix(z3,pi);
   xx:=gl24!DirectSum(DirectSum(xx8,xx8),xx8);
end procedure;
mat(gl24,Sym(8)!(1,2,3,4,5,6,7),~aa);
mat(gl24,Sym(8)!(7,8)(1,6)(2,3)(4,5),~bb);
mat(gl24,Sym(8)!(1,2,4)(3,6,5),~cc);
mat(gl24,Sym(8)!(8,5)(7,4)(1,2)(3,6),~ee);
#sub<gl24|aa,bb>;
#sub<gl24|aa,bb,cc,ee>;
dd8:=DiagonalMatrix(z3,8,[-1,-1,1,-1,1,1,-1,1]);
dd:=gl24!DirectSum(DirectSum(dd8,dd8),dd8);
#sub<gl24|aa,bb,dd>;
gl8:=GeneralLinearGroup(8,z3);
hh8:=gl8![-1,1,0,1,0,0,0,-1,
1,-1,0,1,0,0,0,-1,
0,0,-1,0,1,1,-1,0,
1,1,0,-1,0,0,0,-1,
0,0,1,0,-1,1,-1,0,
0,0,1,0,1,-1,-1,0,
0,0,1,0,1,1,1,0,
1,1,0,1,0,0,0,1];
hh:=gl24!DirectSum(DirectSum(hh8,hh8),hh8);
#sub<gl24|aa,bb,dd,hh>;
mr3:=MatrixRing(z3,3);mr8:=MatrixRing(z3,8);
bk3:=gl24!TensorProduct(mr3![0,1,0,0,0,1,1,0,0],mr8!Id(gl8));
bk2:=gl24!TensorProduct(mr3![1,0,0,0,0,1,0,1,0],mr8!Id(gl8));
testj8:={};
```

273

```
for a in z3 do
  for b in z3 do
    for c in z3 do
      for d in z3 do
        for e in z3 do
          for f in z3 do
          tt:=MatrixRing(z3,8)!
          [a,b,a,b,b,c,a,d,
           b,a,b,b,c,a,a,d,
           a,b,b,c,a,a,b,d,
           b,b,c,a,a,b,a,d,
           b,c,a,a,b,a,b,d,
           c,a,a,b,a,b,b,d,
           a,a,b,a,b,b,c,d,
           f,f,f,f,f,f,f,e];
          if hh8*tt*hh8 eq tt then
             testj8:=testj8 join {tt};
          end if;
end for;end for;end for;
end for;end for;end for;
#testj8;
testj8s:=Setseq(testj8);
mr3:=MatrixRing(z3,3);
mr24:=MatrixRing(z3,24);
for i in [1..9] do
  for j in [1..9] do
  uu:=mr24!TensorProduct(mr3![1,0,0,0,1,0,0,0,1],testj8s[i])
      +TensorProduct(mr3![0,1,1,1,0,1,1,1,0],testj8s[j]);
  if uu in gl24 and Order(gl24!uu) eq 2 and
     Order(bb*uu) eq 12 and Order(dd*uu) eq 3 and dd*uu*hh eq hh*dd*uu then
       i,j,Order(dd*uu),#sub<gl24|aa,bb,hh,uu>;
  end if;
end for;end for;
jj:=gl24!TensorProduct(mr3![1,0,0,0,1,0,0,0,1],testj8s[2]) +
TensorProduct(mr3![0,1,1,1,0,1,1,1,0],testj8s[4]);
jj;
#sub<gl24|aa,bb,cc,dd,hh,jj>;
tt89:={};
for a in z3 do
  for b in z3 do
    for c in z3 do
      for d in z3 do
        for e in z3 do
          for f in z3 do
          tt:=MatrixRing(z3,8)!
          [a,b,b,c,b,c,c,d,
           c,a,b,b,c,b,c,d,
           c,c,a,b,b,c,b,d,
           b,c,c,a,b,b,c,d,
           c,b,c,c,a,b,b,d,
           b,c,b,c,c,a,b,d,
```

```
                  b,b,c,b,c,c,a,d,
                  f,f,f,f,f,f,f,e];
                     if hh8*tt eq tt*hh8 then
                        tt89:=tt89 join {tt};
                     end if;
end for;end for;end for;
end for;end for;end for;
#tt89;
tt89s:=Setseq(tt89);
cn:=func<uu|[uu[8][8],uu[8][7]]>;
[cn(tt89s[i]):i in [1..9]];
tt89ss:=[tt89s[i]:i in [9,3,8,7,6,2,1,5,4]];
zz:=bk3;
ww:=gl24!BlockMatrix(3,3,[tt89ss[i]:i in [6,2,3,9,2,3,1,8,8]]);
Order(ww);
jj*ww*jj eq ww^-1;

v2:=Vector(24,
[z3!0,0,0,0,0,0,0,0,2,0,1,0,0,0,2,2,1,0,2,0,0,0,1,1]);
xx:=(jj*bb)^2*aa;
yy:=-(jj*bb)^2;
tt:=-(yy*ww*yy^2)^-1;
ee2:=tt*xx^2*tt*xx^3*tt^-1*xx^-2*tt^-2*xx^-1*tt^3;
v2*xx eq v2;
v2*ee2 eq v2;
#sub<gl24|xx,ee2>;

co1<x,y,t>:=Group<x,y,t|
x^9=y^3=(y^x*y)^2=(y^(x^3),y)=(x*y^-1)^7=(x^2*y)^4=
t^7=t^(y^x*y)*t=(t,x*y^-1)=((t^x*t)^2,x)=
t*(t^y)^2*t^5*(t^(y^2))^4*t^5*y^2=(x*y*x^2*y^2*t^x)^13=1>;
co2:=sub<co1|x,t*x*t,y^2*x^2*y^2*x*y*x^-2*y,
t*x^2*t*x^3*t^-1*x^-2*t^-2*x^-1*t^3>;
CT2:=CosetTable(co1,co2:Hard:=true, CosetLimit:=40000000);
co1p:=CosetTableToPermutationGroup(co1,CT2);
Degree(co1p);
xxp:=co1p.1;yyp:=co1p.2;ttp:=co1p.3;
[#Fix(xxp),#Fix(yyp*xxp),#Fix(yyp),#Fix(ttp)];
```

# References

Assmus, E. F., and Key, J. D. (1992), *Designs and Their Codes*. Cambridge University Press.

Baker, H. F. (1935), 'Note introductory to Klein's group of order 168', *Proc. Cambridge Phil. Soc.*, **31**, 468–481.

Bannai, E., and Ito, T. (1984), *Algebraic Combinatorics I Association Schemes*. Benjamin/Cummings Publishing Company.

Benson, D. J. (1980), *The Simple Group $J_4$*, PhD. thesis, Cambridge University Press.

Berlekamp, E. R., Conway, J. H., and Guy, R. K. (1982), *Winning Ways for Your Mathematical Plays*, Vol. 2. Academic Press.

Bolt, S. (2002), *Some Applications of Symmetric Generation*, PhD. thesis, University of Birmingham.

Borcherds, R. E. (1998), What Is Moonshine? *Documenta Mathematica*, **ICM**, 607–615.

Borcherds, R. E., Conway, J. H., Queen, L., and Sloane, N. J. A. (1984), 'A Monster Lie algebra', *Adv. Math.*, **53(1)**, 75–79.

Bosma, W., and Cannon, J. (1994), *Handbook of Magma Functions*. University of Sydney.

Bray, J. N. (1997), *Symmetric Presentations of Finite Groups,* PhD. thesis, University of Birmingham.

Bray, J. N., and Curtis, R. T. (2003), 'Monomial modular representations and symmetric generation of the Harada–Norton group', *J. Algebra*, **268**, 723–743.

Bray, J. N. and Curtis, R. T. (2004), 'Double coset enumeration of symmetrically generated groups', *J. Group Theory.*, **7**, 167–185.

Bray, J. N., Holt, D. F., and Roney-Dougal, C. M. (2013), *The Maximal Subgroups of the Low-Dimensional Finite Classical Groups*. London Mathematical Society Lecture Note Series, Vol. 407, Cambridge University Press.

Burness, T. and Giudici, M. (2016), *Classical Groups, Derangements and Primes*. Australian Mathematical Society Lecture Series, Vol. 25, Cambridge University Press.

Cameron, P. J. (1994), *Combinatorics: Topics, Techniques, Algorithms*. Cambridge University Press.

Cameron, P. J. (1999), *Permutation Groups*. LMS Student texts, Vol. 45, Cambridge University Press.

# References

Cameron, P. J., and van Lint, J. H. (1991), *Graphs, Codes, Designs and Their Links*. Cambridge University Press.

Carter, R. W. 1972, reprinted (1989), *Simple Groups of Lie Type*. Wiley.

Chevalley, C. (1955), 'Sur certains groupes simples', *Tôhoku Math. J.*, **7**, 14–66.

Choi, C. (1972), 'On subgroups of $M_{24}$. II: the maximal subgroups of $M_{24}$', *Trans. Am. Math. Soc.*, **167**, 29–47.

Cohn, H., Kumar, A., Miller, S. D., Radchenko, D., and Viazovska, M. (2022), 'Universal optimality of the $E_8$ and Leech lattices and interpolation formulas', *Ann. Math.*, **196**(3), 983–1082.

Collins, M. J. (1990), *Representations and Characters of Finite Groups*. Cambridge Studies in Advanced Mathematics, Vol. 22, Cambridge University Press.

Conway, J. H. (1969a), 'A characterisation of Leech's lattice', *Inventiones Math.*, **7**, 137–142.

Conway, J. H. (1969b), 'A group of order 8,315,553,613,086,720,000', *Bull. Lond. Math. Soc.*, **1**, 79–88.

Conway, J. H. (1971), 'Three lectures on exceptional groups'. Pages 215–247 of: Powell, M. B., and Higman, G. (eds), *Finite Simple Groups*. Academic Press.

Conway, J. H. (1973), 'A construction of the smallest Fischer group'. Pages 27–35 of: Gagen, T., Hale, M. P., and Shult, E. E. (eds.), *Finite Groups '72*. North Holland.

Conway, J. H. (1981), 'The hunting of $J_4$', *Eureka*, **41**, 46–54.

Conway, J. H. (1983), 'The automorphism group of the 26-dimensional even unimodular Lorentz lattice', *J. Algebra.*, **80**, 159–163.

Conway, J. H. (1985), 'A simple construction of the Fischer-Griess Monster group', *Invent. Math.*, **79**(3), 513–540.

Conway, J. H., and Norton, S. P. (1979), 'Monstrous Moonshine', *BLMS*, **11**, 308–339.

Conway, J. H., and Sloane, Neil A. J. (1988), *Sphere-Packing, Lattices and Groups*. Springer-Verlag.

Conway, J. H., Norton, S. P. and Soicher, L.H. (1985), 'The Bimonster, the group $Y_{555}$, and the projective plane of order 3'. Pages 27–50 of: Tangora, M.C. (ed.), *Computers in Algebra (Chicago, Il. 1985)*, Lecture Notes in Pure and Applied Mathematics, Vol. 111. Marcel Dekker (1988).

Conway, J. H., Parker, R. A., and Sloane, N. J. A. (1982), 'The covering radius of the Leech lattice', *Proc. Roy. Soc.*, **380**(1779), 261–290.

Conway, J. H., Curtis, R. T., Norton, S. P., Parker, R. A., and Wilson, R. A. (1985), *An Atlas of Finite Groups*. Clarendon Press.

Curtis, R. T. (1972), *The Mathieu Group $M_{24}$ and Related Topics*. PhD thesis, University of Cambridge.

Curtis, R. T. (1976), 'A new combinatorial approach to $M_{24}$', *Math. Proc. Cambridge Phil. Soc.*, **79**, 25–42.

Curtis, R. T. (1977), 'The maximal subgroups of $M_{24}$', *Math. Proc. Cambridge Phil. Soc.*, **81**, 185–192.

Curtis, R. T. (1984), 'Eight octads suffice', *J. Comb. Theory.*, **36**, 116–123.

Curtis, R. T. (1989), 'Natural constructions of the Mathieu groups', *Math. Proc. Cambridge Phil. Soc.*, **106**, 423–429.

Curtis, R. T. (1990), 'Geometric interpretations of the "natural" constructions of the Mathieu groups', *Math. Proc. Cambridge Phil. Soc.*, **107**, 19–26.

Curtis, R. T. (1993), 'Symmetric generation II: The Janko group $J_1$', *J. London Math. Soc.*, **47**, 294–308.

Curtis, R. T. (1996), 'Monomial modular representations and construction of the Held group', *J. Algebra*, **184**, 1205–1227.

Curtis, R. T. (2007a), 'Construction of a family of Moufang loops', *Math. Proc. Cambridge Phil. Soc.*, **142**(2), 233–237.

Curtis, R. T. (2007b), *Symmetric Generation of Groups*. Encyclopedia of Mathematics and Its Applications, Vol. 111, Cambridge University Press.

Curtis, R. T. (2016), 'The Thompson chain of subgroups of the Conway group $Co_1$ and complete graphs on $n$ vertices', *J. Group Theory.*, **19**(6), 959–982.

Curtis, R. T. (2017), 'Construction of the Thompson Chain of subgroups of the Conway group O and complete graphs on $n$ letters'. Pages 73–90 of: Bhargava, M., Guralnick, R., Hiss, G., Lux, K., and Tiep, P. H. (eds), *Finite Simple Groups: Thirty Years of the Atlas and Beyond*, Contemporary Mathematics, Vol. 694. American Mathematical Society.

Curtis, C. W., and Reiner, I. (1962), *Representation Theory of Finite Groups and Associative Algebras*. John Wiley & Sons.

Dickson, L. E. (1919), 'On quaternions and their generalisation and the history of the eight squares theorem', *Annal. Math.*, **20**, 155–171.

Dickson, L. E. (2007), *Linear Groups, with an Exposition of the Galois Field Theory*. Cosimo Classics.

Dieudonné, J. (1963), *La Géometrie des Groupes Classiques*. 2nd edn. Springer-Verlag.

Graves, J. T. (1844), *On the sum of eight squares theorem and the inderlying system of imaginaries*. In letters to William Rowan Hamilton.

Griess, R. L. (1998), *Twelve Sporadic Groups*. Springer.

Halberstam, H., and Ingram, R. E. (eds). (1967), *The Mathematical Papers of Sir William Rowan Hamilton*, Vol. III, Algebra. Cambridge University Press.

Hall, M., and Wales, D. B. (1969), 'A simple group of order 604,800'. Pages 79–90 of: Brauer, R., and Sah, C.-H. (eds), *The Theory of Finite Groups*. W. A. Benjamin.

Harada, K. (1976), On the simple group F of order $2^{14}.3^6.5^6.7.11.19$. In: *Proceedings of the Conference on Finite Groups (Utah 1975)*. Academic Press.

Held, D. (1969), 'Some simple groups related to $M_{24}$'. Pages 121–124 of: Brauer, R., and Sah, C.-H. (eds), *The Theory of Finite Groups*. W. A. Benjamin.

Hilton, H. (1920), *Plane Algebraic Curves*. Oxford University Press.

Hurwitz, A. (1923), 'Über die Komposition der quadratischen Formen', *Math. Ann.* **88**, 1–25.

Isaacs, I. M. (1976), *Character Theory of Finite Groups*. Pure and Applied Mathematics. Academic Press.

Ivanov, A. A. (2018), *The Mathieu Group $M_{24}$*. Cambridge University Press.

Jacobson, N. (1980), *Basic Algebra II*. W. H. Freeman and Company.

Jacobson, N. (1985), *Basic Algebra I*. 2nd edn. W. H. Freeman and Company.

James, G., and Liebeck, M. (1993), *Representations and Characters of Groups*. Cambridge University Press.

Janko, Z. (1965), 'A new finite simple group with abelian Sylow 2-subgroups', *Proc. Natl. Acad. Sci. USA*, **53**, 675–658.

Janko, Z. (1966), 'A new finite simple group with abelian Sylow 2-subgroups and its characterization', *J. Algebra*, **3**, 147–186.

Janko, Z. (1969), Some new simple groups of finite order. Pages 63–64 of: Brauer, R., and Sah, C.-H. (eds), *The Theory of Finite Groups*. W. A. Benjamin.

Janko, Z. (1976), 'A new finite simple group of order 86, 775, 570, 046, 077, 562, 880, which possesses $M_{24}$ and the full covering group of $M_{22}$ as subgroups', *J. Algebra*, **42**, 564–596.

Jónsson, W. (1972), 'On the Mathieu groups $M_{22}, M_{23}, M_{24}$ and the uniqueness of the associated Steiner systems', *Math Z.*, **125**, 193–214.

King, O. H. (2005), *The Subgroup Structure of Finite Classical Groups in Terms of Geometric Configurations*. London Mathematical Society Lecture Note Series, Vol. 327. Cambridge University Press.

Kleidman, P. B., and Liebeck, M. W. (1990), *The Subgroup Structure of the Finite Classical Groups*. Cambridge University Press.

Klein, F. (1878), 'Uber die transformation siebenter ordnung der elliptischen functionen', *Gesammelte Math. Abhandlungen*, 90–135.

Leavitt, D. W., and Magliveras, S. S. (1984), Pages 337–352 of: *Simple 6 − (33, 6, 36) designs from $P\Gamma L_2(32)$, Computational Group Theory*. Academic Press.

Leech, J. (1956), 'The problem of the 13 spheres', *Math. Gazette.*, **40**, 22–23.

Leech, J. (1964), 'Some sphere packings in higher space', *Canad. J. Math.*, **16**, 657–682.

Lempken, W. (1978), 'A 2-local characterization of the Janko's simple group $J_4$', *J. Algebra*, **55**, 403–445.

Lyons, R. (1972), 'Evidence for a new finite simple group', *J. Algebra*, **20**, 540–569.

MacWilliams, F. J., and Sloane, N. J. A. (1977), *The Theory of Error-Correcting Codes*. North Holland.

Mathieu, E. (1861), 'Memoire sur l'étude des fonctions de plusieurs quantités', *J. Math. Pure Appl.*, **6**, 241–243.

Mathieu, E. (1873), 'Sur les fonctions cinq fois transitive de 24 quantités', *J. Math. Pure Appl.*, **18**, 25–46.

Mordell, L. J. (1969), *Diophantine Equations*. Academic Press.

Neubueser, J., et al. (2006), *Groups, Algorithms and Programming*. Aachen.

Neumaier, A., and Seidel, J. J. (1983), 'Discrete hyperbolic geometry', *Combinatorica*, **3**, 219–237.

Niemeier, H.-V. (1973), 'Definite quadratische Formen der Dimension 24 und Diskriminante 1', *J. Number Theory*, **5**, 142–178.

Norton, S. P. (1975), *F and Other Simple Groups*, PhD thesis, Cambridge University Press.

Norton, S. P. (1980), 'The construction of $J_4$', *Proc. of Symposia in Pure Math.*, **37**, 271–278.

O'Nan, M. E. (1976), 'Some evidence for the existence of a new simple group', *Proc. London Math. Soc.*, **32**, 421–479.

Parker, R. A., Wilson, R. A., Bray, J. N. et al. (1999) *Atlas of Finite Group Representations*. http://brauer.maths.qmul.ac.uk/Atlas/

Praeger, C. E., and Soicher, L. H. (1997), *Low Rank Representations and Graphs for Sporadic Groups*. Australian Mathematical Society Lecture Series B. Cambridge University Press, xi+141.

Ree, R. (1960), 'A family of simple groups associated with the simple Lie algebra (G2)', *Bulletin of the American Math. Soc.*, **66**, 508–510.

Ree, R. (1961), 'A family of simple groups associated with the simple Lie algebra of type (F4)', *Bulletin of the American Math. Soc.*, **67**, 115–116.

Reid, M. A. (2023), *Sextactic Points*. Unpublished response to a question from the author.

Rice, A. (2004), John Thomas Graves (1806–1870). *Oxford Dictionary of National Biography*. Oxford University Press.

Rowley, P. (2005), 'A Monster graph', *Proc. London Mathematical Society (3)*, **90**(1), 42–60.

Rowley, Peter, and Walker, L. (2004a), 'A 11,707,448,673,375 vertex graph related to the Baby Monster I', *J. Combinatorial Theory*, **107**(2), 181–213.

Rowley, P., and Walker, L. (2004b), 'A 11,707,448,673,375 vertex graph related to the Baby Monster II', *J. Combinatorial Theory*, **107**(2), 215–261.

Rowley, P., and Walker, L. (2012), 'A 195,747,435 vertex graph related to the Fischer group $Fi_{23}$, III', *JP Journal of Algebra, Number Theory and Appl.*, **27**, 1–44.

Rowley, P., and Walker, L. (2021), 'The point-line collinearity graph of the $Fi'_{24}$ maximal 2-local geometry - the first three discs', *JANTAA*, **24**(1), 53–109.

Rudvalis, A. (1973), 'A new simple group of order $2^{14}.3^3.5^3.7.13.29$', *Notices American Math. Soc.*, **20**, A–95.

Suzuki, M. (1964), 'On a class of doubly transitive groups', *Ann. Math.*, **79**, 514–589.

Taormina, A., and Wendland, K. (2013), 'The overarching symmetry group of Kummer surfaces in the Mathieu group $M_{24}$', *J. High Energy Physics*, **125**(8), 1–63.

Taormina, A., and Wendland, K. (2015a), 'Symmetry-surfing the moduli space of Kummer K3s', *Proc. Symp. Pure Math.*, **90**, 129–154.

Taormina, A., and Wendland, K. (2015b), 'A twist in the $M_{24}$ moonshine theory', *Confluentes Mathematici*, **7**, 83–113.

Tits, J. (1980a), 'Four presentations of Leech's lattice'. Pages 303–307 of: Collins, M.J. (ed), *Finite Simple Groups II*. Academic Press.

Tits, J. (1980b), 'Quaternions over $\mathbb{Q}(\sqrt{5})$, Leech's lattice and the sporadic group of Hall-Janko', *J. Algebra*, **63**, 56–75.

Todd, J. A. (1966), 'Representations of the Mathieu group $M_{24}$ as a collineation group', *Ann. di Math. Pure ed Appl.* **71**, 199–238.

Todd, J. A. (1970), 'Abstract definitions for the Mathieu groups', *Quart. J. Math. Oxford*, **21**, 421–424.

Weyl, H. (1953), *The Classical Groups, Their Invariants and Representations*. 2nd edn. Princeton Landmarks in Mathematics. Princeton University Press.

Wilson, R. A. (1982), 'The quaternionic lattice for $2G_2(4)$ and its maximal subgroups', *J. Algebra*, **77**, 449–466.

Wilson, R. A. (1983), 'The complex Leech lattice and maximal subgroups of the Suzuki group', *J. Algebra*, **84**, 151–188.

Wilson, R. A. (1986), 'Is $J_1$ a subgroup of the Monster?' *Bull. London Math. Soc.*, **18**, 349–350.

Wilson, R. A. (2009a), *The Finite Simple Groups*. Graduate Texts in Mathematics, Vol. 251, Springer-Verlag.

Wilson, R. A. (2009b), 'Octonions and the Leech lattice', *J. Algebra*, **322**, 2186–2190.

Witt, E. (1938a). 'Die 5-fach transitiven Gruppen von Mathieu', *Abh. Math. Sem. Hamb.*, **12**, 256–265.

Witt, E. (1938b). 'Uber Steinersche Systeme', *Abh. Math. Sem. Hamb.*, **12**, 265–275.

# Index

algebra, division, 251
annihilator, 48
arrangements of $\Omega$
  brick, 90
  MOG, 23
  $PGL_2(11)$, 92
  quadum, 87
Aschbacher, M., 163
ATLAS, xix, 114, 149, 168, 258, 259
Baker, H.F., 197, 244
Bannai, E., 169
Benson, D.J., xvi, 5, 222
Berlekamp, E., 5
Biggs, N., 192
binary Golay code, 30–34, 46
  16-ad, 33
  $T_{12}$, 63
  $U_8$, 63
  dodecad, 33
    intersections, 79
  dual code, 35
  dual space, 47
    Pet$_5$, 94
  dual space $P(\Omega)/\mathscr{C}$, 34
  duum, 77
  hexacode, xvi. *See* hexacode
  line space, 30, 81
  octad, 33
  penumbral dodecads, 63
  point space, 29, 81
  sextet, 19, 35, 36, 47
    coarsening, 34, 226
    even line, 47, 228
    intersections, 36, 222
    odd line, 47
    special hexad, 33

  trio, 34, 47, 80
    intersections, 81, 222
    refinement, 34, 47, 80, 226
  umbral dodecads, 61
  umbral hexad, 59, 63
block design, 16, 28
  $t$-$(v, k, \lambda)$ design, 16
  biplane, 16, 152
blocks, 6
Bolt, S., 222
Borcherds, R.E., 164
Bray, J.N., 153, 219, 222, 223
Burness, T.C., 153
Co$_1$, 131, 234–275
Co$_2$, 138–143, 270
  Magma, 142
Co$_3$, 143–149
  degree 276 action, 146
  index in ·O, 145
Cameron, P.J., 7, 169, 181
Cannon, J., xviii
Capdeboscq, I., 163
Carter, R.W., 159
Cayley, Sir Arthur, 251, 252
Cayley algebra, xv
Chevalley, C., 159
Choi, C., 68
Collins, M., 165
complex numbers, 250
Conway element $\xi_T$, 132, 215, 261
Conway, J.H., xv, xvii, 5, 23, 129, 150, 162, 234
Conway group ·O. *See* ·O
coset enumeration, Todd–Coxeter, 217, 219
Coxeter, H.S.M, 244
Curtis, C.W., 165

281

282    *Index*

Dickson, L.E., 153, 252
Dieudonné, J., 153
double coset enumerator, 219–221
    $2^{12} : M_{24}$, 220
    $\cdot J_4$, 230
    $\cdot O$, 221
Dynkin diagrams, 159

Euler's formula, 192
extension
    non-split, 5
    split, 5
finite field
    $\mathbb{F}_4$, 42, 70
    $\mathbb{F}_9$, 79, 237
    $\mathbb{Z}_p$, 28
    prime subfield, 155
Fischer, B., 161
Four Colour Theorem, 194
four squares theorem, 251

GAP, 4, 65
Giudici, M., 153
Gorenstein, D., 163
graph automorphism, 160
Graves, J.T., 251
Gregory, D., 119
Griess, R.L., 4, 162
group
    3-transposition, 162
    Chevalley, 159
        twisted, 160
    free product, 209
    Hall Janko, 211
    Hessian group, 98
    linear, 151
        $L_2(23)$, 82
        $L_4(2)$, 74
        $L_5(2)$, 75
        affine, 152
        linear fractional maps, 152
        primitive element, 152
        semilinear, 157
        sesquilinear, 157
    non-split, 97
    orthogonal, 159
    permutation action, 165
        $|\mathrm{Fix}_\Omega(g)|$, 165
        adjacency matrices, 169
        Burnside's Lemma, 166
        centralizer algebra, 170
        character value, 165
        eigenvalues, 177
        graph diagrams, 172
        multiplicity-free, 171
        orbital, 169
        permutation character, 165
        rank, 168
        rank 3, 184
        self-paired, 169
    permutation module, 165
    quadruply transitive, 1
    quaternion, 154
    quintuply transitive, 1
    right regular representation, 199, 208
    sharply 5-transitive, 1
    simple, 150–163
    sporadic, 150, 161–162
    symplectic, 158, 159
    Tits, 161
    unitary, 153–157
        $U_3(3)$, 237–242
        conjugate-symmetric, 157
        Hermitian, 157
        isotropic, 155
        order, 154
    unitriangular, 76
Guy, R.K., xvi, 5

Hall, Marshall Jr., 235
Hamilton, Sir William Rowan, 251
Harada, K., 236
Held, D., 236
hexacode, 5, 42–45
    counting octads, 45
    even interpretation, 43
    hexacodewords, 42
    odd interpretation, 43
Hoffman–Singleton graph, 111–114
Holt, D.F., 153
Hurwitz, A., 252

Isaacs, I.M., 165
isotropic, 130
Ito, T., 169

$J_4$, 5, 162, 222–233
    sextet elements, 227
Jónsson, W.J., 36
Jacobson, N., 153
James, G.D., 165
Janko group $J_4$. *See* $J_4$
Janko, Z., 161, 222, 235

King, O.H., 153
Kitten, 101–105
　mnemonic, 103
　vertex, 102
Kleidman, P.B., 153
Klein map, 192
　24 heptagons, 193
Klein quartic, 196
　bitangents, 198
　Hessian, 197
　Hurwitz bound, 197
　points of inflexion, 198
　sextactic points, 198
　triangles of inflexion, 199

$L_4(2)$, 74
lattice, 117
　hexagonal lattice, 118
　integral, 129
　　even, 129
　lattice packing, 118
　Leech. *See* Leech lattice
　square lattice, 117
　unimodular, 129
Leavitt, D.W., 16
Leech lattice, xv, 117–130, 215
　$\Lambda/2\Lambda$, 122
　cross, 124–129
　frame of reference, 124
　over $\mathbb{Z}_3$, 259
　short vectors, 122, 124
　standard cross, 124, 135
　type, 122
　types of cross, 126
Leech, J., 119, 131
Lempken, W., 222
Liebeck, M.W., 153, 165
linear code, 28
　binary, 29
　codewords, 28
　Hamming code, 30
　Hamming distance, 28
　length, 28
　minimal weight, 28
　weight, 28
　weight distribution, 32
linear group
　$L_2(23)$, 17, 46, 58
　linear fractional map, 17
Lorentz
　space-like coordinate, 129
　time-like coordinate, 129
Lyons, R., 162, 163

$M_{12}$, 34, 79, 97–116
　involution
　　$1^4 \cdot 2^4$, 105
　Kitten. *See* Kitten
　natural generators, 200–208
　　algebraic approach, 207
　　combinatorial approach, 200
　　geometric approach, 204
　　regular dodecahedron, 204
　outer automorphism, 114
　Todd triangle, 13
$M_{24}$
　elements
　　$1^6 \cdot 3^6$, 59
　　$1^8 \cdot 2^8$, 58
　　$2^{12}$, 58
　$M_{23}$, 68, 143
　natural generators, 187–200
　　algebraic explanation, 199
　　combinatorial approach, 188
　　geometric approach, 192
　subgroups
　　$\mathrm{Pet}_3$, 53
　　$\mathrm{Pet}_4$, 56
　　$\mathrm{Pet}_5$, 94
　　containment, 93
　　duad stabilizer, 68
　　duum stabilizer, 77, 105
　　monad stabilizer, 68
　　octad stabilizer, 73
　　octern group, 52, 82, 187
　　projective group, 82
　　sextet stabilizer, 71
　　Sylow 2-subgroup, 75
　　$T_{12}$ group, 95
　　triad stabilizer, 69
　　trio stabilizer, 80
　Todd's triangle, 17
Magaard, K., 235
Magliveras, S.S., 16
MAGMA, xviii, 4, 52, 65, 67–69, 71, 73, 75, 76, 79, 113, 135, 137, 142, 144–147, 149, 220, 221, 231, 232, 238, 242, 246, 260, 261
　Todd–Coxeter, 244
matrix
　Kronecker product, 170
　monomial, 165
　tensor product, 170
　unitriangular, 75
McKay, J., xvii, 5, 68, 131, 163
Miracle Octad Generator, 18–27

MOG. *See* Miracle Octad Generator
  $A_9$ elements, 260
  brick group, 262
  bricks, 22, 31
  heavy brick, 24
  hexacode, *See* hexacode
  Kitten, 106
  labelling mnemonic, 23
  standard arrangement, 120
  standard labelling, 23
Monstrous Moonshine, 5, 163
Moufang, R., xv

Neumaier, A., 129
Newton, Sir Isaac, 119
norm
  complex, 250
  quaternion, 251
Norton, S.P., 5, 163, 222, 236

·O, 131–137, 212
  elements
    duum, 137
    involutory, 137, 268
    monomial, 137
    octad, 137
    sextet, 137
    triad, 137, 263
  Magma construction, 135
  order, 135
  transitivity, 133
O'Nan, M.E., 162
octonions, xv, 251

$PGL_2(9)$, 78
$P\Gamma L_2(9)$, 78
Parker, R.A., 129, 222
plane
  affine, 11, 99
    ∞-arrangement, 100
    cross, 101
    perpendicular lines, 100
    quadrangle, 100
    square, 101
    triangle, 100
  Fano, 7
  projective, 7, 15, 162
Praeger, C.E., 177
projective line
  $P_1(7)$, 22
  $P_1(9)$, 12, 98
  $P_1(23)$, 17

quaternions, 251

Ree, R., 160
Reid, M.A., 198
Reiner, I., 165
Roney-Dougal, C.M., 153
Rowley, P.J., 5
Rudvalis, A., 162

$S_6$, 78
  Out $S_6$, 106
  Out $A_6$, 108
  $P\Gamma L_2(9)$, 12, 110
  $PGL_2(9)$, 110
  synthematic total, 108
  syntheme, 108, 226
  triple cover, 60
Seidel, J.J., 129
Sloane, N.J.A., 129
Smith, S., 163
Soicher, L.H., 162, 177
Solomon, R., 163
sphere-packing, 118
  density, 118
  kissing number, 119
  lattice packing, 118
Steinberg, R., 159
Steiner system, 6–16
  extension, 6
  $S(l, m, n)$, 6
  $S(2, q, q^2)$, 11
  $S(2, q+1, q^2+q+1)$, 8
  $S(2,3,9)$
    noughts and crosses, 98
    tic-tac-toe, 11, 98
  $S(2,5,21)$, 16
  $S(3,4,10)$, 12, 98
  $S(3,4,16)$, 14
    translation, 14
  $S(3,4,8)$, 9
  $S(4,5,11)$, 12
  $S(5,6,12)$, 12, 97–105
    hexads, 12
    special hexads, 97
  $S(5,8,24)$, 17–27
    octads, 17
    sextet, 19
    special tetrad, 19, 20, 22
    uniqueness, 37–41
  Steiner triple system STS, 7
String Theory, 5
Suzuki, M., 160, 235
Sylvester, J.J., 108
symmetric difference, 29

## Index

symmetric generation
  doubling up, 236
  The Lemma, 210
  monomial, 210
  progenitor, 210
  progeny, 210
  symmetric generators, 210
synthematic total, 3
syntheme, 3

Taormina, A., 5
tensor product, 262
Thackray, J.G., 222
Thompson chain, 234–275
  $3\,{\cdot}\,Suz$, 248
  $N \cong A_9$, 255
  $G_2(4)$, 247
  $K_n$ for $n > 7$, 253
  conclusion, 257
  the Conway group $Co_1$, 249
  doubling up, 254
  extension to $K_n$, 242
  generator $t_{89}$, 263
  the Hall–Janko group HJ, 246
  important identities, 240
  Leech lattice, 258
  triangle, 236
Thompson, J.G., 153, 234
Tits, J., 161
Todd, J.A., xv, xvii, xix, 13, 61, 68, 244
  Todd's lemma, 19, 38
  Todd's triangle, 17, 36

van Lint, J.H., 181
Viazovska, M., 119
Voyager, 31

Wales, D.B., xvii, 68, 235
Walker, L., 5
Weyl, H., 153
Wilson, R.A., xix, 150, 153, 155, 235
Witt, E., 2, 16, 131

$\mathbb{Z}_p$, 135

Printed in the United States
by Baker & Taylor Publisher Services